Newnes
Electrical Engineer's
Handbook

Newnes
Electrical Engineer's
Handbook

D.F. Warne

Newnes

OXFORD   AUCKLAND   BOSTON   JOHANNESBURG   MELBOURNE   NEW DELHI

Newnes
An imprint of Butterworth-Heinemann
Linacre House, Jordan Hill, Oxford OX2 8DP
225 Wildwood Avenue, Woburn, MA 01801-2041
A division of Reed Educational and Professional Publishing Ltd

 A member of the Reed Elsevier plc group

First published 2000

British Library Cataloguing in Publication Data
A catalogue record for this book is available from the British Library

ISBN 0 7506 4879 1

Library of Congress Cataloguing in Publication Data
A catalogue record for this book is available from the
British Library of Congress

Typeset at Replika Press Pvt Ltd, 100% EOU, Delhi 110 040, India
Printed and bound in Great Britain by Biddles Ltd
www.biddles.co.uk

Contents

Acknowledgements

Preparation of this book seemed at the outset to be a fairly straightforward task. It has, however, proved to be quite different. A structural change from the content of the preceding *Newnes Electrical Pocket Book* was necessary because much of the content and scope of this has been taken to other books in the new series such as the *Newnes Building Services Pocket Book* and the *Newnes Electronics Engineer's Pocket Book*. At the same time, much of what is often regarded as 'mature technology' is changing surprisingly rapidly, and much of the remaining material needed a complete overhaul with some new areas introduced.

Inevitably the chapters have been developed at different paces and the task of finalizing the content of each chapter has been quite difficult because of continual development, especially in the technology and in standards. The forbearance of specialist chapter authors during this process of 'spinning plates' is gratefully acknowledged.

No project of this type goes without some personal and social sacrifice. Without the patience and support of my wife Gill the completion of the book would have been very much more difficult.

D.F. Warne
September 1999

Chapter 1

Introduction

There seems to be a trend in the public perception of engineering and technology that to be able to operate a piece of equipment or a system is to understand how it works. Nothing could be further from the truth. The gap between the ability to operate and a genuine understanding is, if anything, widening because much of the complexity added to modern electrical equipment has the specific aim of making it operable or 'user-friendly' without special training or knowledge.

The need for a basic explanation of principles, leading to a simple description of how various important and common classes of electrical equipment works, has never been stronger. Perhaps more so than in its predecessor, *Newnes Electrical Pocket Book*, an attempt is made to address fundamentals in this book, and the reader is encouraged to follow through any areas of interest using the references at the end of each chapter. More comprehensive coverage of all the subjects covered in this pocket book is available in the *Newnes Electrical Engineer's Reference Book*.

More so now than ever before, the specification and performance of electrical equipment is governed by national and international standards. While it would be inappropriate in a pocket book to cover standards in any detail, a summary of key standards is included for reference purposes at the end of each chapter.

The structure of the book is based around three groups of chapters, which address:

- fundamentals and general material
- the design and operation of the main classes of electrical equipment
- special technologies which apply to a range of equipment

The first group comprises three chapters which set out fundamentals and principles running through all aspects of electrical technology.

The opening chapter deals with fundamentals of electric and magnetic fields and circuits, with energy and power conversion principles.

This is followed by a review of the materials that are so crucial to the design of electrical equipment, and these are grouped into sections on magnetic, insulating and conducting materials. In each of these areas technology is moving ahead rapidly. The great increases in the strength of permanent magnets in the past ten years has done much to make possible the miniaturization of equipment such as the Sony Walkman, and the introduction of so many small motors and actuators in our homes and motor cars. Developments in insulating materials mean that increased reliability and operation at higher temperatures can now readily be achieved. Under the heading of conductors there are continuing advances in superconductors, which are now able to operate in liquid nitrogen, and of course semiconductor development has transformed the way in which equipment can be controlled and the processing power in computers.

Finally in this opening group there is a chapter on measurement and instrumentation.

A classical textbook on electrical measurement would in the past have included sections on moving iron and moving coil instruments, but these have been omitted here in favour of the oscilloscope and sensors which now dominate measurements in most areas.

The following group of nine chapters make up the main core of the book and cover the essential groups of electrical equipment found today in commerce and industry.

The opening five chapters here cover generators, transformers, switchgear, fuses and wire and cables. These are the main technologies for the *production and handling* of electrical energy, from high power and high voltage levels down to the powers and voltages found in the household. Exciting developments in this area include the advances made in high voltage switchgear using SF_6 as an insulating medium, the extension of polymer insulation into high voltage cables and the continuing compaction of miniature and moulded-case circuit breakers. A new section in the wire and cables chapter addresses the growing technology of optical fibre cables; although the main use for this technology is in telecommunications, which is outside the scope of the book, a chapter on wires and cables would not be complete without it and optical fibres have a growing number of applications in electrical engineering.

The following four chapters describe different groups of equipment that *use or store* electrical energy. Probably the most important here is electric motors, since these use almost two-thirds of all electrical energy generated. Static power supplies are also of growing importance in applications such as emergency standby for computers; this technology is now based on power electronics and the opportunity is taken in this chapter to explain the fundamentals of power electronic design and technology. Rotating converters were important for many of the duties now handled by power electronics, but these are now in decline and are not covered here. The range of batteries being developed and appearing in a variety of applications is now very large and this is the subject of a special chapter, which also covers the techniques of battery charging and the emerging related technology of fuel cells. If fuel cells fulfil their promise and start to play a greater part in the generation of electricity in the future then we can expect to see coverage of this area grow and perhaps move to the generation section in future editions. Another major electricity consumer is the range of technologies generally known as electroheat. This covers a spectrum of technologies from arc furnaces through microwave heating to ultraviolet drying techniques which are described in a special chapter.

The final group of three chapters cover subjects that embrace a range of technologies and equipment. There is a chapter on power systems which describes the way in which generators, switchgear, transformers, lines and cables are connected and controlled to transmit and distribute our electrical energy. The privatization of electricity supply in countries across the world has brought great changes in the way in which power systems are operated and these are touched upon here. The second chapter in this group concerns electromagnetic compatibility (emc); with the growing amount of electronic and high-frequency equipment in use today it is imperative that precautions are taken to prevent interference and legislation has been introduced to enforce this prevention. The techniques for tackling this are complex and influence a wide range of equipment. Finally there is a chapter describing the design and use of equipment for operation in hazardous and explosive environments; this again covers a wide range of equipments and there are a number of different classifications of protection.

Chapter 2

Principles of electrical engineering

Dr D.W. Shimmin
University of Liverpool

2.1 Nomenclature and units

This book uses notation in accordance with the current British and International Standards. Units for engineering quantities are printed in upright roman characters, with a space between the numerical value and the unit, but no space between the decimal prefix and the unit, e.g. 275 kV. Compound units have a space, dot or / between the unit elements as appropriate, e.g. 1.5 N m, 300 m/s, or 9.81 $\text{m} \cdot \text{s}^{-2}$. Variable symbols are printed in italic typeface, e.g. *V*. For ac quantities, the instantaneous value is printed in lower case italic, peak value in lower case italic with caret (^), and rms value in upper case, e.g. *i*, *î*, *I*. Symbols for the important electrical quantities with their units are given in **Table 2.1**, and decimal prefix symbols are shown in **Table 2.2**. Graphical symbols for basic electrical engineering components are shown on **Fig. 2.1**.

2.2 Electromagnetic fields

2.2.1 Electric fields

Any object can take an *electric* or *electrostatic charge*. When the object is charged positively, it has a deficit of electrons, and when charged negatively it has an excess of electrons. The electron has the smallest known charge, -1.602×10^{-19} C.

Charged objects produce an electric field. The *electric field strength E* (V/m) at a distance d (m) from an isolated point charge Q (C) in air or a vacuum is given by

$$E = \frac{Q}{4\pi\varepsilon_0 d^2} \tag{2.1}$$

where the *permittivity of free space* $\varepsilon_0 = 8.854 \times 10^{-12}$ F/m. If the charge is inside an insulating material with *relative permittivity* ε_r the electric field strength becomes

$$E = \frac{Q}{4\pi\varepsilon_0 \varepsilon_r d^2} \tag{2.2}$$

Any charged object or particle experiences a force when inside an electric field. The force F (N) experienced by a charge Q (C) in an electric field strength E (V/m) is given by

Table 2.1 Symbols for standard quantities and units

Symbol	Quantity	Unit	Unit symbol
A	geometric area	square metre	m^2
B	magnetic flux density	tesla	T
C	capacitance	farad	F
E	electric field strength	volt per metre	V/m
F	mechanical force	newton	N
F_m	magnetomotive force (mmf)	ampere	A or A · t
G	conductance	siemens	S
H	magnetic field strength	ampere per metre	A/m
I	electric current	ampere	A
J	electric current density	ampere per square metre	A/m^2
J	moment of inertia	kilogram metre squared	kg · m^2
L	self-inductance	henry	H
M	mutual inductance	henry	H
N	number of turns		
P	active or real power	watt	W
Q	electric charge	coulomb	C
Q	reactive power	volt ampere reactive	VAR
R	electrical resistance	ohm	Ω
R_m	reluctance	ampere per weber	A/Wb
S	apparent power	volt ampere	V · A
T	mechanical torque	newton metre	N · m
V	electric potential or voltage	volt	V
W	energy or work	joule	J
X	reactance	ohm	Ω
Y	admittance	siemens	S
Z	impedance	ohm	Ω
f	frequency	hertz	Hz
j	square root of −1		
l	length	metre	m
m	mass	kilogram	kg
n	rotational speed	revolution per minute	rpm
p	number of machine pole pairs		
t	time	second	s
v	linear velocity	metre per second	m/s
ε	permittivity	farad per metre	F/m
η	efficiency		
θ	angle	radian or degree	rad or °
λ	power factor		
Λ	permeance	weber per ampere	Wb/A
μ	permeability	henry per metre	H/m
ρ	resistivity	ohm metre	Ω · m
σ	conductivity	siemens per metre	S/m
ϕ	phase angle	radian	rad
Φ	magnetic flux	weber	Wb
Ψ	magnetic flux linkage	weber or weber-turn	Wb or Wb · t
ω	angular velocity or angular frequency	radian per second	rad/s

Table 2.2 Standard decimal prefix symbols

Prefix	Name	Multiple
T	tera	10^{12}
G	giga	10^{9}
M	mega	10^{6}
k	kilo	10^{3}
d	deci	10^{-1}
c	centi	10^{-2}
m	milli	10^{-3}
μ	micro	10^{-6}
n	nano	10^{-9}
p	pico	10^{-12}
f	femto	10^{-15}

$$F = QE \qquad (2.3)$$

Electric field strength is a vector quantity. The direction of the force on one charge due to the electric field of another is repulsive or attractive. Charges with the same polarity repel; charges with opposite polarities attract.

Work must be done to move charges of the same polarity together. The effort required is described by a *voltage* or *electrostatic potential*. The voltage at a point is defined as the work required to move a unit charge from infinity or from earth. (It is normally assumed that the earth is at zero potential.) Positively charged objects have a positive potential relative to the earth.

If a positively charged object is held some distance above the ground, then the voltage at points between the earth and the object rises with distance from the ground, so that there is a *potential gradient* between the earth and the charged object. There is also an electric field pointing away from the object, towards the ground. The electric field strength is equal to the potential gradient, and opposite in direction.

$$E = -\frac{dV}{dx} \qquad (2.4)$$

2.2.2 Electric currents

Electric charges are static if they are separated by an insulator. If charges are separated by a conductor, they can move giving an electric current. A current of one ampere flows if one coulomb passes along the conductor every second.

$$I = \frac{Q}{t} \qquad (2.5)$$

A given current flowing through a thin wire represents a greater density of current than if it flowed through a thicker wire. The *current density* J (A/m^2) in a wire with cross-section area A (m^2) carrying a current I (A) is given by

$$J = \frac{I}{A} \qquad (2.6)$$

For wires made from most conducting materials, the current flowing through the wire is directly related to the difference in potential between the ends of the wire.

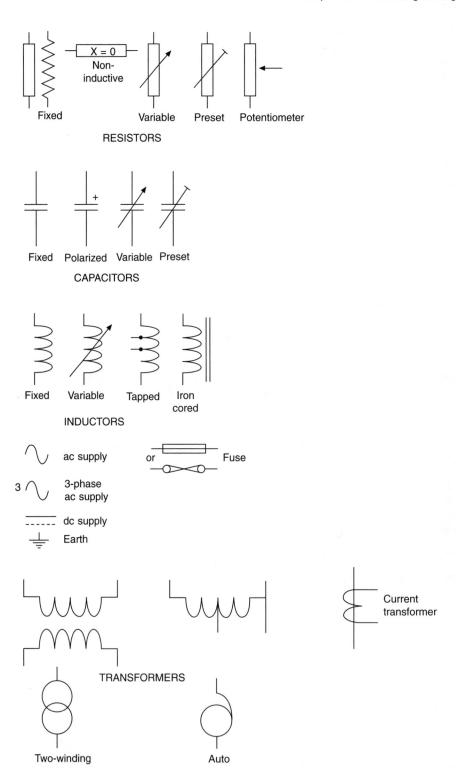

Fig. 2.1 Standard graphical symbols

Ohm's law gives this relationship between the potential difference V (V) and the current I (A) as

$$V = IR \quad \text{or} \quad I = VG \tag{2.7}$$

where R (Ω) is the *resistance*, and G (S) $= 1/R$ is the *conductance* (**Fig. 2.2**). For a wire of length l and cross-section area A, these quantities depend on the *resistivity* ρ ($\Omega \cdot$ m) and *conductivity* σ (S/m) of the material

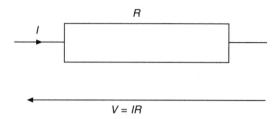

R

I

V = IR

Fig. 2.2 Ohm's law

$$R = \rho \frac{l}{A} \quad \text{and} \quad G = \sigma \frac{A}{l} \tag{2.8}$$

For materials normally described as *conductors* ρ is small, while for *insulators* ρ is large. *Semiconductors* have resistivity in between these extremes, and are usually very dependent on purity and temperature.

In metal conductors, the resistivity increases with temperature approximately linearly:

$$R_T = R_{T_0}(1 + \alpha(T - T_0)) \tag{2.9}$$

for a conductor with resistance R_{T_0} at reference temperature T_0. This is explained in more detail in **section 3.4.1**.

Charges can be stored on conducting objects if the charge is prevented from moving by an insulator. The potential of the charged conductor depends on the *capacitance* C (F) of the metal/insulator object, which is a function of its geometry. The charge is related to the potential by

$$Q = CV \tag{2.10}$$

A simple parallel-plate capacitor, with plate area A, insulator thickness d and relative permittivity ε_r has capacitance

$$C = \frac{\varepsilon_0 \varepsilon_r A}{d} \tag{2.11}$$

2.2.3 Magnetic fields

A flow of current through a wire produces a magnetic field in a circular path around the wire. For a current flowing forwards, the magnetic field follows a clockwise path, as given by the right-hand corkscrew rule (**Fig. 2.3**). The *magnetic field strength* H (A m^{-1}) is a vector quantity whose magnitude at a distance d from a current I is given by

$$H = \frac{I}{2\pi d} \tag{2.12}$$

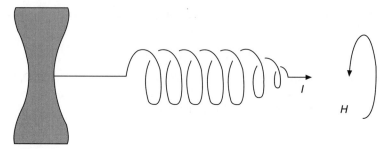

Fig. 2.3 Right-hand corkscrew rule

For a more complicated geometry, *Ampère's law* relates the number of turns N in a coil, each carrying a current I, to the magnetic field strength H and the distance around the lines of magnetic field l.

$$Hl = NI$$

$$= F_m \qquad (2.13)$$

where F_m (ampere-turns) is the *magnetomotive force (mmf)*. This only works for situations where H is uniform along the lines of magnetic field.

The magnetic field produced by a current does not depend on the material surrounding the wire. However, the magnetic force on other conductors is greatly affected by the presence of ferromagnetic materials, such as iron or steel. The magnetic field produces a *magnetic flux density B* (T) in air or vacuum

$$B = \mu_o H \qquad (2.14)$$

where the *permeability of free space* $\mu_o = 4\pi \times 10^{-7}$ H/m. In a ferromagnetic material with *relative permeability* μ_r

$$B = \mu_o \mu_r H \qquad (2.15)$$

A second conductor of length l carrying an electric current I will experience a force F in a magnetic flux density B

$$F = BIl \qquad (2.16)$$

The force is at right angles to both the wire and the magnetic field. Its direction is given by *Fleming's left-hand rule* (**Fig. 2.4**). If the magnetic field is not itself perpendicular to the wire, then the force is reduced; only the component of B at right angles to the wire should be used.

A flow of *magnetic flux* Φ (Wb) is caused by the flux density in a given cross-section area A as

$$\Phi = BA \qquad (2.17)$$

The mmf F_m required to cause a magnetic flux Φ to flow through a region of length l and cross-section area A is given by the *reluctance* R_m (A/Wb) or the *permeance* Λ (Wb/A) of the region

$$F_m = \Phi R_m \quad \text{or} \quad \Phi = \Lambda F_m \qquad (2.18)$$

where

$$R_m = \frac{l}{\mu_o \mu_r A} \qquad (2.19)$$

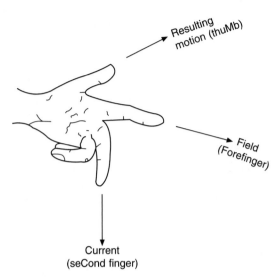

Fig. 2.4 Fleming's left-hand rule

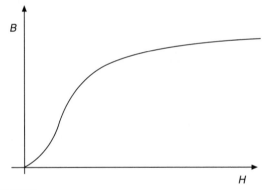

(a) Magnetization curve

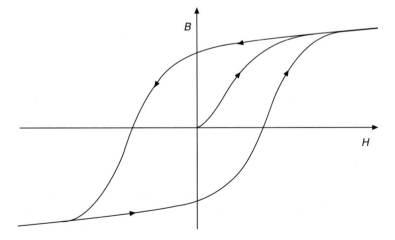

(b) Hysteresis loop

Fig. 2.5 Magnetic characteristics

In ideal materials, the flux density *B* is directly proportional to the magnetic field strength *H*. In ferromagnetic materials the relation between *B* and *H* is non-linear (**Fig. 2.5(a)**), and also depends on the previous magnetic history of the sample. The *magnetization* or *hysteresis* or *BH* loop of the material is followed as the applied magnetic field is changed (**Fig. 2.5(b)**). Energy is dissipated as heat in the material as the operating point is forced around the loop, giving *hysteresis loss* in the material. These concepts are developed further in **section 3.2**.

2.2.4 Electromagnetism

Any change in the magnetic field near a wire generates a voltage in the wire by *electromagnetic induction*. The changing field can be caused by moving the wire in the magnetic field. For a length *l* of wire moving sideways at speed *v* (m/s) across a magnetic flux density *B*, the induced voltage or *electromotive force (emf)* is given by

$$V = Bvl \qquad\qquad (2.20)$$

The direction of the induced voltage is given by *Fleming's right-hand rule* (**Fig. 2.6**). An emf can also be produced by keeping the wire stationary and changing the magnetic field. In either case the induced voltage can be found using *Faraday's law*. If a magnetic flux Φ passes through a coil of *N* turns, the *magnetic flux linkage* Ψ (Wb·t) is

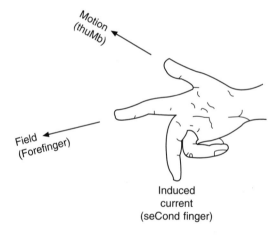

Fig. 2.6 Fleming's right-hand rule

$$\Psi = N\Phi \qquad\qquad (2.21)$$

Faraday's law says that the induced emf is given by

$$V = -\frac{d\Psi}{dt} \qquad\qquad (2.22)$$

The direction of the induced emf is given by *Lenz's law*, which says that the induced voltage is in the direction such that, if the voltage caused a current to flow in the wire, the magnetic field produced by this current would oppose the change in Ψ. The negative sign indicates the opposing nature of the emf.

A current flowing in a simple coil produces a magnetic field. Any change in the current will change the magnetic field, which will in turn induce a *back-emf* in the coil. The *self-inductance* or just *inductance* *L* (H) of the coil relates the induced voltage to the rate of change of current

$$V = L\frac{dI}{dt} \qquad (2.23)$$

Two coils placed close together will interact. The magnetic field of one coil will link with the wire of the second. Changing the current in the primary coil will induce a voltage in the secondary coil, given by the *mutual inductance M* (H)

$$V_2 = M\frac{dI_1}{dt} \qquad (2.24)$$

Placing the coils very close together, on the same former, gives close coupling of the coils. The magnetic flux linking the primary coil nearly all links the secondary coil. The voltages induced in the primary and secondary are each proportional to their number of turns, so that

$$\frac{V_1}{V_2} = \frac{N_1}{N_2} \qquad (2.25)$$

and by conservation of energy, approximately

$$\frac{I_1}{I_2} = \frac{N_2}{N_1} \qquad (2.26)$$

A two-winding transformer consists of two coils wound on the same ferromagnetic core. An autotransformer has only one coil with tapping points. The voltage across each section is proportional to the number of turns in the section. Transformer action is described more fully in **section 6.1**.

2.3 Circuits

2.3.1 DC circuits

DC power is supplied by a battery, dc generator or rectifying power supply from the mains. The power flowing in a dc circuit is the product of the voltage and current

$$P = VI = I^2R = \frac{V^2}{R} \qquad (2.27)$$

Power in a resistor is converted directly into heat.

When two or more resistors are connected in *series,* they carry the same current but their voltages must be added together (**Fig. 2.7**)

$$V = V1 + V2 + V3 \qquad (2.28)$$

As a result, the total resistance is given by

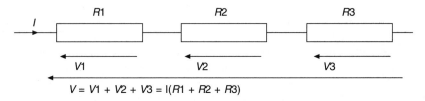

$$V = V1 + V2 + V3 = I(R1 + R2 + R3)$$

Fig. 2.7 Series resistors

$$R = R1 + R2 + R3 \tag{2.29}$$

When two or more resistors are connected in *parallel*, they have the same voltage but their currents must be added together (**Fig. 2.8**)

$$I = I1 + I2 + I3 \tag{2.30}$$

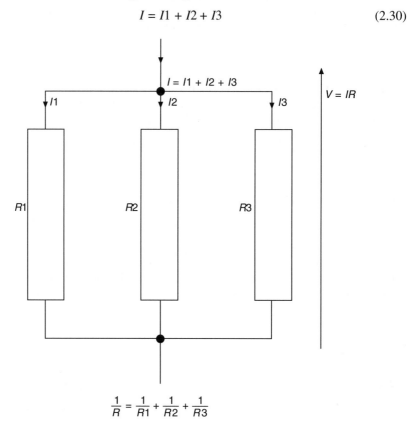

Fig. 2.8 Parallel resistors

The total resistance is given by

$$\frac{1}{R} = \frac{1}{R1} + \frac{1}{R2} + \frac{1}{R3} \tag{2.31}$$

A complicated circuit is made of several components of *branches* connected together at *nodes* forming one or more complete circuits, *loops* or *meshes*. At each node, *Kirchhoff's current law* (**Fig. 2.9(a)**) says that the total current flowing into the node must be balanced by the total current flowing out of the node. In each loop, the sum of all the voltages taken in order around the loop must add to zero, by *Kirchhoff's voltage law* (**Fig. 2.9(b)**). Neither voltage nor current can be lost in a circuit.

DC circuits are made of resistors and voltage or current sources. A circuit with only two connections to the outside world may be internally complicated. However, to the outside world it will behave as if it contains some resistance and possibly a source of voltage or current. The *Thévenin* equivalent circuit consists of a voltage source and a resistor (**Fig. 2.10(a)**), while the *Norton* equivalent circuit consists of a current source and a resistor (**Fig. 2.10(b)**). The resistor equals the internal resistance

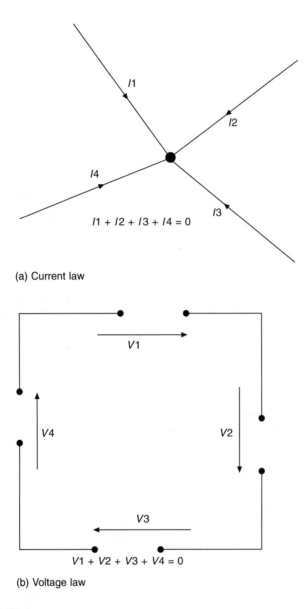

(a) Current law

(b) Voltage law

Fig. 2.9 Kirchhoff's laws

of the circuit, the Thévenin voltage source equals the open-circuit voltage, and the Norton current source is equal to the short-circuit current.

Many circuits contain more than one source of voltage or current. The current flowing in each branch, or the voltage at each node, can be found by considering each source separately and adding the results. During this calculation by *superposition,* all sources except the one being studied must be disabled: voltage sources are short-circuited and current sources are open-circuited. In **Fig. 2.11**, each of the loop currents $I1$ and $I2$ can be found by considering each voltage source separately and adding the results, so that $I1 = I1_a + I1_b$, and $I2 = I2_a + I2_b$.

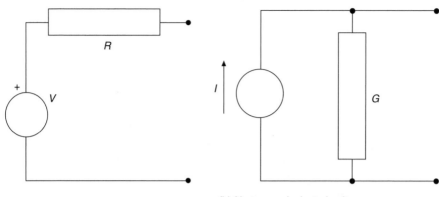

(a) Thévenin equivalent circuit   (b) Norton equivalent circuit

Fig. 2.10 Equivalent circuits

2.3.2 AC circuits

AC power is supplied by the mains from large ac generators or alternators, by a local alternator, or by an electronic synthesis. AC supplies are normally sinusoidal, so that at any instant the voltage is given by

$$v = V_{max} \sin(\omega t - \phi)$$
$$= \hat{v} \sin(\omega t - \phi) \qquad (2.32)$$

V_{max} is the peak voltage or amplitude, ω is the *angular frequency* (rad s^{-1}) and ϕ the *phase angle* (rad). The angular frequency is related to the ordinary *frequency f* (Hz) by

$$\omega = 2\pi f \qquad (2.33)$$

and the *period* is $1/f$. The *peak-to-peak* or *pk–pk* voltage is $2V_{max}$, and the *root mean square* or rms voltage is $V_{max}/\sqrt{2}$. It is conventional for the symbols V and I in ac circuits to refer to the rms values, unless indicated otherwise. AC voltages and currents are shown diagrammatically on a *phasor diagram* (**Fig. 2.12**).

It is convenient to represent ac voltages using complex numbers. A sinusoidal voltage can be written

$$V = V_{max}e^{j\phi} \qquad (2.34)$$

A resistor in an ac circuit behaves the same as in a dc circuit, with the current and voltage in phase and related by the resistance or conductance (**Fig. 2.13**).

The current in an inductor lags the voltage across it by 90° ($\pi/2$ rad) (**Fig. 2.14**). The ac resistance or *reactance X* of an inductor increases with frequency

$$X_L = \omega L \qquad (2.35)$$

The phase shift and reactance are combined in the complex *impedance Z*

$$\frac{V}{I} = Z_L = jX_L = j\omega L \qquad (2.36)$$

Inductors in series behave as resistors in series

$$L_s = L_1 + L_2 + L_3 \qquad (2.37)$$

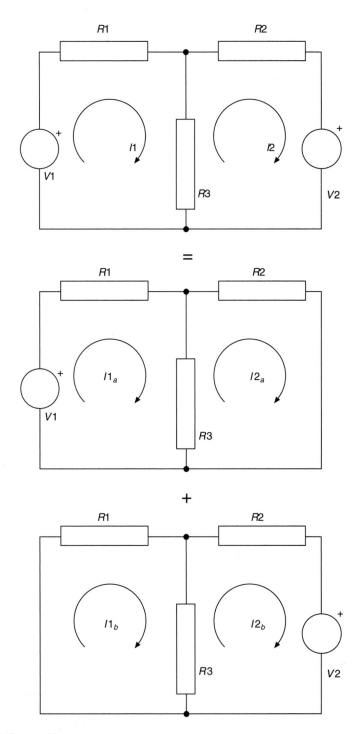

Fig. 2.11 Superposition

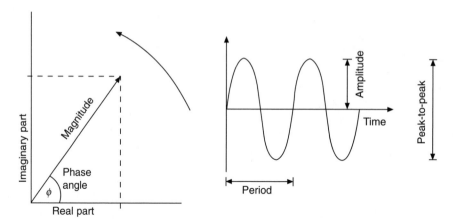

Fig. 2.12 Sinusoidal ac quantities

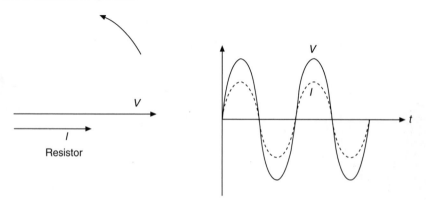

Fig. 2.13 Resistor in an ac circuit

and inductors in parallel behave as resistors in parallel

$$\frac{1}{L_p} = \frac{1}{L_1} + \frac{1}{L_2} + \frac{1}{L_3} \tag{2.38}$$

For a capacitor, the current leads the voltage across it by 90° ($\pi/2$ rad) (**Fig. 2.15**). The reactance decreases with increasing frequency

$$X_c = \frac{1}{\omega C} \tag{2.39}$$

In a capacitor, the current leads the voltage, while in an inductor, the voltage leads the current.

The impedance is given by

$$\frac{V}{I} = Z_c = -jX_c = -\frac{j}{\omega C} = \frac{1}{j\omega C} \tag{2.40}$$

Capacitors in series behave as resistors in parallel (**eqn 2.41**) and capacitors in parallel behave as resistors in series (**eqn 2.42**)

$$\frac{1}{C_s} = \frac{1}{C_1} + \frac{1}{C_2} + \frac{1}{C_3} \tag{2.41}$$

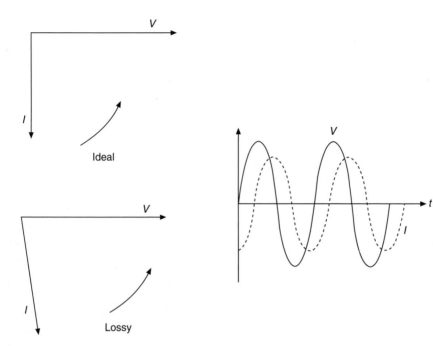

Fig. 2.14 Inductor in an ac circuit

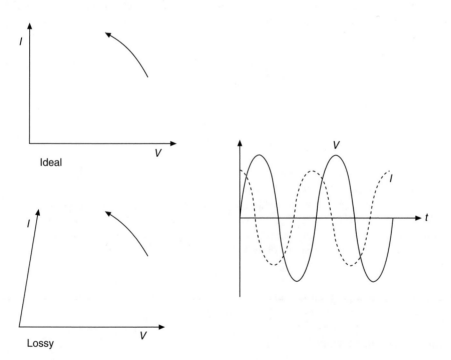

Fig. 2.15 Capacitor in an ac circuit

$$C_p = C_1 + C_2 + C_3 \tag{2.42}$$

The direction of the phase shift in inductors and capacitors is easily remembered by the mnemonic CIVIL (i.e. C-IV, VI-L). Imperfect inductors and capacitors have some inherent resistance, and the phase lead or lag is less than 90°. The difference between the ideal phase angle and the actual angle is called the *loss angle* δ. For a component of reactance X having a series resistance R

$$\tan(\delta) = \frac{R}{X} \tag{2.43}$$

The reciprocal of impedance is *admittance*

$$\frac{I}{V} = Y = \frac{1}{Z} \tag{2.44}$$

Combinations of resistors, capacitors and inductors will have a variation of impedance or admittance with frequency which can be used to select signals at certain frequencies in preference to others. The circuit acts as a *filter*, which can be low-pass, high-pass, band-pass, or band-stop.

An important filter is the *resonant circuit*. A series combination of inductor and capacitor has zero impedance (infinite admittance) at its resonant frequency

$$\omega_0 = 2\pi f_0 = \frac{1}{\sqrt{LC}} \tag{2.45}$$

A parallel combination of inductor and capacitor has infinite impedance (zero admittance) at the same frequency.

In practice a circuit will have some resistance (**Fig. 2.16**), which makes the resonant circuit imperfect. The *quality factor Q* of a series resonant circuit with series resistance R is given by

$$Q = \left(\frac{X_L}{R}\right)_{\omega=\omega_0} = \left(\frac{X_C}{R}\right)_{\omega=\omega_0} \tag{2.46}$$

and for a parallel circuit with shunt resistance R

$$Q = \left(\frac{R}{X_L}\right)_{\omega=\omega_0} = \left(\frac{R}{X_C}\right)_{\omega=\omega_0} \tag{2.47}$$

A filter with a high Q will have a sharper change of impedance with frequency than one with a low Q, and reduced losses.

Mains electricity is generated and distributed using three phases. Three equal voltages are generated at 120° phase intervals. The *phase voltage* V_P is measured with respect to a common star point or neutral point, and the *line voltage* V_L is measured between the separate phases. In magnitude,

$$V_L = \sqrt{3}V_P \tag{2.48}$$

Any three-phase generator, transformer or load can be connected in a *star* or Y configuration, or in a *delta* configuration (**Fig. 2.17**).

2.3.3 Magnetic circuits

A reluctance in a magnetic circuit behaves in the same way as a resistance in a dc electric circuit. Reluctances in series add together

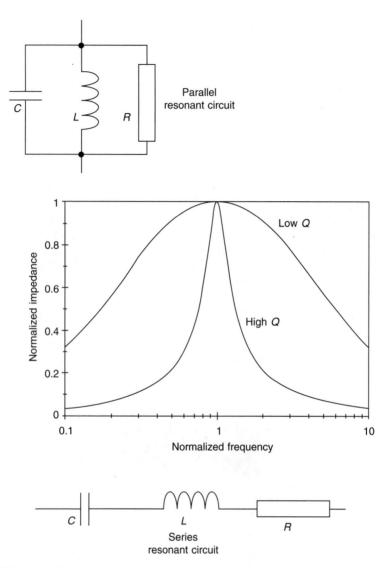

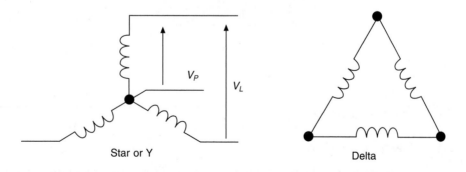

Fig. 2.16 Resonant circuits

Fig. 2.17 Three-phase connections

$$R_m = R_{m_1} + R_{m_2} + R_{m_3}$$

or
$$\frac{1}{\Lambda} = \frac{1}{\Lambda_1} + \frac{1}{\Lambda_2} + \frac{1}{\Lambda_3} \qquad (2.49)$$

and for reluctances in parallel

$$\frac{1}{R_m} = \frac{1}{R_{m_1}} + \frac{1}{R_{m_2}} + \frac{1}{R_{m_3}}$$

or
$$\Lambda = \Lambda_1 + \Lambda_2 + \Lambda_3 \qquad (2.50)$$

Transformers and power reactors may contain no air gap in the magnetic circuit. However, motors and generators always have a small air gap between the rotor and stator. Many reactors also have an air gap to reduce saturation of the ferromagnetic parts. The reluctance of the air gap is in series with the reluctance of the steel rotor and stator. The high relative permeability of steel means that the reluctance of even a small air gap can be much larger than the reluctance of the steel parts of the machine. For a total air gap g (m) in a magnetic circuit, the magnetic flux density B in the air gap is given approximately by

$$B \approx \mu_o \frac{F_m}{g} = \mu_o \frac{N_f I_f}{g} \qquad (2.51)$$

where N_f is the number of series turns on the field winding of the machine, and I_f is the current in the field winding (**Fig. 2.18**).

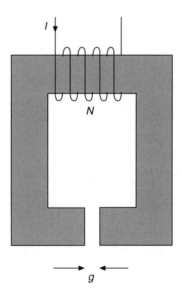

Fig. 2.18 Air-gap magnetic circuit

2.4 Energy and power

2.4.1 Mechanical energy

According to Newton's third law of motion, mechanical force causes movement is a straight line, such that the force F (N) accelerates a mass m (kg) with acceleration a (m/s^2)

$$F = ma \tag{2.52}$$

Rotational movement depends in the same way upon torque T (N m) accelerating a moment of inertia J (kg $\cdot$ m^2) with angular acceleration α (rad/s^2)

$$T = J\alpha \tag{2.53}$$

Movement is often opposed by friction. Friction forces and torques always work against the movement. Friction between dry surfaces has a maximum value, depending on the contact force. Once the driving force exceeds the limiting friction force, the system will move and the friction force stays constant. Viscous damping gives a restraining force which increases with the speed. Friction between lubricated surfaces is mainly a viscous effect.

Objects store potential energy when they are lifted up. The stored energy W (J) of a mass m is proportional to the height h (m) above ground level when the acceleration due to gravity $g = 9.81$ m/s

$$W = mgh \tag{2.54}$$

Moving objects have kinetic energy depending on their linear speed v (m/s) or angular speed ω (rad/s)

$$W = \frac{1}{2}mv^2$$

$$= \frac{1}{2}J\omega^2 \tag{2.55}$$

Mechanical work is done whenever an object is moved a distance x (m) against an opposing force, or through an angle θ (rad) against an opposing torque

$$W = Fx$$

$$= T\theta \tag{2.56}$$

Mechanical power P (W) is the rate of doing work

$$P = Fv$$

$$= T\omega \tag{2.57}$$

2.4.2 Electrical energy

In electrical circuits, electrical potential energy is stored in a capacitance C charged to a voltage V

$$W = \frac{1}{2}CV^2 \tag{2.58}$$

while kinetic energy is stored in an inductance L carrying a current I

$$W = \frac{1}{2}LI^2 \tag{2.59}$$

Capacitors and inductors store electrical energy. Resistors dissipate energy and convert it into heat. The power dissipated and lost to the electrical system in a resistor R has already been shown in **eqn 2.27**.

Electrical and mechanical systems can convert and store energy, but overall the total energy in a system is conserved. The overall energy balance in an electromechanical system can be written electrical energy in + mechanical energy in

> = electrical energy lost in resistance
> + mechanical energy lost in friction
> + magnetic energy lost in steel core
> + increase in stored mechanical energy
> + increase in stored electrical energy   (2.60)

The energy balance is sometimes illustrated in a power flow diagram (**Fig. 2.19**).

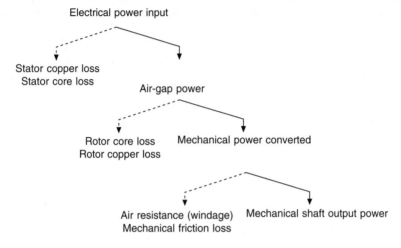

Fig. 2.19 Power flow diagram (example of a typical motor)

The overall efficiency of a system is the ratio of the useful output power to the total input power, in whatever form

$$\eta = \frac{P_{\text{out}}}{P_{\text{in}}} \qquad (2.61)$$

In an ac circuit, the instantaneous power depends on the instantaneous product of voltage and current. For sinusoidal voltage and current waveforms, the *apparent power S* (VA) is the product of the rms voltage and the rms current.

$$S = VI \qquad (2.62)$$

When the voltage waveform leads the current waveform by an angle ϕ, the average, *active* or *real power P*(W) is

$$P = VI \cos(\phi) \qquad (2.63)$$

The factor relating the real power to the apparent power is the *power factor* λ

$$\lambda = \cos(\phi) \qquad (2.64)$$

The real power is converted into heat or mechanical power. In addition there is an oscillating flow of instantaneous power, which is stored and then released each cycle by the capacitance and inductance in the circuit. This imaginary or *reactive power Q* (VAr) is given by

$$Q = VI \sin(\phi) \qquad (2.65)$$

By convention an inductive circuit (where the current lags the voltage) absorbs VAr, while a capacitive circuit (where the current leads the voltage) acts as a source of VAr. The relationship between S, P and Q is

$$S^2 = P^2 + Q^2 \qquad (2.66)$$

In a three-phase circuit, the total real power is the sum of the power flowing into each phase. For a balanced three-phase circuit with phase-neutral voltage V_P and phase current I_P the total power is the sum of the powers in each phase

$$P = 3V_P I_P \cos(\phi) \qquad (2.67)$$

In terms of the line voltage V_L, the real power is

$$P = \sqrt{3}V_L I_P \cos(\phi) \qquad (2.68)$$

Similar relations hold for the reactive and apparent power.

The power in a three-phase circuit can be measured using three wattmeters, one per phase, and the measurements added together (**Fig. 2.20(a)**). If it is known that the

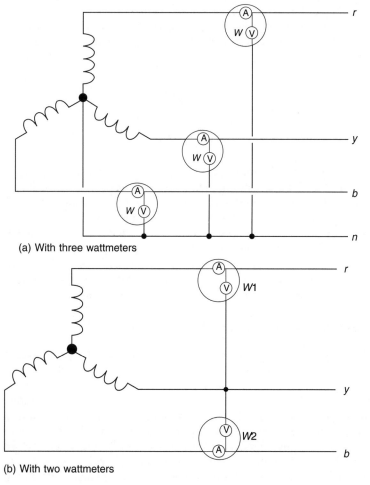

(a) With three wattmeters

(b) With two wattmeters

Fig. 2.20 Three-phase power measurement

load is balanced, with equal current and power in each phase, the measurement can be made with just two wattmeters (**Fig. 2.20(b)**). The total power is then

$$W = W_1 + W_2 \tag{2.69}$$

The two wattmeter arrangement also yields the power factor angle

$$\tan(\phi) = \frac{\sqrt{3}(W_1 - W_2)}{(W_1 + W_2)} \tag{2.70}$$

2.4.3 Per-unit notation

Power systems often involve transformers which step the voltage up or down. Transformers are very efficient, so that the output power from the transformer is only slightly less than the input power. Analysis and design of the circuit is made easier if the circuit values are normalized, such that the normalized values are the same on both sides of the transformers. This is acccomplished using the *per-unit* system. A given section of the power system operates at a certain *base voltage* V_B. Across transformers the voltage steps up or down according to the turns ratio. The base voltage will be different on each side of the transformer. A section of the system is allocated a base apparent power rating or *base VA* $V \cdot A_B$. This will be the same on both sides of the transformer. Combining these base quantities gives a base impedance Z_B and base current I_B

$$Z_B = \frac{V_B^2}{V \cdot A_B}$$

$$I_B = \frac{V \cdot A_B}{V_B} \tag{2.71}$$

The base impedance and base current change across a transformer.

All voltages and impedances in the system are normalized using the appropriate base value. The resulting normalized quantities are per-unit values. Once the circuit has been converted to per-unit values, the transformers have no effect on nominal tap position, and can be replaced by their own equivalent per-unit impedance.

The per-unit values are sometimes quoted as *per cent* values, by multiplying by 100 per cent.

A particular advantage of the per-unit and per cent notation is that the resulting impedances are very similar for equipment of very different sizes.

2.4.4 Energy transformation effects

Most electrical energy is generated by electromagnetic induction. However, electricity can be produced by other means than electromagnetic induction. Batteries use electrochemistry to produce low voltages (typically 1–2 V). An electrolyte is a solution of chemicals in water such that the chemical separates into positively and negatively charged ions when dissolved. The charged ions react with the conducting electrodes and release energy, as well as giving up their charge (**Fig. 2.21**). A fixed *electrode potential* is associated with the reaction at each electrode; the difference between the two electrode potentials drives a current around an external circuit. The electrolyte must be sealed into a safe container to make a suitable battery. 'Dry' cells use an electrolyte in the form of a gel or thick paste. A *primary cell* releases electricity as the chemicals react, and the cell is discarded once all the active chemicals have been used up, or the electrodes have become contaminated. A *secondary cell* uses a reversible chemical reaction, so that it can be recharged to regenerate the active chemicals. The *fuel cell* is a primary cell which is constructed so that the active chemicals (fuel) pass through the cell, and the cell can be used for long periods by replenishing the chemicals. Large batteries consist of cells connected in series or parallel to increase the output voltage or current. The main types of primary, secondary and fuel cell are described in **sections 12.2, 12.3** and **12.4**, respectively.

Electricity can be generated directly from heat. When two different materials are used in an electrical circuit, a small electrochemical voltage (contact potential) is generated at the junction. In most circuits these contact potentials cancel out around the circuit and no current flows. However, the junction potential varies with temperature, so that if one junction is at a different temperature from the others, the contact potentials will not cancel out and the net circuit voltage causes current to flow *(Seebeck effect)*. The available voltage is very small, but can be made more useful by

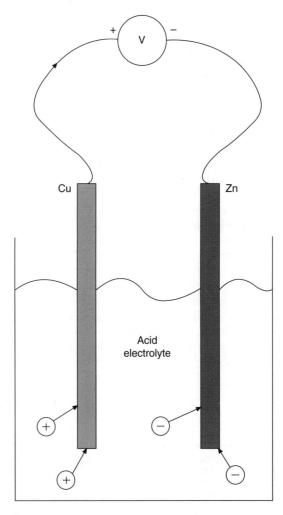

Fig. 2.21 Simple cell

connecting many pairs of hot and cold junctions in series. The *thermocouple* is used mostly for measurement of temperature by this effect, rather than for the generation of electrical power (**Fig. 2.22**). The efficiency of energy conversion is greater with semiconductor junctions, but metal junctions have a more consistent coefficient and are preferred for accurate measurements. The effect can be reversed with suitable materials, so that passing an electric current around the circuit makes one junction hotter and the other colder (*Peltier effect*). Such miniature heat pumps are used for cooling small components.

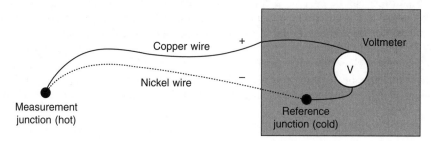

Fig. 2.22 A thermocouple (Seebeck effect)

Certain crystalline chemicals are made from charged ions of different sizes. When a voltage is applied across the crystal, the charged ions move slightly towards the side of opposite polarity, causing a small distortion of the crystal. Conversely, applying a force so as to distort the crystal moves the charged ions and generates a voltage. This *piezoelectric effect* is used to generate high voltages from a small mechanical force, but very little current is available. The use of piezoelectric sensors for the measurement of mechanical pressure and force is described in **section 4.42**. Ferromagnetic materials also distort slightly in a magnetic field. The *magnetostrictive effect* produces low frequency vibration (hum) in ac machines and transformers.

Electricity can be produced directly from light. The *photovoltaic effect* occurs when light falls on suitable materials, releasing electrons from the material and generating electricity. The magnitude of the effect is greater with short wavelength light (blue) than long wavelength light (red), and stops altogether beyond a wavelength threshold. Light falling on small photovoltaic cells is used for light measurement, communications and for proximity sensors. On a larger scale, semiconductor solar cells are being made with usable efficiency for power generation.

Light is produced from electricity in incandescent filament bulbs, by heating a wire to sufficiently high temperature that it glows. Fluorescent lights produce an electrical discharge through a low-pressure gas. The discharge emits ultraviolet radiation, which causes a fluorescent coating on the inside of the tube to glow.

References

2A.   Professional brief edited by Burns, R.W., Dellow, F. and Forbes, R.G., *Symbols and Abbreviations for use in Electrical and Electronic Engineering,* The Institution of Electrical Engineers, 1992.

2B.   BS 3939:1985, *Graphical symbols for electrical power, telecommunications and electronics diagrams,* BSI, 1985.

2C.   BS 5555:1993 (ISO 1000:1992), *SI units and recommendations for the use of their multiples and certain other units,* BSI, 1993.

2D.   BS 5775:1993 (ISO 32:1992), *Quantities, units and symbols, Part 5: electricity and magnetism,* BSI, 1993.

2E. Smith, R.J. and Dorf, R.C., *Circuits, Devices and Systems,* 5th edn, Wiley, 1992, ISBN 0-471-55221-6.

2F. Hughes, E. (revised Smith, I.M.), *Hughes Electrical Technology,* Longman Scientific and Technical, 1995, ISBN 0-582-22696-1.

2G. Bird, J.O., *Higher Electrical Technology,* Butterworth-Heinemann, 1994, ISBN 0-7506-01019.

2H. Breithaupt, J., *Understanding Physics for Advanced Level*, S. Thornes, 1990, ISBN 0-7487-0510-4.

2I. Del Toro, V., *Electrical Engineering Fundamentals,* Prentice-Hall, 1986, ISBN 0-13-247131-0.

Chapter 3

Materials for electrical engineering

Professor A.G. Clegg
Magnet Centre, University of Sunderland

Mr A.G. Whitman
Jones Stroud Insulations

3.1   Introduction

The performance of most types of electrical equipment (as opposed to electronic equipment) relies for their safe and efficient performance on an electrical circuit and the means to keep this circuit isolated from the surrounding materials and environment. Many types of equipment also have a magnetic circuit, which is linked to the electrical circuit by the laws outlined in **Chapter 2**.

The main material characteristics of relevance to electrical engineering are therefore those associated with conductors for the electrical circuit, with the insulation system necessary to isolate this circuit, and with the specialized steels and permanent magnets used for the magnetic circuit.

Although other properties, e.g. mechanical, thermal and chemical, are also relevant, these are often important in specialized cases and coverage of these properties is best left to other books which address these areas more broadly. The scope of this chapter is restricted to the main types and characteristics of conductors, insulation systems and magnetic materials which are used generally in electrical plant and equipment.

3.2   Magnetic materials

All materials have magnetic properties. These characteristic properties may be divided into five groups:

- diamagnetic
- paramagnetic
- ferromagnetic
- antiferromagnetic
- ferrimagnetic

Only the ferromagnetic and ferrimagnetic materials have properties that are useful in practical applications.

Ferromagnetic properties are confined almost entirely to iron, nickel and cobalt and their alloys. The only exceptions are some alloys of manganese and some of the rare earth elements.

Ferrimagnetism is the magnetism of the mixed oxides of the ferromagnetic elements. These are variously called ferrites and garnets. The basic ferrite is magnetite, or Fe_3O_4, which can be written as $FeO \cdot Fe_2O_3$. By substituting for the FeO with other divalent oxides, a wide range of compounds with useful properties can be produced. The main advantage of these materials is that they have high electrical resistivity which minimizes eddy currents when they are used at high frequencies.

The important parameters in magnetic materials can be defined as:

- *permeability* – this is the flux density B per unit of magnetic field H, as defined in **eqs 2.14** and **2.15**. It is usual and more convenient to quote the value of relative permeability μ_r, which is $B/\mu_0 H$. A curve showing the variation of permeability with magnetic field for a ferromagnetic material is given in **Fig. 3.1**. This is derived from the initial magnetization curve and it indicates that the permeability is a variable which is dependent on the magnetic field. The two important values are the *initial permeability*, which is the slope of the magnetization curve at $H = 0$, and the *maximum permeability*, corresponding to the knee of the magnetization curve.

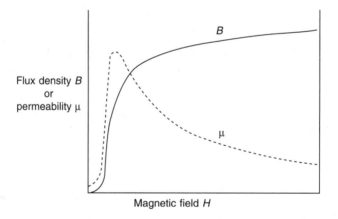

Fig. 3.1 Magnetization and permeability curves

- *saturation* – when sufficient field is applied to a magnetic material it becomes saturated. Any further increase in the field will not increase the magnetization and any increase in the flux density will be due to the added field. The *saturation magnetization* is M_s in A/m and J_s or B_s in tesla.
- *remanence, B_r and coercivity, H_c* – these are the points on the hysteresis loop shown in **Fig. 3.2** at which the field H is zero and the flux density B is zero, respectively. It is assumed that in passing round this loop the material has been saturated. If this is not the case, then an inner loop is traversed with lower values of remanence and coercivity.

Ferromagnetic and ferrimagnetic materials are characterized by moderate to high

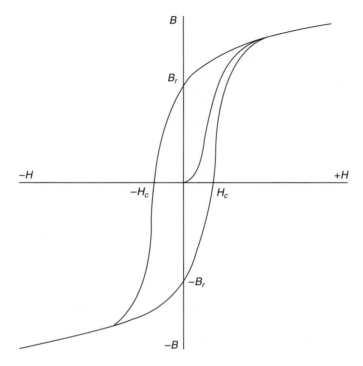

Fig. 3.2 Magnetization and hysteresis curves

permeabilities, as shown in **Table 3.1**. The permeability varies with the applied magnetic field, rising to a maximum at the knee of the *B–H* curve and reducing to a low value at very high fields. They also exhibit magnetic hysteresis whereby the intensity of magnetization of the material varies according to whether the field is being increased in a positive sense or decreased in a negative sense, as shown in **Fig. 3.2**. When the magnetization is cycled continuously round a hysteresis loop, as

Table 3.1 Properties of soft magnetic materials

	Maximum permeability μ $(\times10^{-3})$	Saturation magnetization J_s (T)	Coercivity H_c (A/m)	Curie temperature (°C)	Resistivity $(\Omega \cdot m \times 10^8)$
Sheet steels					
3% Si grain-oriented	90	2.0	6–7	745	48
2.5% Si non-oriented	8	2.0	40	745	44
<0.5% Si non-oriented	8	2.1	50–100	770	12
Low-carbon iron	3–10	2.1	50–120	770	12
Nickel–iron					
78% Ni	250–400	0.8	1.0	350	40
50% Ni	100	1.5–1.6	10	530	60
Amorphous					
Iron-based	35–600	1.3–1.8	1.0–1.6	310–415	120–140
Ferrite					$(\Omega \cdot m)$
MnZn	1.8–3.0	0.45–0.50	15–27	110–220	0.05–10
NiZn	0.12–1.5	0.20–0.35	30–300	250–400	$>10^4$
Garnet	–	0.3–1.7	–	11–280	$>10^6$

for example when the applied field arises from an alternating current, there is an energy loss proportional to the area of the included loop. This is the *hysteresis loss*, and it is measured in joules/m^3. High hysteresis loss is associated with permanent magnetic characteristics exhibited by materials commonly termed '*hard*' magnetic materials, as these often have hard mechanical properties. Those materials with low hysteresis loss are termed '*soft*' and are difficult to magnetize permanently.

Ferromagnetic or ferrimagnetic properties disappear reversibly if the material is heated above the Curie temperature, at which point it becomes *paramagnetic,* that is effectively non-magnetic.

3.2.1 Soft (high-permeability) materials

There is a wide variety of soft magnetic materials for applications from constant dc field through 50 Hz up to microwave frequencies. For the bulk of applications iron, steel or cast iron are used. They have the advantage of low cost and strength, but they should only be used for dc applications since they have low electrical resistivity which would result in eddy currents if used in alternating fields.

In choosing a high-permeability material for a particular application there may be special considerations and the following is a guide to the choice. If the frequency of the applied voltage is 10 kHz or above then ferrites or garnets will normally be used. For constant-field applications mild steel will be used, or low-carbon iron where the highest permeability or lowest coercivity is needed. For 50 Hz power transformers, grain-oriented silicon steel is used but there is now serious competition from amorphous strip and although this is more expensive the core loss is significantly lower than that of silicon steel. For the highest permeability and lowest coercivity in specialist applications up to about 10 kHz nickel iron would be the preferred choice although its cost may be prohibitive.

3.2.1.1 Sheet steels

There are a number of grades of sheet steel and these comprise by far the greatest part of the soft, high-permeability material which is used. These metallic materials have a comparatively low electrical resistivity and they are used in sheet (or lamination) form because this limits the flow of eddy currents and the losses which result. The range of sheet thickness used for 50 or 60 Hz applications is 0.35–0.65 mm for non-oriented materials and 0.13–0.35 mm for grain-oriented silicon steels. For higher frequencies of 400–1000 Hz thicknesses of 0.05–0.20 mm are used.

In grain-oriented steel an increasing level of silicon content reduces the losses but also reduces the permeability; for many applications a 3 per cent silicon content represents a good balance. The effect of the silicon is to increase the electrical resistivity of the steel; not only does this reduce eddy current losses but it also improves the stability of the steel and aids the production of grain orientation. The sheet is subject to cold rolling and a complex annealing treatment to produce the grain orientation in the rolling direction and this also gives improved magnetic permeability in that direction. The properties are further improved at the annealing stage with a glass film on the surface which holds the steel in a state of tension and provides electrical insulation between laminations in a core. A phosphate coating is applied to complete the tensioning and insulation. The grain size in the resulting sheets is comparatively large and the domain boundaries are quite widely spaced. Artificial grain boundaries can be produced by laying down lines of ablated spots on the steel surface; this gives a stress and atomic disruption pattern which pins domain walls and leads to a smaller wall spacing. Various methods have been used to produce this ablating, including spark and laser techniques. All these processes are applied within the steel manufacturer's works and the resulting steel is often referred to as *fully processed*. The main application of this grain-oriented material is in power transformers where low power loss is important, since the transformer is always connected even when its loading is at a minimum.

For rotating machinery, especially motors rated more than about 100 kW, non-oriented steels with a lower silicon content are used. Whilst efficiency remains important in this application, high permeability is now also important in order to minimize magnetizing current and to maximize torque. This material is also used in smaller transformers, chokes for fluorescent tubes, meters and magnetic shielding.

For smaller motors silicon-free non-oriented steels are often used. These are produced by the steel manufacturer with a relatively high carbon content. The carbon makes the sheet sufficiently hard for punching into laminations and after punching the material is decarburized and annealed to increase the grain size. Because of the need for this secondary processing the materials are often known as *semi-processed*. These materials are generally cheaper than silicon steels but in their finished form they have a much higher permeability; in small motors particularly this can be more important than efficiency. Other applications include relays and magnetic clutches.

Silicon steels are also produced in the form of bars, rods or wires for relays, stepping motors and gyroscope housings. The tensile strength of the silicon steels may be improved by the addition of alloying elements such as manganese; this type of material is used in highly stressed parts of the magnetic circuit in high-speed motors or generators.

3.2.1.2 Amorphous alloys

This class of alloys, often called *metallic glasses*, is now an established group of soft magnetic materials. The materials are produced by rapid solidification of the alloy at cooling rates of about a million degrees centigrade per second. The alloys solidify with a glass-like atomic structure which is a non-crystalline frozen liquid. The rapid cooling is achieved by causing the molten alloy to flow through an orifice onto a rapidly rotating water-cooled drum. This can produce sheets as thin as 10 μm and a metre or more wide.

There are two main groups of amorphous alloys. The first is the iron-rich group of magnetic alloys, which have the highest saturation magnetization among the amorphous alloys and are based on inexpensive raw materials. Iron-rich alloys are currently being used in long-term tests in power transformers in the USA. The second is the cobalt-based group, which has very low or zero magnetostriction, leading to the highest permeability and the lowest core loss. Cobalt-based alloys are used for a variety of high-frequency applications including pulsing devices and tape recorder heads, where their mechanical hardness provides excellent wear resistance.

All of these alloys have a resistivity which is higher than that of conventional crystalline electrical steels. Because of this, eddy current losses are minimized both at 50 Hz and at higher frequencies. The alloys have other advantages including flexibility without loss of hardness, high tensile strength and better corrosion resistance than similar crystalline materials.

3.2.1.3   Nickel–iron alloys

The very high magnetic permeability and low coercivity of nickel–iron alloys are
due to two fundamental properties, which are *magnetostriction* and *magnetic anisotropy*.
At a nickel content of about 78 per cent both of these parameters are zero.
Magnetostriction has been briefly referred to in **section 2.4.4**; it is the change of
dimensions in a material due to magnetization, and when this is zero there are no
internal stresses induced during magnetizing. Magnetic anisotropy is the difference
between magnetic behaviour in different directions; when this is zero the magnetization
increases steeply under the influence of a magnetic field, independent of crystal
direction. The alloys with about 78 per cent nickel content are variously called
Mumetal or *Permalloy*.

The commercial alloys in this class have additions of chromium, copper and
molybdenum in order to increase the resistivity and improve the magnetic properties.
Applications include special transformers, circuit breakers, magnetic recording heads
and magnetic shielding.

Fifty per cent nickel–iron alloys have the highest saturation magnetization of this
class of materials, and this results in the best flux-carrying capacity. They have a
higher permeability and better corrosion resistance than silicon iron materials, but
they are more expensive. A wide range of properties can be produced by various
processing techniques. Severe cold reduction produces a cube texture and a square
hysteresis loop in annealed strip. The properties can also be tailored by annealing the
material below the Curie temperature in a magnetic field. Applications for this material
include chokes, relays and small motors.

3.2.1.4   Ferrites and garnets

Ferrites are iron oxide compounds containing one or more other metal oxide. The
important high magnetic permeability materials are manganese zinc ferrite and nickel
zinc ferrite. They are prepared from the constituent oxides in powder form, preferably
of the same particle size intimately mixed. The mixture is fired at about 1000°C and
this is followed by crushing, milling and then pressing of the powder in a die or
extrusion to the required shape. The resulting compact is a black brittle ceramic and
any subsequent machining must be by grinding. The materials may be prepared with
high permeability, and because their high electric resistivity limits eddy currents to
a negligible level they can be used at frequencies up to 20 MHz as the solid core of
inductors or transformers.

A combination of hysteresis, eddy current and residual losses occurs, and these
components may be separately controlled by composition and processing conditions,
taking into account the required permeability and the working frequency.

The saturation flux density of ferrites is relatively low, making them unsuitable
for power applications. Their use is therefore almost entirely in the electronic and
telecommunications industry where they have now largely replaced laminated alloy
and powder cores.

Garnets are used for frequencies of 100 MHz and above. They have resistivities
in excess of 10^8 $\Omega \cdot$ m compared with ferrite resistivity which is up to 10^3 $\Omega \cdot$ m.
Since eddy current losses are limited by resistivity, they are greatly reduced in
garnets. The basic Yttrium–Iron Garnet (YIG) composition is $3Y_2O_3 \cdot 5Fe_2O_3$. This
is modified to obtain improved properties such as very low loss and greater temperature
stability by the addition of other elements including Al and Gd. The materials are

prepared by heating of the mixed oxides under pressure at over 1300°C for up to 10 hours. Garnets are used in microwave circuits in filters, isolators, circulators and mixers.

3.2.1.5   Soft magnetic composites

These composites (SMC) consist of iron or iron alloy powder mixed with a binder and a small amount of lubricant. The composite is pressed into the final shape which can be quite complex. The lubricant reduces the friction during pressing and aids the ejection of the part from the die. After pressing the parts are either cured at 150 to 275°C or heat treated at 500°C.

The isolation of the particles within the binder minimizes the eddy currents and makes the components very suitable for medium frequency applications up to 1 kHz or more. They have the added advantages of being isotropic and can be used for components with quite complex shapes.

3.2.2   Hard (permanent magnet) materials

The key properties of a permanent magnet material are given by the demagnetization curve, which is the hysteresis curve in the second quadrant between B_r and $-H_c$. It can be shown that when a piece of permanent magnet material is a part of a magnetic circuit, the magnetic field generated in a gap in the circuit is proportional to $B \times H \times V$, where B and H represent a point on the demagnetization curve and V is the volume of permanent magnet. To obtain a given field with a minimum volume of magnet the product $B \times H$ must therefore be a maximum, and the $(BH)_{max}$ value is useful in comparing material characteristics.

The original permanent magnet materials were steels, but these have now been superseded by better and more stable materials including *Alnico, ferrites* and *rare earth* alloys. The magnetic properties of all the permanent magnet materials are summarized in **Table 3.2**.

3.2.2.1   Alnico alloys

A wide range of alloys with magnetically useful properties is based on the Al–Ni–Co–Fe system. These alloys are characterized by high remanence, high available energy and moderately high coercivity. They have a low and reversible temperature coefficient of about –0.02 per cent/°C and the widest useful temperature range (up to over 500°C) of any permanent magnet material.

The alloys are produced either by melting or sintering together the constituent elements. Anisotropy is achieved by heating to a high temperature and allowing the material to cool at a controlled rate in a magnetic field in the direction in which the magnets are to be magnetized. The properties are much improved in this direction at the expense of properties in the other directions. This is followed by a tempering treatment in the range 650–550°C. A range of coercivities can be produced by varying the cobalt content. The properties in the preferred direction may be further improved by producing an alloy with columnar crystals.

3.2.2.2   Ferrites

The permanent magnet ferrites are also called ceramics and they are mixtures of ferric oxide and an oxide of a divalent heavy metal, usually barium or strontium. These ferrites are made by mixing together barium or strontium carbonate with iron

Table 3.2 Characteristics of permanent magnet materials

	Reman-ence B_r (T)	Energy product $(BH)_{max}$ (kJ/m³)	Coercivity H_c (kA/m)	Relative recoil permea-bility	Curie tempera-ture (°C)	Resistivity ($\Omega \cdot m \times 10^8$)
Alnico						
Normal anisotropic	1.1–1.3	36–43	46–60	2.6–4.4	800–850	50
High coercivity	0.8–0.9	32–46	95–150	2.0–2.8	800–850	50
Columnar	1.35	60	60–130	1.8	800–850	50
Ferrites (ceramics)						
Barium isotropic	0.22	8	130–155 (a)	1.2	450	10^{12}
Barium anisotropic	0.39	28.5	150 (a)	1.05	450	10^{12}
Strontium anisotropic	0.36–0.43	24–34	240–300 (a)	1.05	450	10^{12}
Bonded ferrite						
Isotropic	0.14	3.2	90 (a)	1.1	450	*
Anisotropic	0.23–0.27	10–14	180 (a)	1.05	450	*
Rare earth						
SmCo₅ sintered	0.9	160	600–660 (b)	1.05	>700	50–60
SmCo₅ bonded	0.5–0.6	56–64	400–460 (b)	1.1	>700	*
Sm₂Co₁₇ sintered	1.1	150–240	600–700 (c)	1.05	>700	75–85
NdFeB sintered	0.9–1.4	160–360	750–920 (d)	1.05	310	140–160
NdFeB bonded	0.25–0.8	10–95	180–460 (e)	1.05	310	*

Intrinsic coercivities: (a) 160–340 kA/m   * Dependent on the resistivity of the polymer bond.
(b) 800–1500 kA/m
(c) 600–1300 kA/m
(d) 950–3000 kA/m
(e) 460–1300 kA/m

oxide. The mixture is fired and the resulting material is milled to a particle size of about 1 µm. The powder is then pressed to the required shape in a die and anisotropic magnets are produced by applying a magnetic field in the pressing direction. The resulting compact is then fired.

3.2.2.3 Rare earth alloys

The $(BH)_{max}$ values that can be achieved with rare earth alloys are 4–6 times greater than those for Alnico or ferrite.

There are three main permanent magnet rare earth alloys, two of which (SmCo₅ and Sm₂Co₁₇) are based on samarium and cobalt, the other being neodymium iron boron (NdFeB). These materials may be produced by alloying the constituent elements together, or more usually by reducing a mixture of the oxides together in a hydrogen atmosphere using calcium as the reducing agent. The alloy is then milled to a particle size of about 10 µm, pressed in a magnetic field and sintered in vacuum.

SmCo₅ was the first alloy to be available, but this has gradually been replaced by Sm₂Co₁₇ because of its lower cost and better temperature stability. The more recently developed NdFeB magnets have the advantage of higher remanence B_r and higher $(BH)_{max}$, and they are lower in cost because the raw materials are cheaper. The disadvantage of NdFeB materials is that they are subject to corrosion and they suffer from a rapid change of magnetic properties (particularly coercivity) with temperature. Corrosion can be prevented by coating the magnets and the properties at elevated temperature may be improved by small additions of other elements.

3.2.2.4   Bonded magnets

Ferrites and rare earth magnets are also produced in bonded forms. The magnet powder particles are mixed with the bond and the resulting compact can be rolled, pressed or injection moulded. For a flexible magnet the bonds may be rubber, or for a rigid magnet they may be nylon, polypropylene, resin or other polymers. Although the magnetic properties are reduced by the bond, they can be easily cut or sliced and in contrast to the sintered magnets they are not subject to cracking or chipping. Rolled or pressed magnets give the best properties as some anisotropy may be induced, but injection moulding is sometimes preferred to produce complex shapes which might even incorporate other components. Injection moulding also produces a precise shape with no waste of material.

3.2.2.5   Applications

Permanent magnets have a very wide range of applications and virtually every part of industry and commerce uses them to some extent. At one time Alnico was the only available high-energy material, but it has gradually been replaced by ferrites and rare earth alloys except in high-stability applications. Ferrites are much cheaper than Alnico but because of their lower flux density and energy product a larger magnet is often required. However, about 70 per cent of magnets used are ferrite. They find bulk applications in loudspeakers, small motors and generators and a wide range of electronic applications. The rare earth alloys are more expensive but despite this and because of their much greater strength they are being used in increasing quantities. They give the opportunity for miniaturization, the 'Walkman' portable stereo player being an example. **Figures 3.3** and **3.4** show the range of applications for permanent magnets in the home and in a car.

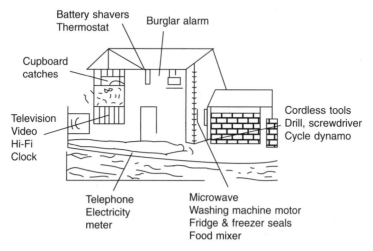

Fig. 3.3 Magnets in the home

3.2.3   Other materials

In addition to the two main groups of soft and hard magnetic materials there are other materials which meet special needs.

The feebly magnetic steels are austenitic, and their virtually non-magnetic properties are achieved by additions of chromium and nickel to a low-carbon steel. To attain a

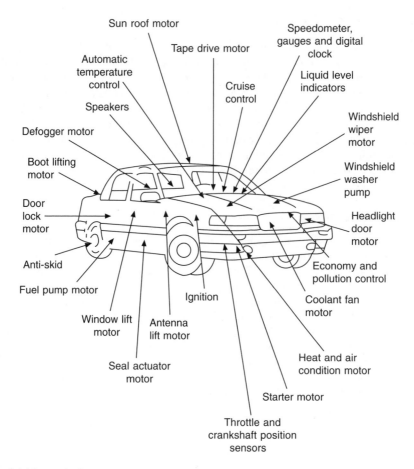

Fig. 3.4 Magnets in the car

relative permeability of 1.05 or less the recommended composition is 18 per cent chromium and 10 per cent nickel, or greater. These steels, which have minimum strength requirements, are used for non-magnetic parts of machinery, for magnetic measuring equipment and for minesweeping equipment, where magnetic flux can actuate magnetic mines.

Magnetic recording makes use of fine magnetic particles which are embedded in the tape or disc. These particles are of metal or of iron or chromium oxide and the choice depends on a compromise between price and quality. The heads used for magnetic recording are usually made from high-permeability ferrites, but amorphous metal is now also being used. Magnetic storage is a rapidly expanding area with higher and higher information densities being achieved.

3.2.4 Standards

The main international standard for magnetic materials is IEC 60404, to which BS 6404 is equivalent. This standard has many parts with specifications for the properties of silicon steels, nickel irons and permanent magnets. Also included are measurement standards for these materials.

3.3   Insulating materials

The reason for using insulating materials is to separate electrically the conducting parts of equipment from each other and from earthed components. Earthed components may include the mechanical casing or structure that is necessary to enable the equipment to be handled and to operate. Whereas the 'active' parts of the equipment play a useful role in its operation, the insulation is in many ways a necessary evil. For example, in an electric motor the copper of the winding and the steel core making up the magnetic circuit are the active components and both contribute to the power output of the motor; the insulation which keeps these two components apart contributes nothing, in fact it takes up valuable space and it may be considered by the designer as not much more than a nuisance.

For these reasons, insulating materials have become a design focus in many types of electrical equipment, with many companies employing specialists in this field and carrying out sophisticated life testing of insulation systems. Such is the importance attached to this field that major international conferences on the subject are held regularly, for instance by the IEEE in USA and by IEE and BEAMA in the UK.

The simplest way to define an insulating material is to state what it is not. It is not a good conductor of electricity and it has a high electrical resistance that decreases with rising temperature, unlike conductors. The following are the most important properties of insulating materials:

- *volume resistivity*, which is also known as specific resistance
- *relative permittivity* (or dielectric constant), which is defined as the ratio of the electric flux density produced in the material to that produced in a vacuum by the same electric field strength. The definitions have been set down in **eqs 2.1** and **2.2.** Relative permittivity can be expressed as the ratio of the capacitance of a capacitor made of that material to that of the same capacitor using a vacuum as its dielectric (see **eqn 2.11**).
- *dielectric loss* (or electrical dissipation factor), which is defined as the ratio of the power loss in a dielectric material to the total power transmitted through it. It is given by the tangent of the loss angle and is commonly known as *tan delta*. Tan delta has been defined in **eqn 2.43**.

The volume resistivity, relative permittivity and tan delta values for a range of insulating materials are shown in **Table 3.3**.

The most important characteristic of an insulating material is its ability to withstand electric stress without breaking down. This ability is sometimes known as its *dielectric strength*, and is usually quoted in kV/mm. Typical values may range from 5 to 100 kV/mm, but it is dependent on a number of other factors which include the speed of application of the electric field, the length of time for which it is applied, temperature and whether ac or dc voltage is used.

Another significant aspect of all insulating materials that dominates the way in which they are categorized is the maximum temperature at which they will perform satisfactorily. Generally speaking, insulating materials deteriorate more quickly at higher temperatures and the deterioration can reach a point at which the insulation ceases to perform its required function. This characteristic is known as ageing, and for each material it has been usual to assign a maximum temperature beyond which it is unwise to operate if a reasonable life is to be achieved. The main gradings or

Table 3.3 Representative properties of typical insulating materials

	Volume resistivity $(\Omega \cdot m)$	Relative permittivity	Tan delta (at 50 Hz)
vacuum	infinity	1.0	0
air	infinity	1.0006	0
mineral insulating oil	10^{11}–10^{13}	2.0–2.5	0.0002
pressboard	10^{8}	3.1	0.013
dry paper	10^{10}	1.9–2.9	0.005
oiled paper	–	2.8–4.0	0.005
porcelain	10^{10}–10^{12}	5.0–7.0	–
E-glass	10^{16}	6.1–6.7	0.002–0.005
polyester resin	10^{14}–10^{16}	2.8–4.1	0.008–0.041
epoxy resin	10^{12}–10^{15}	3.5–4.5	0.01
mica	10^{11}–10^{15}	4.5–7.0	0.0003
micapaper	10^{13}–10^{17}	5.0–8.7	0.0003
PETP film	10^{18}	3.3	0.0025
aramid paper	10^{16}	2.5–3.5	0.005–0.020
epoxy glass laminate	–	4.5–4.7	0.008
silicone glass laminate	–	4.5–6.0	0.003
polystyrene	10^{15}	2.6	0.0002
polyethylene	10^{15}	2.3	0.0001
methyl methacrylate	10^{13}	2.8	0.06
polyvinyl chloride	10^{11}	5.0–7.0	0.1
fused quartz	10^{16}	3.9	–

classes of insulation as defined in IEC 60085 and its UK equivalent BS 2757 are listed in **Table 3.4**. Where a thermal class is used to describe an item of electrical equipment, it normally represents the maximum temperature found within that product under rated load and other conditions. However, not all the insulation is necessarily located at the point of maximum temperature, and insulation with a lower thermal classification may be used in other parts of the equipment.

Table 3.4 Thermal classes for insulation

Thermal class	Operating temperature (°C)
Y	90
A	105
E	120
B	130
F	155
H	180
200	200
220	220
250	250

The ageing of insulation depends not only on the physical and chemical properties of the material and the thermal stress to which it is exposed, but also on the presence and degree of influence of mechanical, electrical and environmental stresses. The processing of the material during manufacture and the way in which it is used in the complete equipment may also significantly affect the ageing process. The definition of a useful lifetime will also vary according to the type and usage of equipment; for instance the running hours of a domestic appliance and a power station generator will be very different over a 25-year period. All of these factors should therefore influence the choice of insulating material for a particular application.

There is therefore a general movement in the development of standards and methods of testing for insulating materials towards the consideration of combinations of materials or *insulating systems*, rather than focusing on individual materials. It is not uncommon to consider life testing in which more than one form of stress is introduced; this is known as *multifunctional* or *multifactor testing*.

Primary insulation is often taken to mean the main insulation, as in the PVC coating on a live conductor or wire. *Secondary* insulation refers to a second 'line of defence' which ensures that even if the primary insulation is damaged, the exposed live component does not cause an outer metal casing to become live. Sleeving is frequently used as a secondary insulation.

Insulating materials may be divided into basic groups which are *solid dielectrics, liquid dielectrics*, *gas* and *vacuum*. Each is covered separately in the following sections.

3.3.1 Solid dielectrics

Solid dielectric insulating materials have in the past (for instance in BS 5691 part 2) been subdivided into three general groups:

- solid insulation of all forms not undergoing a transformation during application
- solid sheet insulation for winding or stacking, obtained by bonding superimposed layers
- insulation which is solid in its final state, but is applied in the form of a liquid or paste for filling, varnishing or bonding

A more convenient and up-to-date way to subdivide this very large group of materials is used by the IEC Technical Committee 15: Insulating Materials:

- inorganic (ceramic and glass) materials
- plastic films
- flexible insulating sleeving
- rigid fibrous reinforced laminates
- resins and varnishes
- pressure-sensitive adhesive tapes
- cellulosic materials
- combined flexible materials
- mica products

This subdivision is organized on the basis of application and is therefore more helpful to the practising engineer. A brief description of each of these classes of material is given in the following sections.

3.3.1.1 Inorganic (ceramic and glass) materials

A major application for materials in this category is in high-voltage overhead lines as suspension insulators (see **Figs 14.2** and **14.8**), or as bushings on high-voltage transformers and switchgear (see **Fig. 6.14**). In either case the material is formed into the well-known series of flanged discs to increase the creepage distance along the surface of the complete insulator. Ceramic materials are used for a number of reasons including:

- ease of production of a wide range of shapes
- good electrical breakdown strength
- retention of insulating characteristics in the event of surface damage

3.3.1.2   Plastic films

Materials such as polyethylene terephthalate (PETP, more commonly referred to as polyester), polycarbonate, polyimide and polyethylene naphthalate (PEN) have been used as films in a variety of applications such as the insulation between foils in capacitors, slot insulation in rotating electrical machines (either by themselves or as a composite with other sheet materials) and more recently as a backing for mica-based products used in the insulation of high-voltage equipment. Plastic films are used in applications requiring dimensional stability, high dielectric strength, moisture resistance and physical toughness.

3.3.1.3   Flexible insulating sleeving

This fulfils a number of requirements including the provision of primary or secondary electrical insulation of component wiring, the protection of cables and components from the deleterious effects of mechanical and thermal damage, and as a rapid and low-cost method of bunching and containing cables. Sleevings may find application in electrical machines, transformers, domestic and heating appliances, light fittings, cable connections, switchgear and as wiring harnesses in domestic appliances and in vehicles. They are used because of their ease of application, flexibility and high dielectric strength; they lend themselves to an extremely wide range of formats including shrink sleeving, expandable constructions and textile-reinforced grades for low and high voltage and temperatures across the range $-70°C$ to $+450°C$.

3.3.1.4   Rigid fibrous reinforced laminates

In the manufacture of most electrical equipment there is a need for items machined out of solid board or in the form of tubes and rods. These items can take the form of densified wood or laminates of paper, woven cotton, glass or polyester, or glass or polyester random mats laminated together with a thermosetting resin which might be phenolic, epoxy, polyester, melamine, silicone or polyimide, depending upon the properties required. Rigid boards are used because they are capable of being machined to size, and they retain their shape and properties during their service life, unlike unseasoned timber which was used in early equipment.

3.3.1.5   Resins and varnishes

In addition to their use in the laminates outlined in **section 3.3.1.4**, a wide range of varnishes and resins are used by themselves in the impregnation and coating of electrical equipment in order to improve its resistance to working conditions, to enhance its electrical characteristics and to increase its working life. At first many resins and varnishes were based on naturally occurring materials such as bitumen, shellac and vegetable oils, but now they are synthetically produced in a comprehensive range of thermoplastic, thermosetting and elastomeric forms. The more common types are phenolic, polyester, epoxy, silicone and polyimide, and these can be formulated to provide the most suitable processing and the required final characteristics. Varnishes and resins are used because of their ability to impregnate, coat and bond basic insulating materials; this assists in the application of the insulating materials and it significantly improves their service life and their ability to withstand dirt and moisture.

3.3.1.6 Pressure-sensitive adhesive tapes

Certain types of pressure-sensitive adhesive (PSA) tapes have become so much a part of modern life that the trade names have been absorbed into everyday language. The usefulness of PSA in short lengths for holding down, sealing or locating applies equally in the field of insulation and a range of tapes has been developed which is based on paper, film or woven glass cloth, coated with suitable adhesive such as rubber, silicone or acrylic.

3.3.1.7 Cellulosic materials

Materials in the form of papers, pressboards and presspapers continue to play a vital role in oil-filled power transformers. Included in this area are other materials which are produced by paper-making techniques, but which use aramid fibres; these materials have found wide application in high-temperature and dry-type transformers as well as in other types of electrical equipment. Cellulose materials are mainly used in conjunction with oil, and it is their porous nature that lends itself to successful use in transformers and cables.

3.3.1.8 Combined flexible materials

In order to produce suitable materials with the required properties such as tear strength, electric strength and thermal resistance at an acceptable price, a range of laminated or combined flexible sheet products has been developed. These employ cellulosic, aramid and glass fleeces as well as other materials, in combination with many of the plastic films already referred to, in a range of forms to suit the application. These products are used in large quantities in low-voltage electric motors.

3.3.1.9 Mica products

Materials based on mica, which is a naturally occurring mineral, play a central part in the design and manufacture of high-voltage rotating machines. Originally the material was in the form of mica splittings, but at present the industry uses predominantly micapaper, which is produced by breaking down the mica into small platelets by chemical or mechanical means, producing a slurry and then feeding this through a traditional paper-making machine. The resulting micapaper, when suitably supported by a woven glass or film backing and impregnated with epoxy or a similar resin, is used to insulate the copper bars which make up the stator winding of the machine. Micapaper is used in the ground insulation of the winding which is shown as part of the stator slot section illustrated in **Fig. 3.5**.

Micapaper is available in a resin-rich form, in which all the necessary resin for consolidation of the insulation around the winding is included within the material. This consolidation is usually carried out in large steam or electrically heated processes into which an insulated coil side or bar can be placed; heat and pressure are then

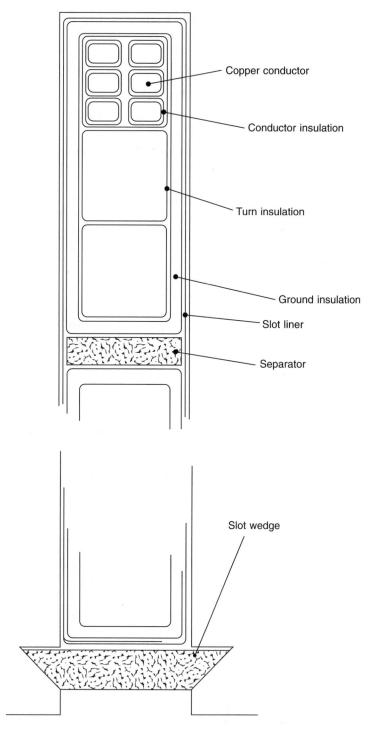

Fig. 3.5 Stator coil section of a high-voltage electrical machine (courtesy of ALSTOM Electrical Machines Ltd)

applied as necessary to cure fully the resin-rich micapaper insulation. Alternatively it is available for use with vacuum pressure impregnation (VPI), in which most of the resin is introduced after winding the machine. The use of VPI eliminates the need to consolidate the insulation in a press; consolidation is achieved either by the use of hydrostatic forces or by ensuring full impregnation of the bars or coils already placed or wound into slots. A large electrical machine stator is shown being lowered into a VPI tank in **Fig. 3.6**.

Fig. 3.6 A large vacuum and pressure impregnation facility capable of treating stators up to 100 MVA (courtesy of ALSTOM Electrical Machines Ltd)

Mica-based products dominate high-voltage insulation systems because of their unique combination of properties, which include:

- high dielectric strength
- low dielectric loss at high frequency
- high surface and volume resistivity
- excellent resistance to corona discharge and electric arc erosion
- temperature capability from –273°C to 1000°C
- flame resistance
- excellent chemical resistance
- high resistance to compressive forces

3.3.1.10   Textile insulation

Although the use of fully varnished fabric is becoming less common, products using glass and polyester-based yarn, and to a lesser extent cotton and rayon, are still in use.

A much larger range of unvarnished narrow-fabric products, more commonly called woven tapes, exist and these use a variety of combinations of different glass and polyester yarns tailored to meet specific applications. Primarily these tapes are used for finishing on top of other insulation such as micapaper, in order to provide a tough outer surface which can readily be coated with a final varnish or paint finish. When manufactured on modern shuttle-less looms they are an economic proposition.

Woven tapes are used because of their ease of application, good conformity and bedding down. An example of the complex shapes that they can be used to cover is shown in **Fig. 3.7**, which illustrates the endwinding of a high-voltage electrical machine.

3.3.1.11   Elastomers and thermoplastics

There is a very wide range of polymeric and rubber-like insulation materials. These have traditionally been dealt with in connection with electric cables and IEC and BSI reflect this by dealing with them separately in Technical Committee 20: Electrical Cables. Some elastomers such as silicone have found application in sleeving, traction systems and increasingly as overhead line insulators, but the bulk of their application continues to be related to cables. The leading materials such as PVC, MDPE, XLPE and EPR are therefore referred to in **Chapter 9**.

3.3.2   Liquid dielectrics

A liquid dielectric remains in the liquid state throughout its working life, unlike resins and varnishes which become solid after processing.

The principal uses of liquid dielectrics are as a filling and cooling medium for transformers, capacitors and rheostats, as an arc-quenching medium in switchgear and as an impregnant of absorbent insulation used mainly in transformers, switchgear, capacitors and cables. The important properties of dielectric liquids are therefore electric strength, viscosity, chemical stability and flashpoint.

Typical materials include highly refined hydrocarbon mineral oils obtained from selected crude petroleum, silicone fluids, synthetic esters and hydrocarbons with high molecular weight. A specially interesting material for cables has been the waxy Mineral Insulating Non-Draining (MIND) compound which has been used in paper-insulated cables; this is described in **section 9.2.3**. A group of polychlorinated biphenyls (PCBs) has been used in transformers, but these materials are now being replaced and carefully disposed of because of their toxic nature and their resistance to biological and chemical degradation.

3.3.3   Gas insulation

Two gases already in common use for insulation are nitrogen and sulphur hexafluoride (SF_6). Nitrogen is used as an insulating medium in some sealed transformers, while SF_6 is finding increasing use in transmission and distribution switchgear because of its insulating properties and its arc-extinguishing capabilities; this is described further in **sections 7.4.1** and **7.5.3(b)**.

Fig. 3.7 Stator endwinding bracing system for a 4-pole, 11 kV, 8 MW induction motor (courtesy of ALSTOM Electrical Machines Ltd)

3.3.4 Vacuum insulation

Vacuum insulation is now used in a range of medium-voltage switchgear. Like SF_6, it has both insulating and arc-extinguishing properties. The action of the vacuum in a circuit breaker is explained in **section 7.4.1**.

3.3.5 Standards

A selection of national and international standards covering the insulation field is given in **Table 3.5**. The majority of the IEC standards with their EN and BSEN equivalents consist of three parts:

Table 3.5 National and international standards relating to insulating materials

IEC	EN/HD	BS	Subject	N. American
60085	HD 566S1	2757	Thermal evaluation and classification of electrical insulation	
60216	HD 611.1S1	5691	Determination of thermal endurance properties of electrical insulating materials	
60243	EN 60243	EN 60243	Methods of test for electric strength of solid insulating materials	
60371	EN 60371	EN 60371	Specifications for insulating materials based on mica	
60394		5689	Varnish fabrics for electrical purposes	
60454	EN 60454	EN 60454	Specifications for pressure-sensitive adhesive tapes for electrical purposes	
60455	EN 60455	EN 60455 5664	Specifications for solventless polymerizable resinous compounds used for electrical insulation	
60464	EN 60464	5629	Specifications for insulating varnishes containing solvent	
60626	EN 60626	EN 60626	Combined flexible materials for electrical insulation	
60641	EN 60641	EN 60641	Specification for pressboard and presspaper for electrical purposes	
60672	EN 60672	EN 60672	Specifications for ceramic and glass insulating materials	
60674	EN 60674	EN 60674	Specifications for plastic films for electrical purposes	
60684	EN 60684	EN 60684	Specifications for flexible insulating sleevings	NEMA VS–1
60819	EN 60819	EN 60819-1 7824	Non-cellulosic papers for electrical purposes	
60893	EN 60893	EN 60893	Specifications for industrial rigid laminated sheets based on thermosetting resins for electrical purposes	NEMA LI–1
61067	EN 61067	EN 61067	Specifications for glass and glass polyester fibre woven tapes	
61068	EN 61068	EN 61068	Polyester fibre woven tapes	
61212	EN 61212	EN 61212 6128	Industrial rigid laminated tubes and rods based on thermosetting resins for electrical purposes	

- Part 1: Definitions, classification and general requirements
- Part 2: Methods of test
- Part 3: Specifications for individual materials

3.4 Conducting materials

Strictly, conducting materials fall into three groups, which are conductors, semiconductors and imperfect insulators. Insulators have been covered in **section 3.3**, so the focus here is on conductors and semiconductors.

3.4.1 Conductors

In general, metals and alloys are conductors of electricity. The conductivity in metals such as copper and aluminium is due to electrons which are attracted to the positive terminal when a voltage is applied. The freedom with which the electrons can move determines the conductivity and resistivity. The restraints on electron movements are impurities, stresses and thermal lattice vibrations, so to obtain the highest conductivity the metal must be very pure and in the annealed state. With increasing temperature the thermal lattice vibrations increase and conductivity is therefore reduced.

The principal materials for commercial application as conductors are the pure metals aluminium and copper, although very widely used are the alloys of these metals, with small additions of other elements to improve their properties for particular applications. **Table 3.6** compares typical values of the key parameters for the two metals.

Table 3.6 Comparison of the typical properties of aluminium and copper

	Aluminium	**Copper**	**Units**
Electrical conductivity at 20°C	37.2×10^6	58.1×10^6	S/m
Electrical resistivity at 20°C	2.69	1.72	$\mu\Omega \cdot$ cm
Density at 20°C	2689.8	8933	kg/m^3
Temperature coefficient of resistance (0–100°C)	4.2×10^{-3}	3.93×10^{-3}	per °C
Thermal conductivity (0–100°C)	238	599	W/m °C
Mean specific heat (0–100°C)	0.909	0.388	J/g°C
Coefficient of linear expansion (0–100°C)	23.5×10^{-6}	16.8×10^{-6}	per °C
Melting point	660	1085	°C

3.4.1.1 Copper and its alloys

Copper has the highest electrical and thermal conductivity of the common industrial metals. It has good mechanical properties, is easy to solder, is readily available and has high scrap value. It is widely used in wire form, and **Table 3.7** gives information for the commonly used wire sizes.

The electrical resistance of copper, as of all other pure metals, varies with temperature. The variation is sufficient to reduce the conductivity of pure copper at 100°C to about 76 per cent of its value at 20°C.

The resistance R_{t1} at temperature t_1 is given by the relationship

$$R_{t1} = R_t[1 + \alpha_t(t_1 - t)] \tag{3.1}$$

where α_t is the constant-mass temperature coefficient of resistance of copper at the reference temperature t (°C). Although resistance may be regarded for practical purposes as a linear function of temperature, the value of the temperature coefficient is not constant, but depends upon the reference temperature according to the law in **eqn 3.2**.

$$\alpha_t = \frac{1}{t + 1/\alpha_0} = \frac{1}{t + 234.45} \tag{3.2}$$

At 20°C, the value of α_{20} which is given by **eqn 3.2** is 0.00393/°C, and this is the value which is adopted by IEC. Multiplier constants and their reciprocals correlating

Table 3.7 IEC and BSI recommended sizes of annealed copper wires

Nominal diameter (mm)	Cross-sectional area (mm²)	Weight (kg/km)	Resistance (Ω/km)	Current rating at 1000 A/in² (A)
5.000	19.64	174.56	.8703	30.43
4.750	17.72	157.54	.9646	27.47
4.500	15.90	141.39	1.0750	24.65
4.000	12.57	111.71	1.3602	19.48
3.750	11.045	98.19	1.5476	17.12
3.550	9.898	87.99	1.7269	15.43
3.350	8.814	78.38	1.9393	13.66
3.150	7.793	69.28	2.193	12.08
3.000	7.069	62.84	2.418	10.96
2.800	6.158	54.74	2.776	9.544
2.650	5.515	49.03	3.099	8.549
2.500	4.909	43.64	3.482	7.609
2.360	4.374	38.89	3.908	6.780
2.240	3.941	35.03	4.338	6.108
2.120	3.530	31.38	4.843	5.471
2.000	3.142	27.93	5.441	4.869
1.900	2.835	25.21	6.029	4.375
1.800	2.545	22.62	6.718	3.944
1.700	2.270	20.18	7.531	3.518
1.600	2.011	17.874	8.502	3.116
1.500	1.767	15.710	9.673	2.739
1.400	1.539	13.685	11.10	2.386
1.320	1.368	12.166	12.49	2.121
1.250	1.227	10.910	13.93	1.902
1.180	1.094	9.722	15.63	1.695
1.120	.9852	8.758	17.35	1.527
1.060	.8825	7.845	19.37	1.368
1.000	.7854	6.982	21.76	1.217
.950	.7088	6.301	24.12	1.099
.900	.6362	5.656	26.87	.9861
.850	.5675	5.045	30.12	.8796
.800	.5027	4.469	34.01	.7791
.750	.4418	3.927	38.69	.6848
.710	.3959	3.520	43.18	.6137
.630	.3117	2.771	54.84	.4832
.560	.2463	2.190	69.40	.3818
.500	.1964	1.746	87.06	.3043
.450	.1590	1.414	107.5	.2465
.400	.1257	1.117	136.0	.1948
.355	.0990	.880	172.7	.1534
.315	.0779	.693	219.3	.1208
.280	.0616	.547	277.6	.0954
.250	.0491	.436	348.2	.0761
.224	.0394	.350	433.8	.0611
.200	.0314	.279	544.1	.0487
.180	.0255	.227	671.8	.0396
.160	.0201	.179	850.2	.0312
.140	.0154	.137	1110	.0239
.125	.0123	.109	1393	.0190
.112	.00985	.0876	1735	.0153
.100	.00785	.0698	2176	.0122
.090	.00636	.0566	2687	.0099
.080	.00503	.0447	3401	.0078
.071	.00396	.0352	4318	.0061
.063	.00312	.0277	5484	.0048

the resistance of copper at a standard temperature with the resistance at other temperatures may be obtained from tables which are included in BS 125, BS 1432–1434 and BS 4109.

Cadmium copper, chromium copper, silver copper, tellurium copper and sulphur copper find wide application in the electrical industry where high conductivity is required. The key physical properties of these alloys are shown in **Table 3.8**. It can be seen that some of the alloys are deoxidized and some are 'tough pitch' (oxygen containing) or deoxidized. Tough pitch coppers and alloys become embrittled at elevated temperatures in a reducing atmosphere, and where such conditions are likely to be met, oxygen-free or deoxidized materials should be used.

Cadmium copper has greater strength than ordinary copper under both static and alternating stresses and it has better resistance to wear. It is particularly suitable for the contact wires in electric railways, tramways, trolley buses, gantry cranes and similar equipment, and is also used in overhead telecommunications lines and transmission lines of long span. It retains its hardness and strength in temperatures at which high conductivity copper would soften, and is used in electrode holders for spot and seam welding of steel; it has also been used in commutator bars for certain types of motor. Because it has a comparatively high elastic limit in the work-hardened condition, it is also used in small springs required to carry current, and is used as thin hard-rolled strip for reinforcing the lead sheaths of cables which operate under internal pressure. Castings of cadmium copper have some application in switchgear components and in the secondaries of transformers for welding machines. Cadmium copper can be soft soldered, silver soldered and brazed in the same way as ordinary copper, although special fluxes are required under certain conditions, and these should contain fluorides; being a deoxidized material there is no risk of embrittlement by reducing gases during such processes.

Chromium copper is particularly suitable for high-strength applications such as spot and seam types of welding electrodes. Strip and, to a lesser extent, wire are used for light springs which carry current. In its heat-treated state, the material can be used at temperatures up to 350°C without risk of deterioration of properties, and it is used

Table 3.8 Selected physical properties of copper alloys

Property	Cadmium copper	Chromium copper	Silver copper	Tellurium copper	Sulphur copper
Content	0.7–1.0% cadmium	0.4–0.8% chromium	0.03–0.1% silver	0.3–0.7% tellurium	0.3–0.6% sulphur
Tough pitch (oxygen-containing) or deoxidized	deoxidized	deoxidized	tough pitch or deoxidized	tough pitch or deoxidized	deoxidized
Modulus of elasticity (10^9 N m^{-2})	132	108	118	118*	118
Resistivity at 20°C (10^{-8} $\Omega \cdot$ m)					
annealed	2.2–1.9	–	1.74–1.71	1.76**	1.81
solution heat treated	–	4.9	–	–	–
precipitation hardened	–	2.3–2.0	–	–	–
cold worked	2.3–2.0	–	1.78	1.80	1.85

*Tough pitch.
**Solution heat treated or annealed.

for commutator segments in rotating machines where the temperatures are higher than normal. In the solution heat-treated condition, chromium copper is soft and can be machined; in the hardened state it is not difficult to cut but it is not free-machining like leaded brass or tellurium copper. Joining methods similar to cadmium copper are applicable, and chromium copper can be welded using gas-shielded arcs.

Silver copper has the same electrical conductivity as ordinary high-conductivity copper, but its softening temperature, after hardening by cold work, is much higher and its resistance to creep at moderately elevated temperatures is enhanced. Because its outstanding properties are in the work-hardened state it is rarely required in the annealed condition. Its principal uses are in electrical machines which operate at high temperatures or are exposed to high temperatures in manufacture. Examples of the latter are soft soldering or stoving of insulating materials. Silver copper is available in hard-drawn or rolled rods and sections, especially those designed for commutator segments, rotor bars and similar applications. Silver copper can be soft soldered, silver soldered, brazed or welded without difficulty but the temperatures involved in all these processes are sufficient to anneal the material if in the cold-worked condition. Because the tough pitch material contains oxygen as dispersed particles of cuprous oxide, it is also important to avoid heating it to brazing and welding temperatures in a reducing atmosphere. In the work-hardened state silver copper is not free-cutting, but it is not difficult to machine.

Tellurium copper offers free-machining, high electrical conductivity, retention of work hardening at moderately elevated temperatures and good corrosion resistance. It is unsuitable for most types of welding, but gas-shielded arc welding and resistance welding can be effected with care. A typical application is magnetron bodies, which are often machined from solid. Tellurium copper can be soft soldered, silver soldered and brazed without difficulty. For tough pitch, brazing should be done in an inert or slightly oxidizing atmosphere since reducing atmospheres are conducive to embrittlement. Deoxidized tellurium copper is not subject to embrittlement.

Sulphur copper is free-machining and does not have the tendency of tellurium copper to form coarse stringers in the structure which can affect accuracy and finish. It has greater resistance to softening than high-conductivity copper at moderately high temperatures and gives good corrosion resistance. Sulphur copper has applications in all machined parts requiring high electrical conductivity, such as contacts and connectors; its joining characteristics are similar to those of tellurium copper. It is deoxidized with a controlled amount of phosphorus and therefore does not suffer from hydrogen embrittlement in normal torch brazing, but long exposure to a reducing atmosphere can result in loss of sulphur and consequent embrittlement.

3.4.1.2 Aluminium and its alloys

For many years aluminium has been used as a conductor in most branches of electrical engineering. Several aluminium alloys are also good conductors, combining strength with acceptable conductivity. Aluminium is less dense and cheaper than copper, and its price is not subject to the same wide fluctuations as copper. World production of aluminium has steadily increased over recent years to overtake that of copper, which it has replaced in many electrical applications.

There are two specifications for aluminium, one for *pure metal grade 1E* and the other for a *heat-treatable alloy 91E*. Grade 1E is available in a number of forms which are extruded tube (E1E), solid conductor (C1E), wire (G1E) and rolled strip (D1E). The heat-treatable alloy, which has moderate strength and a conductivity approaching that of aluminium, is available in tubes and sections (E91E).

The main application areas are:

Busbars. Although aluminium has been used as busbars for many years, only recently has it been accepted generally. The electricity supply industry has now adopted aluminium busbars as standard in 400 kV substations, and they are also used widely in switchgear, plating shops, rising mains and in the UK aluminium smelting plants. Sometimes the busbars are tin-plated in applications where joints have to be opened and remade frequently.

Cable. The use of aluminium in wires and cables is described at length in **Chapter 9**. Aluminium is used extensively in cables rated up to 11 kV and house wiring cable above 2.5 mm² is also available with aluminium conductor.

Overhead lines. The Aluminium Conductor Steel Reinforced (ACSR) conductor referred to in **section 14.3** and **Fig. 14.3** is the standard adopted throughout the world, although in the USA Aluminium Conductor Aluminium alloy wire Reinforced (ACAR) is rapidly gaining acceptance; it offers freedom from bimetallic corrosion and improved conductance for a given cross-section.

Motors. The use of aluminium in cage rotors of induction motors is described in **Chapter 10**. Motor frames are often die-cast or extruded from aluminium, and shaft-driven cooling fans are sometimes of cast aluminium.

Foil windings are suitable for transformers, reactors and solenoids. They offer a better space factor than a wire-wound copper coil, the aluminium conductor occupying about 90 per cent of the space, compared to 60 per cent occupied by copper. Heat transfer is aided by the improved space factor and the reduced insulation that is needed in foil windings, and efficient radial heat transfer ensures an even temperature gradient. Windings of transformers are described in more depth in **section 6.2.2**.

Heating elements have been developed in aluminium but they are not widely used at present. Applications include foil film wallpaper, curing concrete and possibly soil warming.

Heat sinks are an ideal application for aluminium because of its high thermal conductivity and the ease of extrusion or casting into solid or hollow shapes with integral fins. They are used in a variety of applications such as semiconductor devices and transformer tanks. The low weight of aluminium heat sinks make them ideal for pole-mounted transformers and there is the added advantage that the material does not react with transformer oil to form a sludge.

3.4.1.3 Resistance alloys

Many alloys with high resistivity have been developed, the two main applications being resistors and heating elements.

Alloys for standard resistors are required to have a low temperature coefficient of resistivity in the region of room temperature. The traditionally used alloy is Manganin, but this has increasingly been replaced by Ni–Cr–Al alloys with the trade names

Karma and Evanohm. The resistivity of these alloys is about 1.3 $\mu\Omega \cdot$ m and the temperature coefficient is $\pm 0.5 \times 10^{-5}/°C$. For lower-precision applications copper–nickel alloys are used, but these have a lower resistivity and a relatively high thermo emf against copper.

For heating elements in electric fires, storage heaters and industrial and laboratory furnaces there is a considerable range of alloys available. A considerable resistivity is required from the alloy in order to limit the bulk of wire required, and the temperature coefficient of resistivity must be small so that the current remains reasonably constant with a constant applied voltage. Ni–Cr alloys are used for temperatures up to 1100°C, and Cr–Fe–Al alloys are used up to 1400°C. Ceramic rods are used for higher temperatures and silicon carbide may be used up to 1600°C. For even higher temperatures, the cermets $MoSi_2$ and Zircothal are used. The maximum temperatures at which the materials may be used depend on the type of atmosphere.

3.4.2 Semiconductors

A semiconductor is able at room temperature to conduct electricity more readily than an insulator but less readily than a conductor. At low temperatures, pure semiconductors behave like insulators. When the temperature of a semiconductor is increased, or when it is illuminated, electrons are energized and these become conduction electrons. Deficiencies or '*holes*' are left behind; these are said to be carriers of positive electricity. The resulting conduction is called *intrinsic conduction*.

The common semiconductors include elements such as silicon, germanium and selenium and compounds such as indium arsenide and gallium antimonide. Germanium was originally used for the manufacture of semiconductor devices, but because of resistivity problems and difficulty of supply it was replaced by silicon which is now the dominant material for production of all active devices such as diodes, bipolar transistors, MOSFETs, thryistors and IGBTs.

Both silicon and germanium are group IV elements of the periodic table, having four electrons in their outer orbit. This results in a diamond-type crystal giving a tight bond of the electrons. **Figure 3.8** shows the atoms in a silicon crystal. Each atom is

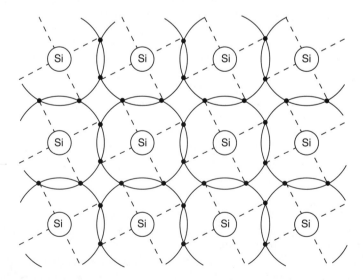

Fig. 3.8 Atoms in a silicon crystal

surrounded by eight electrons, four its own and four from neighbouring atoms; this is the maximum number in an orbit and it results in a strong equilibrium. It is for this reason that pure crystals of silicon and germanium are not good conductors at low temperature.

3.4.2.1 Impurity effects and doping

The conductivity of group IV semiconductors like silicon can be greatly increased by the addition of small amounts of elements from group V (such as phosphorus, arsenic or tin) or group III (such as boron, aluminium, gallium or indium).

Phosphorus has five electrons in its outer shell and when an atom of phosphorus replaces an atom of silicon it generates a *free electron*, as shown in **Fig. 3.9**. This is called *doping*. The extra electrons are very mobile; when a voltage is applied they move very easily and a current passes. If 10^{16} phosphorus atoms/cm^3 are added to a pure crystal, the electron concentration is greatly increased and the conductivity is increased by a factor of about a million. The impurities are called *donor* atoms and the material is an *impurity semiconductor*. This is called an *n-type semiconductor*, and *n* represents the excess of free electron carriers.

If the material is doped with group III atoms such as indium, then a similar effect occurs. This is shown in **Fig. 3.10**. The missing electron forms a '*hole*' in the structure which acts as a positive carrier. This structure is known as a *p-type semiconductor* and *p* represents the excess of positive carriers. The impurities are called *acceptor* atoms.

A single crystal containing both n-type and p-type regions can be prepared by introducing the donor and acceptor impurities into molten silicon at different stages of the crystal formation. The resultant crystal has two distinct regions of p-type and n-type material, and the boundary joining the two areas is known as a *p–n junction*. Such a junction may also be produced by placing a piece of donor impurity material against the surface of a p-type crystal or a piece of acceptor impurity material against an n-type crystal, and applying heat to diffuse the impurity atoms through the outer layer. When an external voltage is applied, the n–p junction acts as a rectifier, permitting current to flow in only one direction. If the p-type region is connected to the positive terminal of a battery and the n-type to the negative terminal, a large current flows through the material across the junction, but when the battery is connected in the opposite manner, no current flows. This characteristic is shown in **Fig. 3.11**.

3.4.2.2 The transistor

Many types of device can be built with quite elaborate combinations and constructions based around the n–p and p–n junction. Further information on these devices may be found in **reference 3F**.

Possibly the most important single device is the transistor, in which a combination of two or more junctions may be used to achieve amplification. One type, known as the n–p–n junction transistor, consists of a very thin layer of p-type material between two sections of n-type material, arranged in a circuit shown in **Fig. 3.12**. The n-type material at the left of the diagram is the *emitter* element of the transistor, constituting the electron source. To permit the flow of current across the n–p junction, the emitter has a small negative voltage with respect to the p-type layer, or *base* component, that controls the electron flow. The n-type material in the output circuit serves as the *collector* element, which has a large positive voltage with respect to the base in order to prevent reverse current flow. Electrons moving from the emitter enter the base, are

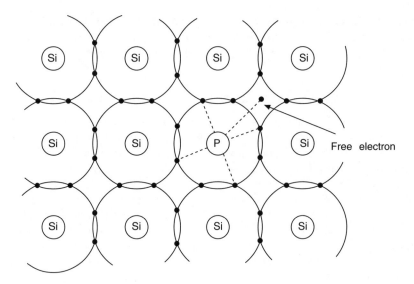

Fig. 3.9 Representation of an n-type semiconductor

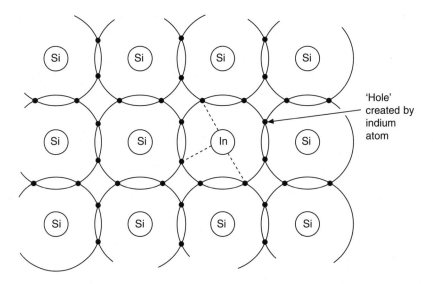

Fig. 3.10 Representation of a p-type semiconductor

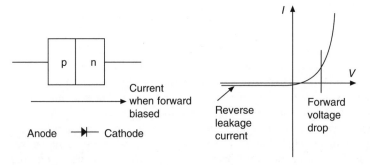

Fig. 3.11 Construction, symbol and characteristic of a semiconductor diode

attracted to the positively charged collector, and flow through the output circuit. The input impedance between the collector and base is low, whereas the output impedance between collector and base is high. Therefore, small changes in the voltage of the base cause large changes in the voltage drop across the collector resistance, making this type of transistor an effective amplifier.

Similar in operation to the n–p–n type is the p–n–p junction transistor also shown in **Fig. 3.12**. This also has two junctions and is equivalent to a triode vacuum tube.

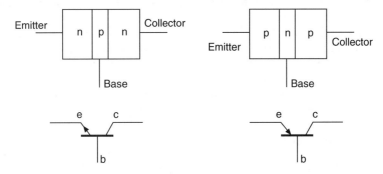

Fig. 3.12 Construction and symbols for n–p–n and p–n–p transistors

Other types, such as the n–p–n–p junction transistor, provide greater amplification than these two-junction transistors.

3.4.2.3 Printed circuits and integrated circuits

A *printed circuit* is an electrical circuit made by printing and bonding conducting material as a network of fine threads on a thin ceramic or polymer insulating sheet. This replaces the wiring used in conventional circuits. Other elements such as transistors, resistors and capacitors can be deposited onto the same base as the printed circuit.

An *integrated circuit* is effectively a combination of many printed circuits. It is formed as a single unit by diffusing impurities into single-crystal silicon, which then serves as a semiconductor material, or by etching the silicon by means of electron beams. Several hundred Integrated Circuits (ICs) are made at a time on a thin wafer several centimetres in diameter, and the wafer is subsequently sliced into individual ICs called *chips*.

In *large-scale integration* (LSI), several thousand circuit elements such as resistors and transistors are combined in a 5 mm square area of silicon no more than 0.5 mm thick. Over 200 such circuits can be arrayed on a silicon wafer 100 mm in diameter.

In *very large-scale integration* (VLSI), hundreds of thousands of circuit elements fit onto a single silicon chip. Individual circuit elements on a chip are interconnected by thin metal or semiconductor films which are insulated from the rest of the circuit by thin dielectric layers. This is achieved by the formation of a silicon dioxide layer on the silicon wafer surface, silicon dioxide being an excellent dielectric. *Metal oxide semiconductor field effect transistors* (MOSFETs) are made using this technique. These transistors are used for high-frequency switching applications and for random access memories in computers. They have very high speed and low power consumption.

3.4.2.4 The microprocessor

The microprocessor is a single chip of silicon which has the ability to control processes.

It can form the central processing unit (CPU) of a small computer and it can be used in a wide range of other applications. A microprocessor may incorporate from a thousand up to several hundred thousand elements. It typically contains a *read-only memory* (ROM), that is a memory that can be read repeatedly but cannot be changed, but it may also have some *random-access memory* (RAM) for holding transient data. Also present in a microprocessor are registers for holding computing instructions, for holding the 'address' of each instruction in turn and for holding data, and a logic unit. Interfaces for connecting with external memories and other systems are included as required.

The microprocessors used in personal computers (PCs) have been the subject of intensive development during the last decade. The speed of operation is usually defined as a frequency and chips with frequencies of 667 MHz or higher are now available; this corresponds to an individual operation time of 1.5 nanoseconds. The amount of information that can be transferred in parallel and held in registers is known as a *bit*, and 64-bit processors are now available.

3.4.3 Superconductors

The ideal superconducting state is characterized by two fundamental properties, which are the disappearance of resistance when the temperature is reduced to a critical value, and the expulsion of any magnetic flux in the material when the critical temperature (T_c) is reached. Superconductivity was first discovered in mercury in 1911. Other elements have subsequently been found to exhibit superconductivity and theories have been developed to explain the phenomenon. The critical temperatures for these materials were typically about 10K ($-263°C$) which meant that they had to be cooled with liquid helium at 4K. In general these materials have been of academic interest only because they could only support a low current density in a low magnetic field without losing their superconducting properties.

In the 1950s a new class of materials was discovered. These are metallic alloys, the most important being niobium titanium and niobium tin. The highest critical temperatures achieved by these materials is 23.2K and they can be used to produce magnetic flux densities of over 15 T. The main commercial application for these low-T_c superconductors is for magnets in medical imaging equipment which requires the high fields to excite magnetic resonance in nuclei of hydrogen and other elements. The magnet or solenoid of the magnetic resonance imaging (MRI) unit has an internal diameter of about 1.2 m and the patient to be examined is put into this aperture. The image from the resonance test shows unexpected concentrations of fluids or tissue and enables a diagnosis. Superconducting magnets producing high magnetic fields are also used in magnetic research and in high-energy physics research; other applications such as dc motors and generators, levitated trains, cables and ac switches have been explored but the complexity and high cost of providing liquid helium has prevented commercial development.

In late 1986 a ceramic material LaBaCuO was discovered to be superconducting at 35K and in 1987 the material YBaCuO was found to have a critical temperature of 92K. Since that time the critical temperatures of new materials has progressively increased to over 130K. Examples of these are BiSrCaCuO (with a T_c of 106K), ThBaCaCuO (T_c of 125K) and HgBaCaCuO (T_c of 133K). The enormous significance of these discoveries is that the materials will be superconducting in liquid nitrogen, which has a boiling point of 77K and is much easier and cheaper to provide than helium.

The consequence of these *high-temperature superconductors* has been an unprecedented upsurge of research activity throughout the world. Much of this work is directed towards finding materials with even higher T_c values and towards establishing viable production methods. These new materials are brittle and unless they are in the form of very thin films they will only operate up to current densities of about $0.1-1$ A/mm^2. The indications are, however, that they will operate in flux densities exceeding 50 T.

Attention is now being turned to applications for the materials which are available. Their brittleness rules out their direct use as conventional wires in magnets or cables, although a prototype cable carrying 2300 A over 1 m has been made. In this case the superconducting ceramic oxide was deposited onto flexible tape, and this technique leads to the possible use of cable for winding solenoids to produce high magnetic fields. Other applications which are already in use are superconducting quantum interference device (SQUID) magnetometers which are used for detecting very small currents in the human body to detect and pinpoint defects. The materials are also in use in transmission aerials, in which low noise at low temperatures is an advantage.

3.5 Standards

Each country has in the past had its own standards for materials. Over the past 20 years or so there has been a movement towards international standards which for electrical materials are produced by IEC (International Electrotechnical Commission). When an IEC standard is produced the member countries copy this standard and

Table 3.9 Standards for conducting materials

	BS	IEC	Subject of standard
Copper	1432 1433 1434	60356	Specifications for copper for electrical purposes.
	4109		Specifications for copper for electrical purposes. Wire for general electrical purposes and for insulated cables and flexible cords.
	7884		Specifications for copper and copper–cadmium stranded conductors for overhead electric traction and power transmission systems.
Aluminium	215 pts1&2	60207	Aluminium conductors.
	2627		Wrought aluminium for electrical purposes – wire
	2897		Wrought aluminium for electrical purposes – strip with drawn or rolled edges.
	2898		Wrought aluminium for electrical purposes – bars, tubes and sections.
	3242	60208	Aluminium alloy stranded conductors for overhead power transmission.
	3988	60121	Wrought aluminium for electrical purposes – solid conductors for insulated cables.
	6360	60228 60228A	Specifications for conductors in insulated cables and cords.
Semiconductors	4727 pt1 group 05	60050 (521)	Semiconductor terminology.
	6493	60747	This has many parts with specifications for diodes, transistors, thyristors and integrated circuits.

issue it under their own covers. This now applies to BSI, ASTM, DIN and the French standards organization. Where appropriate, standards are also issued as applying to all the European Union.

Leading standards for electrical materials are shown in **Table 3.9**.

References

3A.   Jones, G.R., Laughton M.A. and Say, M.G. (eds), *Electrical Engineer's Reference Book* (15th edn), Butterworth-Heinemann, 1993.

3B.   Brandes, E.A. and Brook, G.R. *Smithells Metals Reference Book* (7th edn), Butterworth-Heinemann, 1992.

3C.   McCaig, M. and Clegg, A.G. *Permanent Magnets in Theory and Practice* (2nd edn), Pentech and Wiley, 1987.

3D.   Sillars, R.W. *Electrical Insulating Materials and Their Application*, Peter Peregrinus, 1973.

3E.   Tillar Shugg, W. *Handbook of Electrical and Electronic Insulating Materials*, Van Nostrand, 1986.

3F.   Brindley, K. *Newnes Electronics Engineer's Pocket Book*, Butterworth-Heinemann, 1993.

3G.   Mayo, William E. *Processing and Applications of High T_c Superconductors*, Metallurgical Society Inc., 1988.

3H.   Warnes, L.A.A. *Electronic and Electrical Engineering*, Macmillan, 1994 (good chapters on semiconductors).

Chapter 4

Measurement and instrumentation

Dr M.J. Cunningham
Department of Computer Science, University of Manchester

4.1   Introduction

Measurement is an activity which pervades almost all human endeavours. In industrial countries about 5 per cent of the gross national product is devoted to making measurements and this figure is comparable with the expenditure on the health services in the UK. Three major areas of the use of measurement and instrumentation are in product manufacture, in the process industries and in the exchange of commodities. Measurement by means of instruments takes place at all stages of the manufacturing process from the validation of bought-in components, through the assessment of the process, in safety tests and environmental tests and in the final check of conformance with the customer's specification. Measurement is the arbiter of quality – the ability to meet the customer's needs.

Instruments pervade the process industries where flow, temperature, pressure and liquid level are very commonly measured quantities. The ability to control these quantities depends on the ability to measure the quantities satisfactorily. Thus measurement, often by means of a sensor, is found in every control loop of the process.

When commodities change hands, from pears to petroleum, acceptable measurement is the dispenser of fair play. A traditional role of government is to ensure the availability of an accepted system of measurement for trade throughout the country.

When some property of an object is measured a number is assigned to the property in terms of the agreed unit value in such a way that the number faithfully reflects the characteristics of the property. This is a definition of measurement. In producing a number which reflects the property, the quantity (how much?) of the quality (of what sort?) is obtained. In order that the number does reflect the property in question, it is essential that the influence of other qualities is kept to a minimum. An ideal measurement would therefore be one for which the result is only affected by the size of the quantity to be measured and by no other. Much effort in designing and using instruments goes into meeting this condition of satisfactory measurement as closely as possible.

It is clear from the above that the subject of measurement and instrumentation is a vast one. Restricting the subject to electrical engineering is in fact not much of a restriction as very many non-electrical measurements are converted into electrical measurements via suitable sensors, owing to the accuracy and convenience of electrical measurement. It is therefore impossible to cover this subject completely, nor is it necessary. By describing the major instruments – the Cathode Ray Oscilloscope (CRO), the Digital Storage Oscilloscope (DSO), the frequency counter, the Digital Voltmeter

(DVM), the Digital Multimeter (DMM) and some important sensors – the principles and practice of measurement and instrumentation can be outlined.

4.2 The oscilloscope

The most commonly used instrument in electronic and electrical engineering is the cathode ray oscilloscope or its digital counterpart, the digital storage oscilloscope. It is difficult to envisage design, development, testing or maintenance without the oscilloscope. The oscilloscope makes visible the waveforms in the circuits of interest and is therefore the means whereby the engineer can bring his knowledge, experience and creativity to bear on the situation.

4.2.1 The cathode ray oscilloscope

The CRO is an instrument which presents to the engineer a 'graph' of the voltage signal at a point in a circuit against time, as shown in **Fig. 4.1**. This is a very helpful form of presentation for the engineer since this is the way in which signals are conceived. It is possible to measure both time and amplitude information with such a display. The display can be used quantitatively to investigate, for example, waveform distortion. How this is achieved can be seen by a consideration of the construction of the CRO.

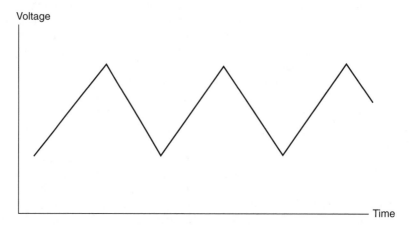

Fig. 4.1 The CRO display

Figure 4.2 shows a simplified diagram of a Cathode Ray Tube (CRT). The CRT is at the heart of the CRO. It consists of an indirectly heated cathode which is coated with a material capable of giving a large number of electrons when heated. The electrons are attracted away from the vicinity of the cathode by the anode which is at a high positive voltage. By careful design the electrons pass along the tube and hit the phosphor coating inside the CRT screen. Phosphor is the name of a family of materials that possess fluorescence and phosphorescence. *Fluorescence* is the property of a material to give off light when struck by the fast moving electrons. *Phosphorescence* is the property of a material to keep giving off light for some time after having been struck by the electrons. The time for which the light is given out is the *persistence* of the phosphor. CROs are supplied with either general purpose or long persistence phosphors. General purpose phosphors have persistences of the order of tens of

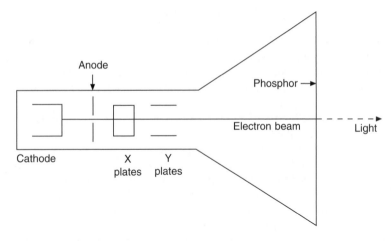

Fig. 4.2 Cathode ray tube construction

milliseconds and long persistence phosphors have persistences of the order of a few seconds. A long persistence phosphor is advantageous for a very low frequency waveform, perhaps less than 1 Hz, as the whole waveform can be seen by means of the long period of phosphorescence. However, long persistence phosphors are easily damaged by, for example, allowing the electron beam to be stationary for some time. Most CROs therefore have general purpose phosphors.

The deflection of the electron beam in order to display the applied signal is achieved very directly in the CRO by applying a scaled version of the signal to the deflection plates. The relationship between the deflection and the applied voltage is a linear one and so the interpretation of the resulting display is an easy matter.

In order for the display on the CRO screen to look like a graph of the signal it is necessary to have uniformly increasing horizontal deflection as well as the vertical signal deflection described. A linearly rising voltage applied to the X deflection plates achieves this. To obtain an easily viewed display of a repetitive signal, it is usual for internal circuitry to produce a repeated sweep from left to right. The waveform which achieves this is called a saw-tooth waveform and is shown in **Fig. 4.3.**

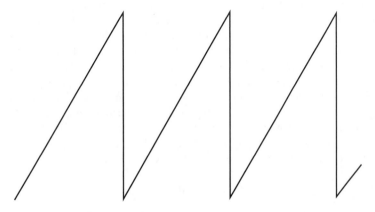

Fig. 4.3 The saw-tooth waveform

4.2.2   The steady display of a repetitive signal

To obtain a usable display, it is clearly necessary for the frequencies of the timebase and the input frequency to be simply related, such as $1:1$, $1:2$, $1:3$ etc. This presents some difficulty since the frequency of the input waveform will probably be unknown until viewed on the CRO. To overcome this, a *variable timebase frequency control* is provided on the CRO front panel. When this is adjusted to bring the timebase frequency close to a simple relationship with that of the input frequency, internal circuitry locks the timebase frequency to the exact relationship. The display on the screen will therefore be that of one period of the input signal for a $1:1$ ratio, two periods of the input signal for a $2:1$ ratio and so on. Thus a steady display of the input signal is provided for signals over a very wide frequency range. Parameters of interest such as period and amplitude can be measured conveniently. An example of a $1:1$ ratio is shown in **Fig. 4.4**. This is called the synchronized method of display.

It often happens, however, that the engineer would like to look at only a small part of the period of a repetitive waveform. **Figure 4.5** shows a sinusoidal type input, but with a ripple on each maximum positive excursion. The synchronized method for signal display shown earlier will not allow the detailed examination of the ripple area of interest, but the trigger method will. In the internal trigger method, a variable dc voltage is set by a control on the front panel. This trigger level voltage is compared with the input signal and one sweep of the timebase is enabled when the input signal exceeds the trigger level. The timebase runs for a time set by the front panel and is in no way related to the value of the input signal. After the sweep, the spot is blanked, flies back and waits for the next occasion when the input exceeds the trigger level. At this time another sweep occurs and so the display is that of the repetition of the same portion of the waveform again and again. By suitable adjustment of the trigger level and the timebase sweep rate, a small detail can be investigated easily as shown in **Fig. 4.5**.

4.2.3   The display of one signal against another – Lissajous figures

A CRO can be configured to display one signal against another externally provided signal rather than against the internally provided timebase signal. The ensuing displays are called Lissajous figures after the nineteenth-century Frenchman who investigated the effect of the combination of simple harmonic motions. The method can be very useful for measuring phase shift in a circuit (the concept of phase angle in an ac circuit has been explained in **section 2.3.2**). **Figure 4.6** shows a linear circuit supplied by a sinusoidal voltage source. The output must also be sinusoidal, but the amplitude and phase will be different from that of the applied input. If the input signal is displayed on the horizontal axis and the output signal is displayed on the vertical axis, then a display such as **Fig. 4.7** occurs. The display can be used to measure the gain, which is the maximum to minimum vertical excursion divided by the maximum to minimum horizontal excursion, y_g/x_g, taking into account the scaling of the x and y controls. More useful, however, is the phase measurement capability. If the horizontal deflection signal is $v_x = V_x \sin \omega t$ and the vertical deflection signal is $v_y = V_y \sin(\omega t + \phi)$ then the phase shift ϕ can be found by the following:

at
$$t = 0, v_x = 0$$

$$v_y = V_y \sin(0 + \phi) = V_o$$

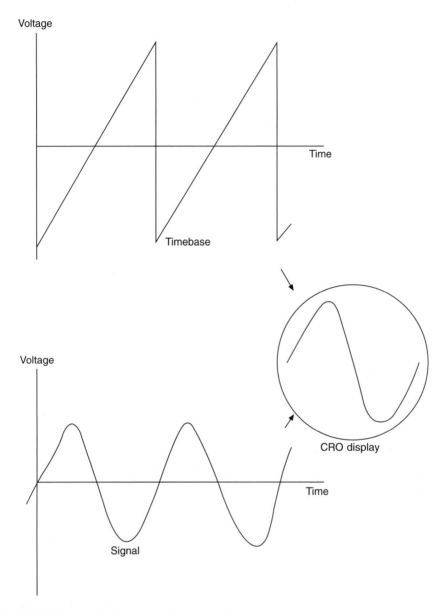

Fig. 4.4 CRO display for 1 : 1 ratio of timebase and input frequency

therefore
$$\sin \phi = \frac{V_o}{V_y}$$

$$\phi = \sin^{-1} = \frac{V_o}{V_y}$$

Expressed in words the phase shift between two signals (such as the input and the output in **Fig. 4.6**) can be measured by taking the ratio of the maximum y excursion and the intercept on the x axis and then performing the inverse sine operation. In this

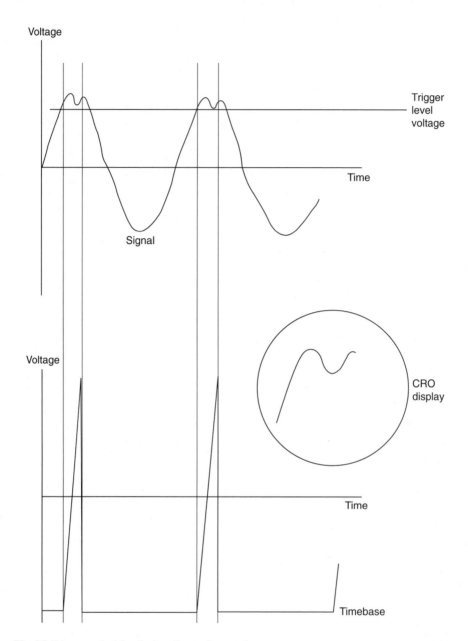

Fig. 4.5 Trigger method for display of part of a waveform

way it is easy to observe the phase shift in, for example, an amplifier and to explore how the phase shift changes with frequency.

4.2.4 The display of two signals against time

Very often it is helpful to be able to observe the waveform and make measurements at two points in a circuit simultaneously. This allows the assessment of the performance of the circuit, for example the distortion introduced by an audio amplifier. In order

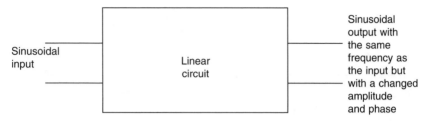

Fig. 4.6 A linear circuit

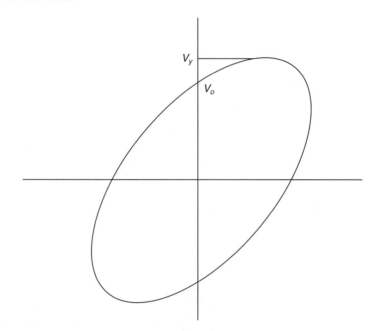

Fig. 4.7 A Lissajous figure

to do this without significantly increasing the cost of the CRO, then one electron beam must write both signals. Provided this is carefully done, the eye does not see this, but sees two traces simultaneously.

There are two ways of achieving this and most CROs provide both, often automatically switching between the two to suit the frequency of the signal. The two ways are called the *alternate method* and the *chop method*. The effect of the alternate method is shown in **Fig. 4.8**. In the alternate method, a sweep of the timebase is traced with alternately one input then the other connected to the Y deflection plates. The two inputs y_1 and y_2 have their own gain controls and Y deflection controls. This method works well provided the input signal frequency is not too low. For low frequencies, it becomes evident to the eye that the display consists of one trace followed by another.

For low frequencies the chop method is appropriate, and the effect of this is shown in **Fig. 4.9**. An internal oscillator running at about 500 kHz is used to switch the two inputs in small sections to the deflection plates. For low-frequency signals, there are very many sections for a sweep, for example 5000 for a 100 Hz signal, and the display appears continuous. At high frequencies, the sectional nature becomes very evident to the eye and the alternate method is used.

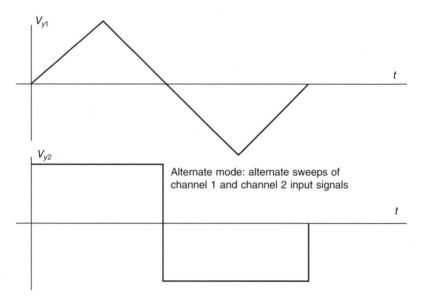

Fig. 4.8 The alternate mode

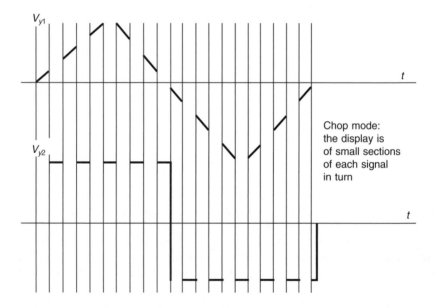

Fig. 4.9 Display of two *y* inputs in the chop mode

4.2.5 The ac/dc switch

There are applications where it is desirable to be able to measure the characteristics of a small varying quantity superimposed on a large dc value. A typical case would be the assessment of residual ripple on a dc power supply. The CRO has an ac/dc switch on the front panel and in the dc position, the displayed waveform would look like **Fig. 4.10(a)**. If the sensitivity of the CRO is increased by changing the gain switch, the trace would disappear from the screen. In order to be able to view an enlarged version of the ac component, the switch should be set in the ac position. This introduces a logical capacitor in series with the signal, blocking the dc component. The display then looks like **Fig. 4.10(b)**. It is possible in this position to increase the sensitivity as desired to view the ac quantity. If the displayed signal is very low in frequency, say a few Hz, much of the signal will also be effectively blocked by the capacitor in the ac position and a much reduced amplitude of display will result. For this reason the ac/dc switch should be kept in the dc position for general use.

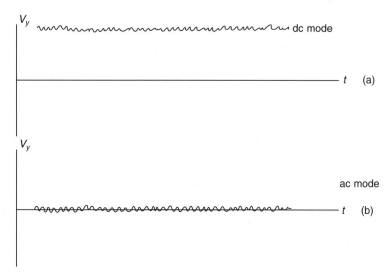

Fig. 4.10 The effect of the ac/dc switch

4.2.6 The loading effect of the CRO

In common with all instruments, the CRO affects the condition of the circuit to which it is connected. This always happens and it is important to discover whether this effect is significant or not. It is called the *loading effect* of the instrument. A manufacturer will state, usually by the Y input socket, the effective resistance and capacitance of the CRO. The effect of the CRO on the circuit is as if it were a parallel resistor and capacitor, shown in **Fig. 4.11**, with typical values for resistance and capacitance shown. As an example a simple potential divider circuit is shown in **Fig. 4.11**. Before connecting the CRO, the voltage across the lower 1 MΩ resistor is 6 V pk–pk by symmetry. When the CRO is connected in the case of low frequencies for which the capacitive effect of the CRO can be neglected, the 1 MΩ of the CRO reduces the lower resistance to a combined resistance of 1/2 MΩ. This means that the voltage across the lower resistance will fall to 4 V pk–pk and this is what a perfectly accurate CRO will read. Of course the voltage will return to 6 V pk–pk when the CRO is removed. The effect of the CRO on the circuit is clearly very significant and the effect becomes worse at high frequencies, where the effect of the capacitance of the CRO is to lower further the impedance of the instrument.

If the characteristics of the circuit and the CRO are known, as in the situation shown in **Fig. 4.11**, then it is possible to calculate the effect of the loading and correct for it. Usually, however, the CRO is the means of investigating the circuit and no such previous knowledge will be available. The loading effect will only be serious where the impedance of the circuit approaches the impedance of the CRO and it can

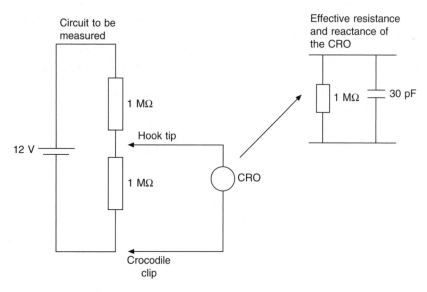

Fig. 4.11 CRO loading effect

usually be neglected if the circuit impedance is not greater than one-tenth of the CRO impedance. There are many cases where this condition is not met or it is not known if it is met. The solution is to use a suitable probe with the CRO.

4.2.7 CRO probes

In the body of a typical CRO probe is mounted a switch labelled ×1 and ×10. In the ×1 position the signal is passed directly to the CRO. In the ×10 position a 9 MΩ resistor and variable parallel capacitor are switched in series with the signal lead, as shown in **Fig. 4.12**. At low frequencies, where the effect of the two capacitances can be neglected, the effect of the ×10 switch is to increase the effective CRO resistance to 10 MΩ thus making the loading effect much less significant. At high frequency, the effect of the probe capacitor reduces the effective capacitance of the instrument.

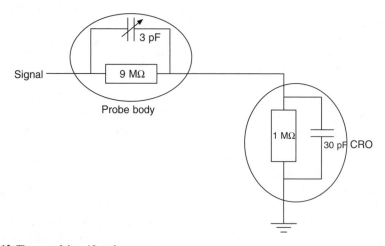

Fig. 4.12 The use of the ×10 probe

Cable capacitance makes the effect less predictable than for resistance, but it will also significantly reduce the loading effect at high frequencies.

The effect of the ×10 probe will also be to drop nine-tenths of the signal voltage across the probe and so only one-tenth will be presented to the CRO. The amplitude readings of the CRO must be multiplied by 10 to give the signal amplitude, hence the name of the probe.

Other more specialist probes are available to further reduce the loading effect of the CRO, but it is found that for many cases the ×10 probe gives a sufficient reduction.

4.2.8   CRO bandwidth and the measurement of rise time

Probably the most important parameter of the CRO is the *bandwidth*. The bandwidth gives the maximum frequency for which a faithful display of amplitude takes palce. If a sinusoidal signal above the bandwidth frequency is applied to the CRO, then the displayed amplitude will be smaller than would be the case for the same size input at a lower frequency. This is illustrated in **Fig. 4.13**, and it is clear that the instrument calibration can only be used for sinusoids below the bandwidth frequency. This is not the most significant effect of the bandwidth of the CRO. It has a direct bearing on the fastest rising edge of a rapidly changing voltage which can faithfully be displayed on the CRO. The display and measurement of such edges are very important in assessing the digital signals that occur in, for example, computers. A CRO cannot respond instantaneously to a very sudden change in voltage. If a very sudden change is applied, the response of the CRO will look like **Fig. 4.14**. The response is usually described by the rise time, the time taken for the response to change from 10 per cent to 90 per cent of the final value. It is found that there is a simple relation between the rise time of the CRO and the CRO bandwidth, which is given by

$$\text{Bandwidth} \times \text{Rise time} = 0.35$$

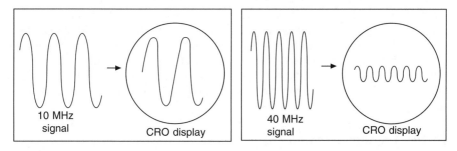

Fig. 4.13 Effect of a CRO bandwidth of 20 MHz

For example, a 20 MHz CRO will have a rise time of $0.35/(20 \times 10^6)$ seconds, or 17.5 ns. This means that if the instrument is presented with a voltage waveform with rise times faster than this value, the rise time that will be displayed is the CRO rise time. In order for the effect of the properties of the CRO not to significantly intrude on the display, the fastest rise time that can be displayed should be not less than eight times the CRO rise time. From this relation, the required CRO bandwidth can be calculated in order to display a given signal rise time. For example, if the signal rise time is 80 ns, then the CRO rise time must be 10 ns in order not to change the display significantly. A 10 ns rise time corresponds to a CRO bandwidth of $0.35/10^{-8}$ Hz, or 35 MHz. This procedure allows the necessary bandwidth to be calculated for a given

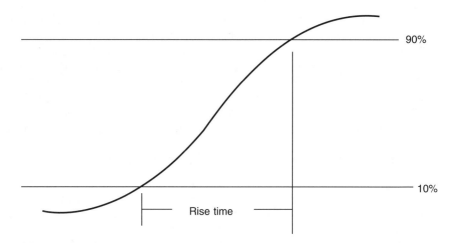

Fig. 4.14 Response of a CRO to a very fast change in voltage

rise time and checked against the instrument in use, and it can establish whether a new instrument is needed.

Since fast rising voltages are commonly found in digital circuits, the need for high bandwidth CROs is clear.

4.2.9   The high bandwidth CRO

A high bandwidth CRO is significantly more expensive than a general purpose CRO. There are several reasons for this. In the first place, as the spot moves over the surface very rapidly at high frequencies, the emitted light reduces. To compensate for this the electron beam velocity is increased, but this tends to reduce the sensitivity of the instrument and so extra care is required with the electronics. As the electron beam passes through the deflection plates the forces on the electrons change significantly at high frequencies. This again reduces the sensitivity. This is overcome by splitting the Y deflection plates into a number of small sections and introducing time delays between each. This ensures that the signal on each plate remains closely the same as the electron beam passes down the tube. It makes the CRT much more expensive, however.

4.2.10   The digital storage oscilloscope

The digital storage oscilloscope solves some of the problems of the CRO at high frequencies by means of storing the signal and then displaying at an effective lower frequency. This allows a cheaper CRT to be used. The block diagram of a DSO Y deflection system is shown in **Fig. 4.15**. The Analogue to Digital (A/D) converter gives a digital version of the voltage input suitable for storing in memory. For the DSO, the CRT ceases to be the critical component at high frequencies. The critical component for the DSO is the A/D converter. This element samples the input waveform at regular intervals and converts to a digital equivalent. If the sampling rate is not high enough or the levels of quantization not fine enough, then this will reflect on the accuracy of the quantities measured. An illustration of what happens on a leading voltage edge with a DSO is shown in **Fig. 4.16**. The samples are shown by crosses. This means that the analogue voltage to be measured is converted into numbers at the

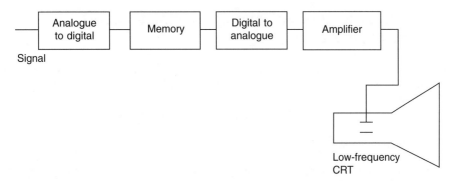

Fig. 4.15 Y-deflection system in a digital storage oscilloscope

times shown. The fastest rising edge which can be displayed is when there is one sample on the rising edge. The effective rise time of the DSO is therefore given by 1.6 sample intervals, this is the time from 10 per cent to 90 per cent at the final value. This represents the DSO rise time and a signal should have a slower rise time than this, say six times slower. The fastest acceptable signal rise time is therefore 6×1.6 sampling intervals, or 10/sampling rate.

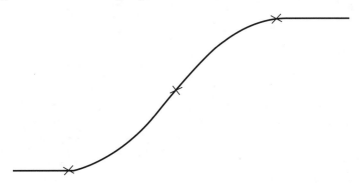

Fig. 4.16 DSO rise time

The sampling rate of the DSO is therefore a key parameter of the instrument and is quoted as, for example, 500 M samples per second. Such an instrument is capable of faithfully displaying a signal rise time of 20 ns, and this roughly corresponds to a 140 MHz bandwidth CRO.

The other key parameter of the A/D converter is the fineness of quantization, or the number of bits of the A/D. If the number of bits is not sufficiently high, then significant information is lost in the A/D process. It is for this reason that a signal displayed on a cheap DSO often appears to have a better signal to noise ratio than for a CRO. The DSO is merely not responding to the small voltage differences of the noise component.

If a good A/D converter is used, then the DSO will give a faithful display of an applied signal over a wide range of amplitudes sand frequencies. The DSO has other advantages than cheaper CRT construction. The information in digital form can easily be manipulated to find, for example, average values and rms values. The information can be easily directly communicated to a computer and a hard copy of the display is fairly easily made available.

4.3   Digital frequency counters, voltmeters and multimeters

Digital multimeters, voltmeters and frequency counters are the next most common instruments used by the electronic and electrical engineer after the CRO and DSO. Since they develop one from the other, the frequency counter will be described first, then the digital voltmeter (DVM) and then the digital multimeter (DMM).

4.3.1   The frequency counter

The ability to measure accurately the frequency of a signal is a common requirement, for example when designing a filter circuit. This can be achieved by counting the number of periods of the signal that occur in one second, which is the frequency of the signal. The basis of the instrument shown in **Fig. 4.17**.

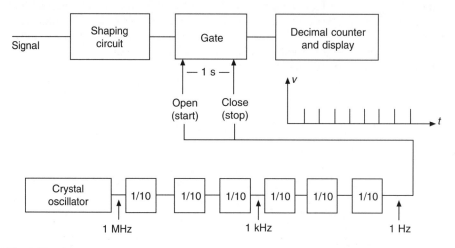

Fig. 4.17 A frequency counter

   The signal is converted into voltage pulses at the same frequency by the shaping circuits, the output from which passes through a gate, is converted into a decimal number, and then shown on a suitable display such as liquid crystal or light emitting diode. The gate is opened for 1 second by arranging that the signal which opens the gate and the signal which closes the gate are derived from a quartz crystal oscillator. This arrangement works very well for high-frequency signals, since the number of pulses counted is very high, but not so well for a low-frequency signal, for example one with a frequency of 10 Hz.

   For such a signal, the ±1 error, which occurs for all digital instruments, is a major source of uncertainty in the measurement results. When the number of pulses from the shaping circuit is measured, the gating time is not synchronized with the signal. It is therefore possible to just miss pulses or just include pulses at either end of the gating period, as shown in **Fig. 4.18**. The measured result is therefore 9 Hz and 11 Hz respectively, whereas 10 Hz would be a more representative value. Because of the ever present ±1 error in digital instruments, it is important to ensure that the number of pulses counted is very high so that the effect of the ± error becomes very small.

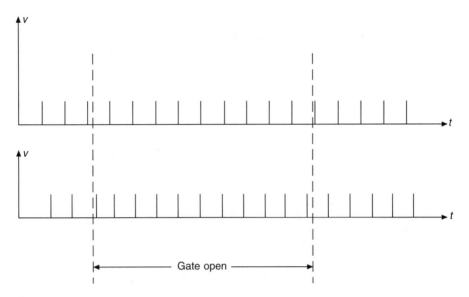

Fig. 4.18 The ±1 error

4.3.1.1 The measurement of low frequencies

For a low-frequency signal, the above condition of ensuring many counted pulses is difficult to achieve. It would be possible to wait for 10 seconds instead of 1 second for the result, which would help a little in reducing the effect of the ±1 error, but this would still not be acceptable, neither would waiting for 100 seconds. The preferred solution is to swap the signal and crystal oscillator output around as shown in **Fig. 4.19**. The signal now opens the gate for exactly one period. The decimal counter will accumulate the number of 1 µs pulses from the crystal oscillator in one period of the signal and the instrument therefore measures the period of the signal. It is clear that the decimal counter will collect a large number of pulses for a low-frequency signal and the ±1 error will have relatively little effect on the period measurement.

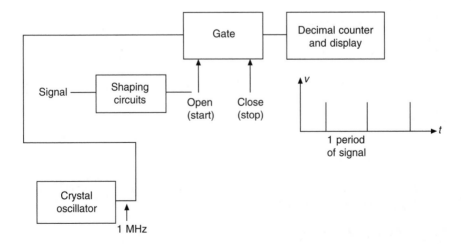

Fig. 4.19 The period counter

Therefore for a high-frequency signal, frequency measurement should be used and for a low-frequency signal, period measurement should be used. In the case illustrated, it makes no difference for a 1 kHz signal which method is used.

4.3.1.2   The accuracy of the frequency counter

The potential accuracy of the frequency counter is very high since essentially counting is involved. The key component for the accuracy of the instrument is the crystal oscillator.

The frequency of a crystal depends on many things but usually the temperature, the supply voltage and its age are the three dominant factors. By selecting a crystal cut at a very carefully specified angle to the crystallographic axis, called the AT cut, a quartz crystal with a very low temperature coefficient can be produced. For instruments for which very high accuracy is specified, this crystal is also enclosed in a temperature controlled box. This is more expensive, but gives a very stable frequency with changes in ambient temperature.

The dc supply to the oscillator needs to be carefully controlled in order to avoid frequency changes owing to this cause. A typical figure would be a change of 1 part in 10^9 for a ±10 per cent change in power supply voltage.

Quartz crystals change their frequency with time after manufacture. In common with many products, the change is most rapid initially and then settles down to a smaller rate of change. A typical characteristic is shown in **Fig. 4.20**. From this it can be seen that the initial rate of change might well be significant and recalibration is required more frequently when the instrument is new. This is described in more detail in **section 4.5**.

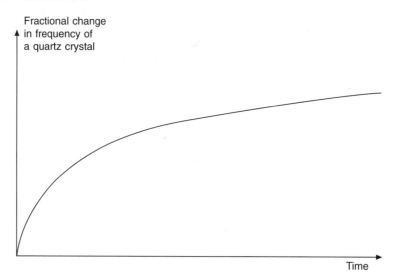

Fig. 4.20 Change in frequency of a quartz crystal with time

With reasonable care with these matters it is possible to produce a very accurate frequency counter for a modest cost.

4.3.2   The digital voltmeter

As frequency and time interval can be measured accurately easily and cheaply, it is attractive to make a digital voltmeter (DVM) which converts the dc voltage to be measured to a corresponding time interval or frequency. There are several methods for achieving this and, by way of illustration, the dual slope DVM will be described.

The block diagram for the dual slope DVM and the voltage waveforms are shown in **Figs 4.21** and **4.22** respectively. At the beginning of the cycle, the unknown voltage is connected to the integrator. The output of an integrator for a dc input is a ramp voltage the slope of which is proportional to the unknown voltage. After one period of the mains supply (20 ms for 50 Hz) the integrator input is switched to an internal stable reference voltage of opposite polarity from that of the voltage to be measured. The output of the integrator then ramps down with a constant slope in all cases. At the same moment as the integrator input is switched, the gate in a timing set-up is opened. This allows pulses from the crystal oscillator to the decimal counter. This continues until the instant at which the output of the integrator returns to the common voltage, when the gate is closed. The number of pulses counted is directly proportional to the unknown voltage and the display is suitably scaled in volts.

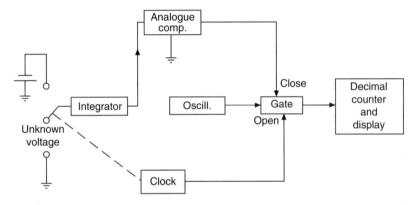

Fig. 4.21 Dual slope DVM

This method for voltage measurement has several attractions. First, it can be manufactured cheaply. Also mains frequency interference, which must always be present, is eliminated by the integration over one period of the mains. The fact that two slopes are involved means that parameters of the integrator, which change with ambient conditions such as temperature, affect both slopes equally and cancel. Reasonable accuracy and noise immunity can therefore be achieved by this method. The accuracy of the digital voltmeter is that of the frequency counter plus the contributions of the extra components, particularly the voltage source and range defining resistors.

4.3.3 The digital multimeter

The digital multimeter (DMM) is an extension to the digital voltmeter. Just as the digital voltmeter turns the voltage to be measured into a time interval, so the digital multimeter turns the quantity to be measured into a voltage.

Since the digital voltmeter contains a frequency counter, it is possible to provide frequency measurement capabilities directly with modest performance. DC current measurement can be obtained by passing the current to be measured through a resistor and then measuring the voltage developed by the digital voltmeter. The digital multimeter contains a chain of stable resistors which can be selected for

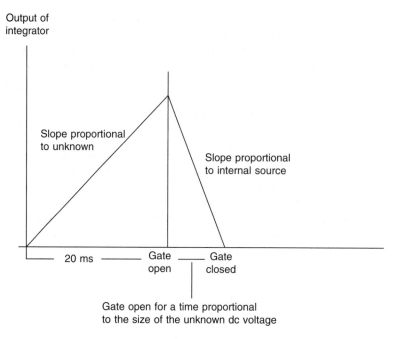

Fig. 4.22 Timing diagram for the dual slope DVM

various current ranges. The accuracy on the current range will clearly not be as high as for dc voltage. AC voltage can be measured by converting to an equivalent dc voltage. The cheaper schemes rely on rectification and averaging and the calibrations only hold for a sinusoidal signal. More generally used is the true rms converter which gives the rms voltage of the measured quantity regardless of its waveform. The instrument with true rms capability tends to be more expensive.

Resistance measurement can be added in the digital multimeter. The most elegant way of doing this for a dual slope measurement is to pass the same current through the resistor to be measured and a stable reference resistor. The dual slope method can be involved and the voltage across the unknown resistor applied to the integrator of **Fig. 4.21** for the first of the slopes and the voltage across the reference resistor applied to the integrator for the second of the slopes. The time for which the gate is open will be proportional to the unknown resistor and the display can be scaled in ohms. This method is insensitive to changes in the value of current flowing through the resistors, but the stability of the reference resistor is clearly important for the accuracy of the DMM in this mode.

4.4 Sensors

A sensor is a device which converts the quantity to be measured into another quantity which is easier to present as a number, to manipulate or to display. The human body of course employs sensors, the eye converting optical information, the ear sound information and so on into internal information suitable for further processing. The range of industrial sensors is vast since almost any physical effect relating two quantities can potentially be used as the basis of a sensor. Most conveniently, the output quantity is an electrical one since this makes available the array of tools of signal conditioning and computation.

Sensors can be categorized in a number of ways. Sensors are described as *active* when energy conversion takes place, such as the piezoelectric sensor converting pressure information into electrical information. Sensors are described as *passive* when external power is required, such as the rotary potentiometer converting angle into slider voltage when an electrical current flows.

Sensors can be categorized according to the quantity which is changed by the input quantity, such as resistive sensors for measuring temperature of pressure or light intensity or strain or magnetic field.

As always in measurement and instrumentation, the acceptable sensor must give information about the quantity to be measured and be as little sensitive as possible to all other influence quantities. This is by no means easy, since the list of resistive sensors above indicates that, for example, a strain gauge will also change its resistance with temperature, pressure, light intensity, magnetic field and so on. Careful design and intelligence in use is necessary for reliable measurement. The sensor must also be reliable, cheap, not unduly disturb the system to be measured and preferably not involve a major education programme on the part of the potential users.

To illustrate from among the vast list of sensors available, three common and typical examples will be described – the strain gauge, the piezoelectric accelerometer and the Linear Variable Differential Transformer (LVDT).

4.4.1 The strain gauge

A strain gauge is a sensor which converts change in length into a change in resistance and hence voltage. By carefully attaching the strain gauge to the surface of an object, dimensional changes in the object can be monitored and measured. A typical simple strain gauge is shown in **Fig. 4.23**. It consists of a metal foil formed into a grid structure by a photo-etching process situated on a resin film. The gauge is designed to have much more conductor along the axis to be measured than at right angles. As the gauge is strained, the resistance changes. The resistance is given by $R = \rho L/A$ where ρ is the resistivity of the material, L is the length and A the cross-sectional area. All three of these change with strain. For commonly used material the gauge factor, the fractional change in resistance caused by a fractional change in length, is close to 2. Alloys are selected with a low temperature coefficient of resistance and a low temperature coefficient of linear expansion.

Fig. 4.23 The strain gauge

If the strain gauge is connected in the Wheatstone bridge circuit shown in **Fig. 4.24**, then the output will change as the strain changes. Influence quantities such as temperature will also produce changes in the output. A useful method for significantly reducing these effects is to mount a second gauge in a suitable position. For the measurement of the strain of a cantilever, shown in **Fig. 4.25**, then the two gauges will experience dimensional changes which are equal but of opposite sign (one

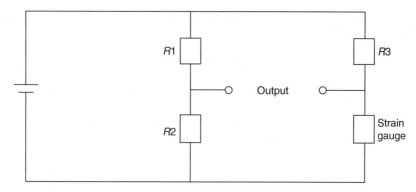

Fig. 4.24 Strain gauge in a Wheatstone bridge circuit

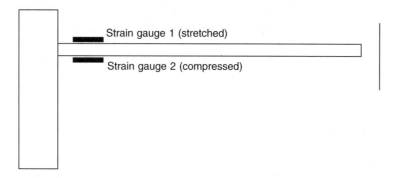

Fig. 4.25 Strain gauges mounted on a cantilever

stretched and the other compressed) but equal changes in resistance owing to influence quantities such as temperature. By connecting the gauges to adjacent arms of the Wheatstone bridge circuit, then the unwanted effects tend to cancel, whilst the effect of the change in resistance with strain is doubled.

4.4.2 The piezoelectric accelerometer

Changes in the vibrations of machinery give information about likely future failure of the equipment, and measurement of vibration using an accelerometer can therefore be used to inform maintenance actions in a cost effective manner. A common way of achieving this is by a piezoelectric accelerometer, shown in **Fig. 4.26**. As its name implies, a piezoelectric element converts pressure into an electrical quantity and vice versa. Usually in accelerometers a manufactured ceramic piezoelectric material such as PZT (lead zirconate titanate) is used. Acceleration causes the piezoelectric element to be compressed, thus giving an electrical output which can be used to monitor the acceleration. In a commercial device, care is taken to avoid sensitivity to acceleration in other axes. This is an active device and the output signal in electrical form is generated from the input signal in mechanical form. These types of device are also known as *transducers*. The piezoelectric accelerometer is usually operated well below its resonance frequency where the output varies little with frequency. There is no dc output from these devices, but when used with a charge amplifier good low-frequency performance can be achieved.

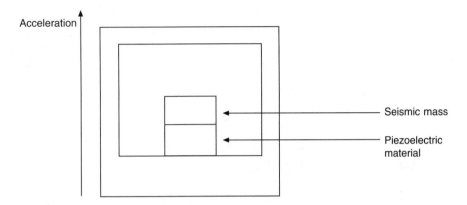

Fig. 4.26 The piezoelectric accelerometer

4.4.3   The linear variable differential transformer

The LVDT has found widespread acceptance as an industrial displacement sensor.

The construction of the LVDT is shown in **Fig. 4.27**. The sensor is a transformer that has a primary and two identical secondary windings which are connected in series opposition. The effect of this is that the output voltage is the difference between the two second voltages. A ferromagnetic rod made from a nickel–iron alloy moves inside the coils and as it moves the output voltage in one coil goes up and the other down. About the centre position, the phase of the signal reverses. LVDTs are made commercially with movement ranges from ±100 μm to ±25 cm.

The LVDT has some attractive features for industrial use. Since there is no mechanical connection between the core and the rest of the instrument friction is low, tolerance

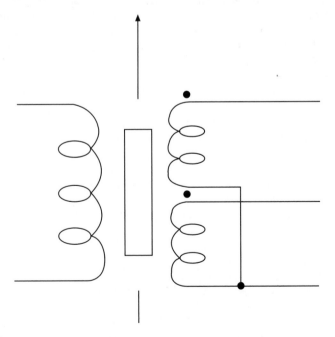

Fig. 4.27 The LVDT

to mechanical overload and shock is high and reliability is high. The windings can be set in epoxy resin and the sensor housed in a metal shield thus enabling the LVDT to be used in hostile environments. There is electrical isolation between the primary and secondary windings thus allowing flexibility in the choice of the common connection.

Because of these attractive features the LVDT is commonly the preferred method for industrial displacement measurement.

4.5   The international measurement system, traceability and calibration

Measurement is a comparison process and all useful measurements are ultimately related to the International Measurement System, the SI system of units. The base units of the SI system are the metre, kilogram, second, ampere, kelvin, candela and mole. All other measurement units are derived from these, but they are often given special names for ease in use. For example, the volt is the kg $m^2s^{-3}A^{-1}$ in terms of the base units, but fortunately the name volt is used instead.

The international community establishes the base units for comparison purposes. This provision is not well known by most of the people who benefit from it, but it is very important that the provision is there.

The process of relating the reading used say in manufacture to assess the performance of a product to the international measurement standards is called *traceability*. Traceability is the unbroken chain of comparisons performed in an acceptable way linking the measurement with the national or international measurement standards. The comparison with a more accurate instrument is termed *calibration*. The user gains traceability by having his instrument calibrated by an accredited calibration laboratory. This calibration laboratory is responsible for the other comparisons back to national or international measurement standards. In the UK, calibration laboratories demonstrating traceability are accredited by UKAS (United Kingdom Accreditation Service). The various accredited calibration laboratories offer their services on a competitive basis related to price, speed of service and a collect and delivery system.

Clearly a calibration is only valid at the instant of calibration and even then the measurement will be subject to the uncertainty associated with the calibration laboratory for that parameter and range. Following calibration, the stability of the instrument's readings is important. The reading of an instrument for the same quantity will change with time or drift and will also be affected by a number of *influence quantities*. An influence quantity is a quantity which is not the subject of the measurement but nevertheless changes the result of the measurement. Common influence quantities are temperature, humidity, power supply voltage and a range of radiations such as electromagnetic interference. It is usually possible to correct for these influence quantities. In some cases the manufacturer will give, for example, the temperature coefficient of the instrument. By means of measuring the difference between the temperature of the room in use from the temperature at which calibration was performed, often 23°C, the correction for temperature can be applied. These corrections are never completely perfect and it is necessary to estimate the residual uncertainty in the measurement from all such causes.

The change in the instrument reading for the same applied quantity with time necessitates regular calibration. As noted when discussing frequency counters, the change in an instrument reading is greatest when new. This means that recalibration is necessary more frequently initially. Depending on use, it might be necessary to recalibrate after 90 days and then once a year.

Many manufacturing and other organizations are obtaining registration to the international quality system standard ISO 9000. In order to obtain such registration, it is necessary to demonstrate the traceability of measurements affecting product quality, that the uncertainty of measurements is suitable and that a systematic procedure is in place for the regular calibration of instruments used.

References

4A.   Gillies, R.B. *Instrumentation and Measurements for Electronic Technicians,* Merrill, 1988.
4B.   Cooper, W.D. and Helfrick, A.D. *Electronic Instrumentation and Measurement Techniques,* Prentice Hall, 1988.
4C.   Bentley, J.P. *Principles of Measurement Systems*, Longman, 1995.
4D.   Pallas-Areny, R. and Webster, J.F. *Sensors and Signal Conditioning,* Wiley, 1991.
4E.   Cunningham, M.J. Measurement Errors and Instrument Inaccuracies, *J. Phys. E.,* 1981, Vol. 14, pp. 901–8.

Chapter 5

Generators

Dr G.W. McLean
Generac Corporation

5.1 Introduction

Throughout the world, whether in underdeveloped areas or in highly developed countries, there is a need for generators in many different applications.

In addition to the underlying need for a public supply of electricity, there are a number of situations in which independent supplies are needed. The applications for generators are categorized broadly:

- public supply networks in which a number of high-power generator sets may operate in parallel
- private or independent generators which may run in parallel with the public supply or isolated from it. Examples of this include:
  - *peak shaving* to reduce the maximum demand of electricity by a user; this can avoid large financial penalties during times of generally high demand on the system
  - *standby emergency generators* to protect the supply to critical circuits such as hospitals or water supplies
  - *temporary supplies* which are needed by the construction industry, or in cases of breakdown
  - *combined heat and power* where the waste heat from the generator engine is used for other purposes such as building heating
- portable supplies, often trailer-mounted, where no alternative supply is available

5.2 Main generator types

The two main types of generator are '*turbo*' or *cylindrical-rotor* and *salient-pole* generators. Both these types are *synchronous* machines in which the rotor turns in exact synchronism with the rotating magnetic field in the stator. Since most generators are in this class, it forms the basis of most of the chapter.

The largest generators used in major power stations are usually turbo-generators. They operate at high speeds and are usually directly coupled to a steam or gas turbine. The general construction of a turbo-generator is shown in **Fig. 5.1**. The rotor is made from solid steel for strength, and embedded in slots within the rotor are the field or excitation windings. The outer stator also contains windings which are located in slots, this is again for mechanical strength and so that the teeth between the slots form

(a)

(b)

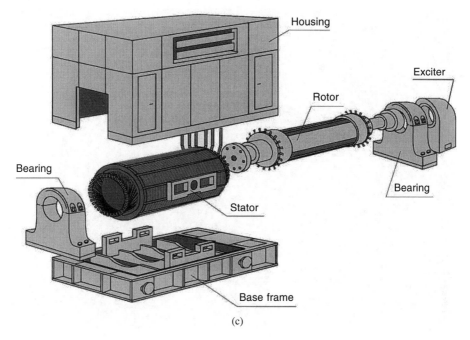

Fig. 5.1 Turbo-generator construction: (a) stator, (b) rotor, (c) assembly (courtesy of ABB ALSTOM Power)

a good magnetic path. Most of the constructional features are very specialized, such as hydrogen cooling instead of air, and direct water cooling inside the stator windings, so only passing reference is made to this class of machine in the following descriptions.

More commonly used in smaller and medium power ranges is the salient-pole generator. An example is shown in **Fig. 5.2**. Here, the rotor windings are wound around the poles which project from the centre of the rotor. The stator construction is similar in form to the turbo-generator stator shown in **Fig. 5.1**.

Less commonly used are *induction generators* and *inductor generators*.

Induction generators have a simple form of rotor construction shown in **Fig. 5.3**, in which aluminium bars are cast into a stack of laminations. These aluminium bars require no insulation and the rotor is therefore much cheaper to manufacture and much more reliable than the generators shown in **Figs 5.1** and **5.2**. The machine has characteristics which suit wind turbines very well, and they also provide a low-cost alternative for small portable generators. The basic action of the induction generator will be described later in this chapter, but both the construction and the operation of the machine are very similar to the induction motor described in **Chapter 10**.

Inductor generators use solid steel rotors cut with slots to produce a flux pulsation in the stator as the rotor turns. These machines are usually used at high speed for specialized applications requiring high frequency.

5.3   Principles of operation

5.3.1   No-load operation

The basic operation of all these generator types can be explained using two simple

Fig. 5.2 Salient-pole generator construction: Top: stator, bottom: rotor (courtesy of Generac Corporation)

Fig. 5.3 Induction generator construction (courtesy of Invensys Brook Crompton)

rules, the first for magnetic circuits and the second for the voltage induced in a
conductor when subjected to a varying magnetic field.

The means of producing a magnetic field using a current in an electric circuit have
been explained in **section 2.2.3**, and **eqs 2.13** and **2.18** have shown that the flux Φ
in a magnetic circuit which has a reluctance R_m is the result of a magneto-motive
force (mmf) F_m, which itself is the result of a current I flowing in a coil of N turns.

$$\Phi = F_m/R_m \tag{5.1}$$

and

$$F_m = IN \tag{5.2}$$

The main magnetic and electrical parts of a salient-pole generator are shown
diagrammatically in **Fig. 5.4**. In **Fig. 5.4(a)**, dc current is supplied to the rotor coils
through brushes and sliprings. The product of the current I and the coil turns N
results in mmf F_m as in **eqn 5.2**, and this acts on the reluctance of the magnetic circuit
to produce a magnetic flux, the path of which is shown by the broken lines in **Fig.
5.4(b)**. As the rotor turns, the flux pattern created by the mmf F_m turns with it; this
is illustrated by the second plot of magnetic flux in **Fig. 5.4(b)**. In **section 2.2.3** it has
also been explained that when a magnetic flux Φ passes through a magnetic circuit
with a cross-section A, the resulting flux density B is given by

$$B = \Phi/A \tag{5.3}$$

Figure 5.4(a) also shows a stator with a single coil with an axial length l. As the rotor
turns its magnetic flux crosses this stator coil with a velocity v. It has been explained
in **section 2.2.4** that an electromotive force (emf) V will be generated, where

$$V = Bvl \tag{5.4}$$

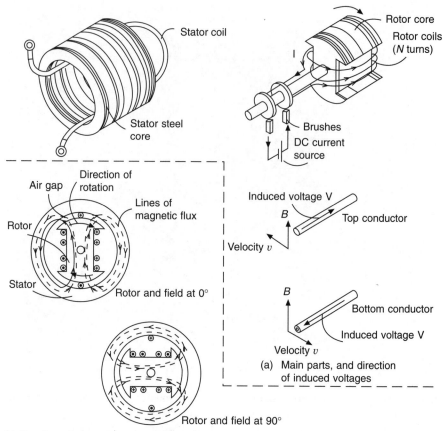

Fig. 5.4 Principles of generator operation

The direction of the voltage is given by Fleming's right-hand rule, as shown in **Fig. 2.6**.

Figure 5.4(b) shows that as the magnetic field rotates, the flux density at the stator coil changes. When the pole face is next to the coil, the air gap flux density B is at its highest, and B falls to zero when the pole is 90° away from the coil. The induced emf or voltage V therefore varies with time (**Fig. 5.5**) in the same pattern as the flux density varies around the rotor periphery. The waveform is repeated for each revolution of the rotor; if the rotor speed is 3000 rpm (or 50 rev/s) then the voltage will pass through 50 cycles/s (or 50 Hz). This is the way in which the frequency of the electricity supply from the generator is established. The case shown in **Fig. 5.4** is a 2-pole rotor, but if a 4-pole rotor were run at 1500 rpm, although the speed is lower, the number of voltage alternations within a revolution is doubled, and a frequency of 50 Hz would also result. The general rule relating the synchronous speed n_s (rpm), number of poles p and the generated frequency f (Hz) is given by

$$f = n_s p / 120 \qquad (5.5)$$

The simple voltage output shown in **Fig. 5.5** can be delivered to the point of use (the 'load') with a pair of wires and this form of supply is known as single phase. If more

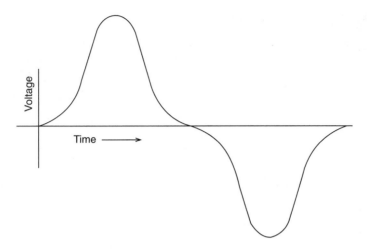

Fig. 5.5 Induced voltage waveform

coils are added to the stator shown in **Fig. 5.4(a)**, and if these are equally spaced, then a three-phase output as shown in **Fig. 5.6** can be generated. The three phases are conventionally labelled 'U', 'V' and 'W'. The positive voltage peaks occur equally spaced, one-third of a cycle apart from each other. The nature of single-phase and three-phase circuits has been explained in **Chapter 2**. The three coils either supply three separate loads, as shown in **Fig. 5.7(a)** for three electric heating elements, or more usually they are arranged in either 'star' or 'delta' in a conventional three-phase circuit (**Fig. 5.7(b)**).

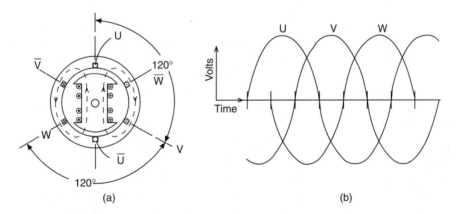

Fig. 5.6 Three-phase generation

In a practical generator the stator windings are embedded in slots, the induced voltage remaining the same as if the winding is in the gap as shown in **Fig. 5.4(b)**. Also, in a practical machine there will be more than the six slots shown in **Fig. 5.6(a)**. This is arranged by splitting the simple coils shown into several subcoils which occupy separate slots, each phase still being connected together to form a continuous winding. **Figures 5.1** and **5.2** show the complexity that results in a complete stator winding.

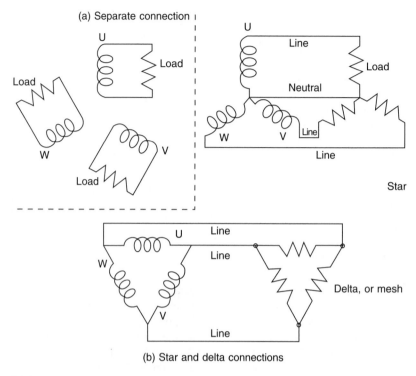

(b) Star and delta connections

Fig. 5.7 Three-phase connections

5.3.2   The effect of load

In the circuits shown in **Fig. 5.7** currents flow in each phase, and these currents will
have a waveform similar to the voltage waveform shown in **Fig. 5.6(b)**. The concept
of a phase shift between voltage and current in ac circuits which contain inductance
or capacitance has been explained in **section 2.3.2**. If an inductive or capacitive load
is connected, then the current waveforms will respectively 'lag' or 'lead' the voltage
waveforms by 90°. For the inductive load case shown in **Fig. 5.6(a)**, the current in
the U phase will be zero, but current will be flowing in V and W phases. It can be
seen that the lines of magnetic flux now enclose not only the rotor excitation current,
but also the stator currents in the V and W phases. **Equations 5.1** and **5.2** show that
the flux is the result of the mmf acting on a magnetic circuit, but it can now be seen
that the mmf is a combination of the ampere-turns from the rotor and the stator
winding. If I_r, I_s, N_r, and N_s are the currents and turns in the stator and rotor windings
respectively then **eqns 5.1** and **5.2** combine to give

$$\Phi = (I_r N_r + I_s N_s)/R_m \tag{5.6}$$

In **Fig. 5.6** a cross is used to indicate a current flow into the page, and a dot shows
current flowing out of the page. It is seen that the stator currents oppose the excitation
current in the rotor and their effect is to reduce the flux, with a corresponding
reduction in the generated voltage. This demagnetizing effect is called '*armature
reaction*'; it is the way in which Lenz's law (**section 2.2.4**) operates in a generator.
   The underlined currents in **eqn 5.6** indicate that these are vectors and a vector
addition is necessary. The armature reaction effect therefore depends on the extent to

which the stator currents lag or lead the voltages (often called the 'phase' or 'phase angle'). If, for example, the generator load is capacitive, the currents will lead the voltages by 90°, and they will be opposite in direction to that shown in **Fig. 5.6** for an inductive load. The ampere-turns of stator and rotor windings will add in this case and the flux and the generated voltage will be higher. In the case of a resistive load, the ampere-turns of the stator will act at 90° to the rotor poles, tending to concentrate the flux towards the trailing edge of the pole and producing magnetic saturation here when large stator currents flow; this reduces the flux and the output voltage, but not so much as in the inductive load case.

The output voltage is influenced not only by armature reaction, but also by voltage drop within the stator winding. This voltage drop is partly due to the internal resistance of the winding, and partly due to flux which links the stator winding but not the rotor winding; this flux is known as '*leakage flux*' and it appears in the stator electrical circuit as a *leakage inductance*, which also creates a voltage drop. The phase angle between stator currents and voltages will affect this voltage drop, producing a greater drop at lagging currents, and a negative drop (an increase) in voltage at leading currents.

In order to maintain a constant output voltage it is therefore necessary to change the excitation current in the rotor to compensate for the load conditions. The variation in rotor current to do this is shown in **Fig. 5.8**.

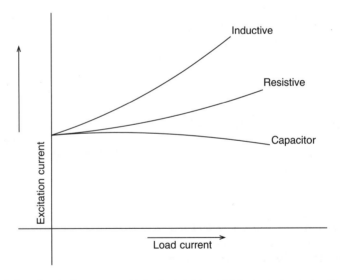

Fig. 5.8 Variation of excitation current with load current to maintain constant output voltage

5.3.3 Damping of transients

Transient changes in stator load result in a change of flux in the rotor pole, and if it can be arranged that this flux change induces a voltage and a flow of current in the pole face, this current will oppose the change in stator flux. To achieve this, it is normal to insert into the pole face a set of aluminium or copper 'damper' bars, connected at either end by a ring or end plate to form a conducting cage in the pole faces.

The *damper cage* has a considerable influence on the transient current flow in the stator, particularly in the case of a short circuit. In addition, if the load in the three phases is unbalanced, the induced currents in the damper cage will act to reduce distortion of the waveform and to reduce asymmetry in the output phase voltages. A single-phase generator represents a severe case of asymmetry, and this requires very careful damper cage design because of the high induced currents.

The cage also helps to damp mechanical oscillations of the rotor speed about the synchronous speed when the generator is connected in parallel with other machines. These oscillations might otherwise become unstable, leading to the poles 'slipping' in relation to the frequency set by other generators, and resulting in a loss of synchronism. Such a condition would be detected immediately by the generator protection circuits and the generator would then be isolated from the network.

5.3.4 Voltage waveform

The specified voltage waveform for a generator is usually a sine wave with minimum distortion. A sine wave supply has advantages for many loads because it minimizes the losses in the equipment; this is especially the case with motors and transformers. The voltage waveform of a practical generator usually contains some distortion or harmonics, as shown in **Fig. 5.9**. The distorted waveform shown in **Fig. 5.9(a)** can be represented as a series of harmonics, consisting of the fundamental required frequency and a series of higher frequencies which are multiples of the fundamental frequency.

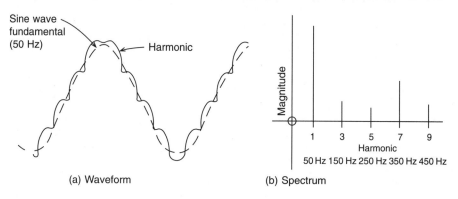

(a) Waveform   (b) Spectrum

Fig. 5.9 Harmonic distortion

The harmonic distortion is calculated using a Fourier analysis or other means of obtaining the spectrum of harmonics. An example of a spectrum is shown in **Fig. 5.9(b)**. Distortion is defined by a '*distortion factor*', where

$$\text{Distortion factor} = (\Sigma V_n^2)^{1/2}/V_1 \qquad (5.7)$$

In **eqn 5.7**, V_n is the magnitude of the nth harmonic, V_1 being the magnitude of the fundamental.

There are several ways in which a generator can be designed to produce minimum distortion factor.

The higher frequency '*slot harmonics*' are due to distortions in the air gap flux density wave, these being created by the stator slot openings. The distortions can be reduced by skewing the stator slots so that they are no longer parallel to the rotor shaft, but form part of a helix. The slots are often skewed by an amount close to the pitch between one stator slot and the next.

A second step is to use more than one stator slot per pole for each phase winding;

Fig. 5.10 shows a winding with three stator slots per pole for each phase. This distributes the effect of the winding better and reduces the harmonics.

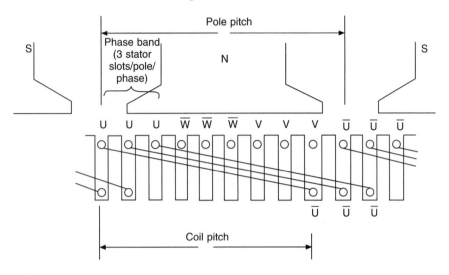

Fig. 5.10 Slot and winding layout

Harmonics can also be reduced by '*short-pitching*' the stator winding as shown in **Fig. 5.10**. Except for the smallest machines, most generators have a double layer winding in which the one side of a coil is laid into a slot above the return side of a different coil. The simplest (fully pitched) winding has all the coil sides of one phase in slots above the return sides of coils in that same phase, but by displacing one layer of the winding with respect to the other, the harmonics are reduced.

A fourth technique is to shape the face of the rotor pole so that the air gap between rotor and stator is larger at the tips. This prevents a 'flat-topped' shape to the flux wave in the air gap and therefore reduces voltage distortion.

Finally, correct spacing of the damper bars in the pole face and proper choice of the arc length of the pole face reduces the high frequency harmonics which are produced by currents induced in the bars by the passage of the stator slots.

5.4   The Automatic Voltage Regulator (AVR)

While some small generators have an inherent ability to produce reasonably constant voltage as the load varies, it is clear from the previous explanations that some form of automatic voltage control is required in the usual form of generator. This control is referred to as an automatic voltage regulator, or AVR, and it is based on a closed-loop control principle.

The basis of this closed-loop control is shown in **Fig. 5.11**. The output voltage is converted, usually through a transformer or resistor network, to a low voltage dc signal, and this 'feedback' signal is subtracted from a fixed 'reference' voltage to produce an 'error' signal.

The error signal is processed by a 'compensator' before being amplified to drive the rotor excitation current. The change in rotor excitation current produces a variation in output voltage, closing the control loop. If the 'gain' of the control loop is large

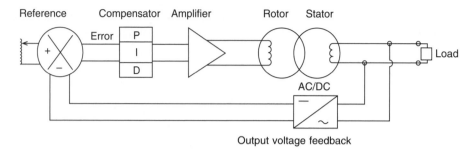

Fig. 5.11 Closed-loop voltage control

enough then only a small error is required to produce the necessary change in excitation current and the output current will therefore remain substantially constant.

A high gain can lead to instability in the circuit, with oscillations in the output voltage; the purpose of the compensating circuit is to enable small errors to be handled in a stable way. The most common form of compensator is a PID circuit in which the error is amplified proportionately (P), integrated (I) and differentiated (D) in three parallel circuits before being added together. Many AVRs have adjustment potentiometers which allow the gains of each channel to be varied in order to achieve the best performance. The integral term enables compensator output to be achieved at zero error, and this produces the minimum error in output voltage.

The layout of a commercial AVR is shown in **Fig. 5.12**.

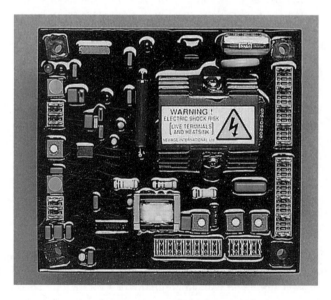

Fig. 5.12 Layout of an automatic voltage regulator (AVR) (courtesy of Newage International)

Many AVRs are now offered with digital circuitry. The principle of the feedback loop remains the same, but the feedback signal is converted to digital form using an analogue to digital converter. The calculations are performed digitally in a microprocessor and the output is on or off, using pulse width modulation (PWM) to vary the average level of dc supplied to the rotor excitation winding. Alternatively, the phase angle of a thyristor bridge can be used to vary the output level; this is known as phase-angle control.

The continuous improvement in power electronic controls and processor power is bringing further advances in voltage and speed control, with more flexible protection of the generator and its connected circuits. An example of recent developments is the variable-speed constant-frequency generator from Generac Corporation; this is illustrated in **Fig. 5.13**. This consists of an variable-speed engine-driven permanent-magnet generator feeding a power electronic frequency-changer circuit, the output being at constant 50 Hz or 60 Hz frequency. A microprocessor is used to control the switching of the output devices and to regulate the engine speed depending upon the load applied to the generator. At low power demand the engine speed is reduced to minimize noise, increase efficiency and extend life expectancy. The result is a saving in the volume and weight of the generator.

Fig. 5.13 Variable-speed constant-frequency generator (courtesy of Generac Corporation)

5.5 Brushless excitation

Although some generators are still produced with brushes and sliprings to provide the rotor current as illustrated in **Fig. 5.4**, most now have a brushless excitation system. The two main techniques for synchronous generators are the separate exciter and capacitor excitation and these are described in the following sections. Also included for convenience here is a brief description of induction generators, since these also provide a brushless system.

5.5.1 Separate exciter

The most common way of supplying dc current to the rotor winding without brushes
and sliprings is shown in **Fig. 5.14**.

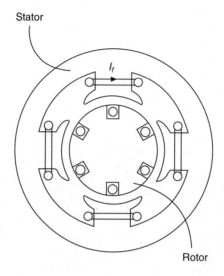

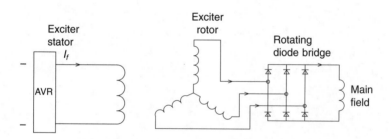

Fig. 5.14 Separate brushless exciter

The output of the AVR drives a dc current I_f through the poles of the *exciter*, which
are mounted in a stator frame. The stator then produces a stationary flux which
induces a voltage in the rotor winding as it turns. **Figure 5.15** shows that the exciter
rotor is mounted on the same shaft as the main generator. The ac voltage produced
by the rotor winding of the exciter is converted to dc by a bridge rectifier which is
also mounted on the rotor shaft. This rectifier unit is shown clearly at the end of the
shaft in **Fig. 5.18**. The dc output of the rectifier is connected to the main rotor
windings by conductors laid in a slot along the rotor shaft. The inductance of the
main generator rotor coils is usually sufficient to smooth out the ripple in the bridge
rectifier output.

The power supply to the AVR is either provided by a separate excitation winding
in the main generator stator, or by a small permanent-magnet generator mounted on

Fig. 5.15 Cut-away section of an ac generator (courtesy of Newage International)

the shaft of the main generator. This is often referred to as a '*pilot exciter*'. The advantage of the pilot exciter is that the generator has a source of power available once the shaft is turning; the voltage supplied to the AVR is completely independent of generator load and there is no reliance on residual magnetism to start the self-excitation process. The first method relies on residual flux in the magnetic circuit of the main generator to start self-excitation.

The pilot exciter also enables the generator to supply current to the connected network even when a short-circuit occurs, enabling the high current to be detected by protection relays which will then disconnect the faulty circuit. If the AVR is supplied from an excitation winding in the main generator stator, the supply voltage is very small when the stator windings experience a short-circuit, and the AVR is unable to drive an adequate rotor excitation current.

One manufacturer uses two excitation windings to provide a voltage from the AVR in short-circuit conditions, so that sufficient current is supplied into the fault to trip the protection system. During a short circuit the air gap flux density in these machines shows a pronounced third harmonic component. This component induces voltage in coils of one of the excitation windings, which are one-third pitched and this therefore delivers a voltage to the AVR under short-circuit conditions. The second excitation winding is fundamental-pitched and provides the major drive for the AVR under normal operating conditions. It is claimed that the performance of this system is comparable to a machine using a permanent-magnet exciter.

Another method used to provide voltage to the AVR under short-circuit conditions is a series transformer driven by the generator output current.

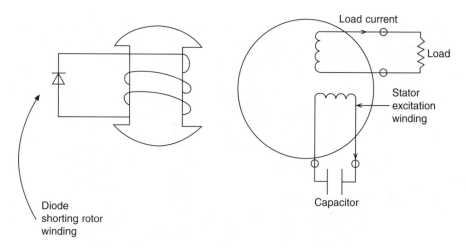

Fig. 5.16 Capacitor excitation

5.5.2 Capacitor excitation

The use of this technique is usually restricted to single-phase generators with a rated output less than 10 kW.

A separate excitation winding in the stator has a capacitor connected directly across its output as shown in **Fig. 5.16**. The rotor is usually of salient-pole construction as described previously, but in this case the rotor winding is shorted through a diode. During starting, residual flux in the rotor body induces a small voltage in the stator excitation winding and a current flows through the capacitor. This current produces two waves of magnetic flux around the air gap of the generator. One wave travels in the same direction as the rotor, to create the armature reaction described in **section 5.3.2**. The second wave travels in a direction opposite to the rotor, and induces a voltage in the rotor windings at twice output frequency. The current circulated in the rotor windings by this induced voltage is rectified by the diode to produce a dc current. This dc current increases the magnetic flux in the machine, which in turn drives more current through the stator excitation winding, which in turn produces more rotor current. This self-excitation process continues until the flux reaches a point at which the magnetic circuit is saturated, and a stable voltage results. The process also produces an inherent AVR action, since any load current in the output stator winding induces more rotor current to offset the armature reaction effect.

5.5.3 Induction generator

The principles and construction of the cage induction motor are explained in **Chapter 10**. If a three-phase motor of this type is energized, it will accelerate as a motor up to a speed near its synchronous speed. As shown in **eqn 5.5**, the synchronous speed for a 4-pole machine operated at 50 Hz would be 1500 rpm. If the machine is driven faster than the synchronous speed by an engine or other prime mover, the machine torque reverses and electrical power is delivered into the connected circuit.

A simple form of wind turbine generator uses an induction machine driven by the wind turbine. The induction machine is first connected to the three-phase supply, and acting as a motor it accelerates the turbine up to near the synchronous speed. At this point, the torque delivered by the wind turbine is sufficient to accelerate the unit further, the speed exceeds the synchronous speed and the induction machine becomes a generator.

It is also possible to operate an induction machine as a generator where there is no separate mains supply available. It is necessary in this case to self-excite the machine, and this is done by connecting capacitors across the stator winding as shown in **Fig. 5.17(a)**. The leading current circulating through the capacitor and the winding produces a travelling wave of mmf acting on the magnetic circuit of the machine. This travelling wave induces currents in the rotor cage which in turn produce the travelling flux wave necessary to induce the stator voltage. For this purpose, some machines have a excitation winding in the stator which is separate from the main stator output winding. **Figure 5.17(b)** shows a single-phase version of the capacitor excitation circuit.

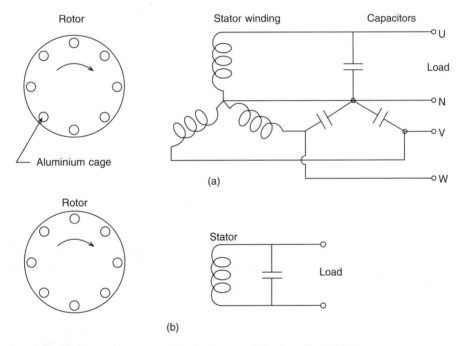

Fig. 5.17 Self-excitation of an induction generator: (a) three phase (b) single phase

In small sizes, the induction generator can provide a low-cost alternative to the synchronous generator, but it has relatively poor performance when supplying a low power factor load.

5.6 Construction

5.6.1 Stator

The stator core is constructed from a stack of thin steel sheets or laminations which are stamped to shape and insulated electrically from each other, either by a thin coating or by an oxide layer which is produced during heat treatment. The steel used has a small silicon content as described in **Chapter 3**; this increases the resistivity of the steel and therefore reduces the losses due to eddy currents. The steel is carefully processed in order to minimize the hysteresis losses because the whole stator core is subjected to alternating magnetic flux. In large turbo-generators the core is built up in segments and grain-oriented steel is used to reduce the losses further.

The stator windings are located in axial slots in the stator core which are formed by the shape of the laminations. Except in high voltage machines, the individual coils of the winding are wound with copper wire covered with a layer of polyester/polyamide enamel which is about 0.05 mm thick. The slots are lined with a tough insulating sheet, usually about 0.25–0.5 mm thick; a popular material is a laminate of Mylar and Nomex. The coils are impregnated when in place with a resin to give the winding mechanical strength as well as to improve the heat transfer from the copper to the cooling air. Windings operating at different voltages, such as the three phases, have a further sheet of insulating material separating them in the end-winding region. A general guide to the insulation materials and processes in use is given in **Chapter 3**.

The mounting of the stator core in its frame differs according to the size of the machine, but in the majority of medium-sized generators the arrangement is as shown in **Fig. 5.15**. At either end of the frame are bearing housings which locate the rotor shaft. These housings or end-bells are cast in smaller and medium size machines, and fabricated in larger sizes. The generator is often mounted directly onto the engine, and in this case it is usual to eliminate the drive-end bearing and to use the rear bearing of the engine to locate the generator shaft.

5.6.2 Rotor

It has already been noted that the construction of a turbo-generator is very specialized and the rotor for these machines is not dealt with here. However, even within the class of salient-pole generators, quite different forms of rotor construction are used, depending upon size.

Generators rated up to about 500 kW use rotor laminations which are stamped in one piece. In larger machines the poles are made separately from stacks of laminations, and each pole is keyed using a dovetail arrangement onto a spider which is mounted on the rotor shaft. In large high-speed machines the poles can be made from solid steel for extra strength and to reduce mechanical distortion; these solid poles are screwed to the shaft, as shown in the large 4-pole machine in **Fig. 5.18**.

The nature of the rotor coils also depends upon the size of the machine. Because the ratio of surface area to volume is larger in the coils of small generators, these are easier to cool. Generators rated above about 25 kW therefore use a 'layer-wound' coil in which each layer of the coil fits exactly into the grooves formed by the layer below; this is illustrated in **Fig. 5.19**. Rectangular cross-section wire can be used to minimize the coil cross-section. The simplest and cheapest way to make the coils, often used in smaller machines, is to wind them in a semi-random way as shown in **Fig. 5.20**. In either case, the coils are impregnated after winding like the stator windings to give extra mechanical strength and to improve the heat transfer by removing air voids within the coil.

The coils are under considerable centrifugal stress when the rotor turns at full speed, and they are usually restrained at both ends of the pole by bars, and by wedges in the interpole spaces, as shown in **Fig. 5.18**.

5.6.3 Cooling

Adequate cooling is a vital part of the design and performance of a generator. Forced

Fig. 5.18 Large salient-pole rotor (courtesy of Brush Electrical Machines)

cooling is needed because of the high loss densities that are necessary to make economic use of the magnetic and electrical materials in the generator.

The most critical areas in the machine are the windings, and particularly the rotor winding. It is explained in **Chapter 3** that the life expectancy of an insulation system decreases rapidly if its operating temperature exceeds recommended temperatures. It is crucial for reliability therefore that the cooling system is designed to maintain the winding temperatures within these recommended limits. As shown in **Table 3.4**, insulation materials and systems are defined by a series of letters according to their temperature capability. As improved insulation with higher temperature capability has become available, this has been adopted in generator windings and so the usual class of insulation has progressed from class A (40–50 years ago) through class E and class B to class F and class H, the latter two being the systems generally in use at present. Class H materials are available and proven, and this system is becoming increasingly accepted. An important part of the generator testing process is to ensure that the cooling system maintains the winding temperature within specified limits, and this is explained later in **section 5.8**.

In turbo-generators it has already been noted that very complex systems using hydrogen and de-ionized water within the stator coils are used. The cooling of small and medium-sized machines is achieved by a flow of air driven around the machine by a rotor-mounted fan.

A typical arrangement is shown in **Fig. 5.15**. In this case the cooling air is drawn in through ducts; it is then drawn through the air gap of the machine and ducts around the back of the stator core before reaching the centrifugal fan which then expels the

Fig. 5.19 Layer-wound rotor (courtesy of Generac Corporation)

air from the machine. There are many variations to the cooling system, particularly for larger generators. In some machines there is a closed circulated air path cooled by a secondary heat exchanger which rejects the heat to the outside atmosphere. This results in a large and often rectangular generator enclosure as shown in **Fig. 5.21**.

5.7   Rating and specification

In order that a generator can be selected to suit a particular application, manufacturers issue specification data. This can be used and interpreted according to the following sections.

5.7.1   Rated output

The key aspect of the specification is the rated output of the generator, which is normally expressed in terms of the apparent power (VA, kVA or MVA) when supplying the maximum load at the rated power factor, assumed to be 0.8. The rated output is usually based upon continuous operation in a maximum ambient temperature of

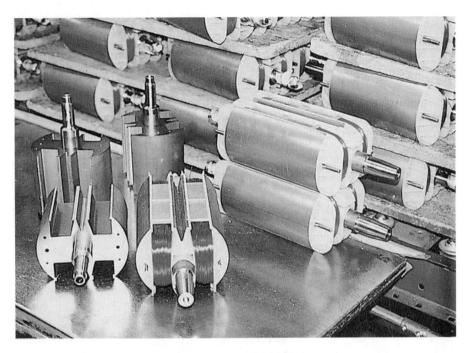

Fig. 5.20 Random-wound rotors (courtesy of Generac Corporation)

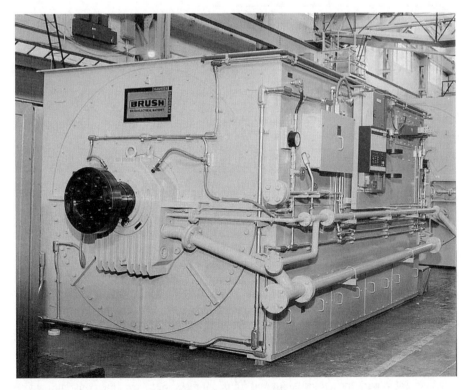

Fig. 5.21 Closed air circuit cooled generator (courtesy of Brush Electrical Machines)

40°C. If the machine has a special short-time rating the nameplate should state the time limits of operation.

The rated output from a given size of generator is related to the size and speed of the machine:

$$\text{Rated output power} = K \times D_g^2 \times L_c \times n_s \qquad (5.8)$$

where D_g is the stator bore diameter, L_c is the stator core length and n_s is the speed. K is a design constant which is proportional to the product of the air gap flux density and the stator current density.

The rated output of a machine is reduced in ambient temperatures exceeding 40°C and at altitudes above sea level exceeding 1000 m. The latter is because the air density is decreased, and its ability to cool the machine is reduced. Derating factors are applied for these conditions and typical values are summarized in **Tables 5.1** and **5.2**.

Table 5.1 Typical derating factors for class H insulation in ambient temperatures above 40°C

Ambient temperature (°C)	Derating factor
40	1.00
45	0.97
50	0.94
55	0.91
60	0.88

Table 5.2 Typical altitude derating factors

Altitude (m)	Derating factor
1000	1.00
1500	0.97
2000	0.97
2500	0.94
3000	0.87
3500	0.82
4000	0.77

5.7.2 Reactances

The generator can be characterized by several reactances, each being useful in working out the performance and protection requirements under different circumstances. These include the *synchronous reactance*, the *transient reactance*, the *subtransient reactance*, the *Potier reactance* and the *negative-* and *zero-sequence reactances*.

The subtransient reactance represents the output impedance of the generator within the first few cycles after a short-circuit occurs at the generator terminals. It is used for selecting protective relays for the connected load circuit. The lower the value of the subtransient reactance, the more onerous is the protection requirement.

Transient reactance represents the impedance of the machine over a slightly longer period, and is relevant to the performance of the generator and its AVR under changing load conditions. A low transient reactance is beneficial in responding to load changes.

Associated with the subtransient and transient reactances are time constants which define the rate of decay of these reactances.

The negative and positive sequence reactances influence the performance of the generator when supplying an unbalanced three-phase load.

5.7.3 Main items of specification

In summary, the following items are important when considering an application and specifying the appropriate generator:

- rated output, expressed as apparent power (VA)
- cooling air requirement (m^3/s)
- moment of inertia of the rotor
- generator weight
- efficiency at full, 3/4, 1/2 and 1/4 load
- stator winding resistance
- reactances and time constants, as listed in **section 5.7.2**
- maximum short-circuit current delivered by the generator

Typical values for a 200 kW generator are shown in **Table 5.3**.

Table 5.3 Typical parameters for a 200 kW generator

Line–line voltage, frequency		400 V, 50 Hz
Cooling air flow		0.4 m^3/s
Rotor moment of inertia		1.9 kg m^2
Weight		600 kg
Efficiency	full load	93%
	3/4 full load	94%
	1/2 full load	94%
	1/4 full load	93%
Stator resistance		0.025 Ω
Rotor resistance		1.9 Ω
Reactances	synchronous, X_d	1.9 pu
	transient, X_d^1	0.2 pu
	subtransient, X_d^{11}	0.1 pu
	positive sequence	0.07 pu
	negative sequence	0.1 pu
	zero sequence	0.07 pu
Time constants	transient, T_{d0}^1	0.8 s
	subtransient, T_d^{11}	0.01 s
Short-circuit current		2.5 pu

5.8 Testing

Type tests are performed by manufacturers in order to confirm that a design meets its specifications, and production tests are done in order to check that each machine as it is manufactured conforms to performance and safety standards.

These tests usually include:

- *full-load tests* to measure the temperature rise of the machine windings and insulation. The temperature rise is calculated from the change in resistance of the stator windings.
- *tests to determine the excitation current* required to deliver a given output

voltage. These are done for open-circuit conditions and also for various load currents and power factors. The resulting curves are usually known as 'saturation' curves.

- *short-circuit tests* to determine the current that can be driven by the generator into a short-circuit fault in the connected load
- *transient short-circuit tests* to determine the subtransient and transient reactances and time constants
- *overspeed tests* to confirm that the rotor does not distort or disintegrate. This test is normally performed at 150 per cent of rated speed, at full rated temperature.

The results from these tests are used to calculate the data described in **section 5.7**, an example of which has been shown in **Table 5.3**.

It is now necessary for any generator manufactured or imported into the European Union that relevant EU Directives are met through certification. Strictly it is the manufacturer of the generator set, including the engine and all controls, that is responsible for this certification, but many manufacturers of generators and AVRs will assist in the tests.

5.9 Standards

The leading international, regional and national standards adopted by users and suppliers of generators for manufacturing and testing are shown in **Table 5.4**.

Table 5.4 International, regional and national standards relating to generators

IEC/ ISO	EN/HD	BS	Subject of standard	N. American
IEC 34-1	EN 60034-1	4999-101	Ratings and performance	
IEC 34-2A		4999-102	Losses and efficiency	
IEC 34-4	EN 60034-4	4999-104	Synchronous machine quantities	
IEC 34-6	EN 60034-6	4999-106	Methods of cooling	
IEC 34-8	HD 53.8	4999-108	Terminal markings	
		4999-140	Voltage regulation, parallel operation	
IEC 34-14/ ISO 2373		4999-142	Vibration	
IEC 529	EN 60529	5490	Degrees of protection	
ISO 8528-3		7698-3	Generators for generating sets	
ISO 8528-8		7698-8	Low power generating sets	
IEC 335-1	EN 60335-1	3456-1	Safety of electrical appliances	
	EN 60742	3535-1	Isolating transformers	
			Motors and generators	NEMA MG 1
			Cylindrical-rotor synchronous generators	IEEE C50.13
			GT-driven cyl-rotor synch generators	IEEE C50.14

Chapter 6

Transformers

Professor D.J. Allan
ALSTOM T&D – Transformers

6.1 Principles of operation

In simple form, a transformer consists of two windings connected by a magnetic core. One winding is connected to a power supply and the other to a load. The circuit containing the load may operate at a voltage which differs widely from the supply voltage, and the supply voltage is modified through the transformer to match the load voltage.

In a practical transformer there may be more than two windings as well as the magnetic core, and there is the need for an insulation system and leads and bushings to allow connection to different circuits. Larger units are housed within a tank for protection and to contain oil for insulation and cooling.

In the simplest case, with no load current flowing, the transformer can be represented by two windings on a common core, as shown in **Fig. 6.1**. It has been explained in **Chapter 2** (**eqns 2.25** and **2.26**) that the input and output voltages and currents in a transformer are related by the number of turns in these two windings, which are usually called the *primary* and *secondary* windings. These equations are repeated here for convenience.

$$V_1/V_2 = N_1/N_2 \qquad (6.1)$$

$$I_1/I_2 = N_2/N_1 \qquad (6.2)$$

The magnetic flux density in the core is determined by the voltage per turn:

$$V_1/N_1 = 4.44 f B_m A \qquad (6.3)$$

In the no-load case, a small current I_0 flows to supply the magnetomotive force which drives the magnetic flux around the transformer core; this current lags the primary voltage by almost 90°. I_0 is limited in magnitude by the effective resistance (R_c) and reactance (X_c) of the magnetizing circuit, as shown in **Fig. 6.2**. The magnetizing current is typically 2 per cent to 5 per cent of the full load current and it has a power factor in the range 0.1 to 0.2.

When the transformer is loaded, there is an internal voltage drop due to the current flowing through each winding. The voltage drops due to the primary and secondary winding resistances (R_1 and R_2) are in phase with the winding voltage, and the voltage drops due to the primary and secondary winding leakage reactances (X_1 and X_2) lag the winding voltage by 90°. The leakage reactances represent those parts of the transformer flux which do not link both windings; they exist due to the flow of

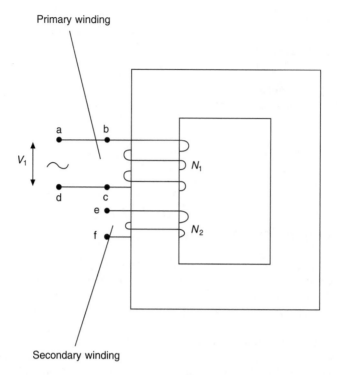

Fig. 6.1 Simple transformer circuit

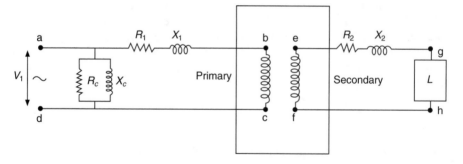

Fig. 6.2 Equivalent circuit of a transformer

opposing currents in each winding and they are affected strongly by the winding geometry.

The current flow and voltage drops within the windings can be calculated using the equivalent circuit shown in **Fig. 6.2**. This circuit is valid for frequencies up to 2 kHz. A vectorial representation of the voltages and currents is shown in **Fig. 6.3** for the case of a load L with a power factor angle Φ.

The decrease in output voltage when a transformer is on load is known as *regulation*. The output voltage is less than the open-circuit voltage calculated according to **eqn 6.1** because of voltage drops within the winding when load current flows through the resistive and reactive components shown in **Fig. 6.2**. The resistive drops are usually much smaller than the reactive voltage drops, especially in large transformers, so the impedance Z of the transformer is predominantly reactive. Regulation is usually

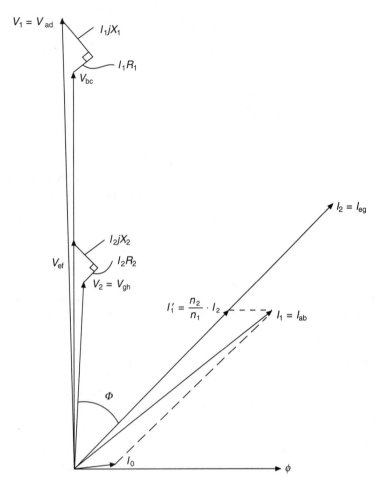

Fig. 6.3 Vector diagram of a transformer on load

expressed as a percentage value relating the vector addition of the internal voltage drops (shown in **Fig. 6.3**) to the applied voltage.

For many applications, the equivalent circuit shown in **Fig. 6.2** can be simplified by ignoring R_c and X_c and referring R_2, X_2 and L to the primary side as shown in **Fig. 6.4**, where:

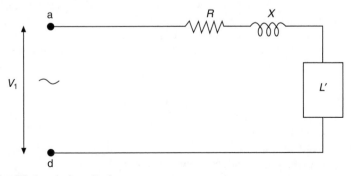

Fig. 6.4 Simplified equivalent circuit

$$R = R_1 + R_2(N_1/N_2)^2 \tag{6.4}$$

$$X = X_1 + X_2(N_1/N_2)^2 \tag{6.5}$$

and

$$L' = L(N_1/N_2)^2 \tag{6.6}$$

The impedance of the transformer is given by:

$$Z = (R^2 + X^2)^{1/2} \tag{6.7}$$

Values for R_1 and R_2 can be established by measuring the resistance of the windings. The value of X is determined by calculation or by derivation from the total impedance Z, which can be measured with one winding of the transformer short-circuited. Z is given by

$$Z = V/I \tag{6.8}$$

where V is the voltage necessary to circulate the full-load current I in the windings under short-circuit conditions. When V is expressed as a percentage of rated voltage, this gives Z as a percentage value referred to rated power.

When the transformer is energized, but without a load applied, the no-load power loss is due to the magnetic characteristics of the core material used and the flow of eddy currents in the core laminations. The loss due to magnetizing current flowing in the winding is small and can be ignored.

When the transformer is loaded, the no-load loss is combined with a larger component of loss due to the flow of load current through the winding resistance. Additional losses on load are due to eddy currents flowing within the conductors and to circulating currents which flow in metallic structural parts of the transformer. The circulating currents are induced by leakage flux which is generated by the load current flowing in the windings and they are load dependent. These additional losses are known as *stray losses*.

The efficiency of a transformer is expressed as:

$$\text{Efficiency} = (\text{output/input}) \times 100\%$$

$$= (\text{input} - \text{losses})/\text{input} \times 100\%$$

$$= [1 - (\text{losses/input})] \times 100\% \tag{6.9}$$

where input, output and losses are all expressed in units of power.

The total losses consist of the no-load loss (or iron loss) which is constant with voltage and the load loss (or copper loss) which is proportional to the square of load current. Total losses are usually less than 2 per cent for distribution transformers, and they may be as low as 0.5 per cent in very large transformers.

No-load and load losses are often specified as target values by the user for larger transformers, or they may be evaluated by the *capitalization* of losses. The capitalization formula is of the type:

$$C_c = C_T + A \times P_0 + B \times P_K \tag{6.10}$$

where C_c = capitalized cost
C_T = tendered price
A = capitalization rate for no-load loss (£/kW)
P_0 = guaranteed no-load loss (kW)

B = capitalization rate for load loss (£/kW)
P_K = guaranteed load loss (kW)

6.2 Main features of construction

6.2.1 The core

The magnetic circuit in a transformer is built from sheets or laminations of electrical steel. Hysteresis loss is controlled by selecting a material with appropriate characteristics, often with large grain size. Eddy current loss is controlled by increasing the resistivity of the material using the silicon content, and by rolling it into a very thin sheet. The power loss characteristics of available materials are shown in **Fig. 6.5**. Generally speaking the steels with lower power loss are more expensive, so that the 0.3 mm thick CGO steel shown is the cheapest of the range shown in **Fig. 6.5**, and the very low loss 0.23 mm thick Hi-B material with domain control techniques used to modify the apparent crystal size is the most expensive. New materials with thicknesses of 0.1 mm are available on a laboratory basis, but the use of such thin materials presents some production problems. The 25 μm thick amorphous ribbon shown with very low loss is made using a rapid cooling technique; it can be used in very low loss transformers up to a few MVA and a core with losses of only one-sixth of the conventional 0.3 mm CGO material is possible. The general principles applying to electrical steels are discussed in **Chapter 3**.

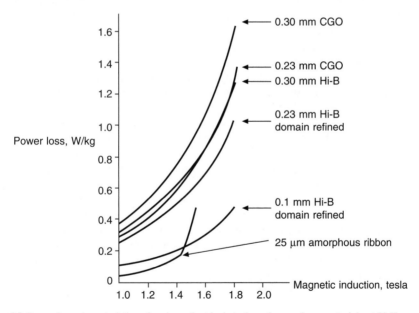

Fig. 6.5 Power loss characteristics of various electrical steels and amorphous materials at 50 Hz

The audible sound radiated by a transformer is generated by magnetostrictive deformation of the core and by electromagnetic forces in the windings, tank walls and magnetic shields. The dominant sound is generated by longitudinal vibrations in the core laminations, which are induced by the magnetic field. The amplitude of the vibrations depends upon the flux density in the core and on the magnetic properties of the core steel. In a large power transformer operating at high flux density the audible sound level can exceed 100 dB(A) and it may be necessary to use high quality core material, improved core-joint techniques and perhaps external cladding or a sound enclosure to reduce the sound to an acceptable level.

In small power transformers, the core may take the form of a continuous strip of steel wound into a coil. The windings may be formed directly onto this core using toroidal winding machines, or the core may be cut to allow preformed windings to be fitted, and reinterlaced with the windings in place.

Where transformer weight is critical, it may be advantageous to operate a local power system at high frequency. **Equation 6.3** relates voltage per turn to frequency and core cross-section. If the system is designed to operate at 400 Hz (a typical requirement for aircraft), it can be seen that the core cross-section will be only 12.5 per cent of that which would be necessary at 50 Hz. The clear advantage in cost and weight must be balanced against higher core losses at the increased frequency (although these can be reduced by using thinner laminations) and against higher load losses caused by high-frequency currents in the windings.

A longer-established form of construction uses the stacked-core technique. For transformers which are rated in hundreds of watts the laminations may be stamped in E and I shapes, or in C and I shapes, then built into cores and assembled round the windings. At higher ratings in kVA or MVA, the usual construction is to cut laminations to length, and to assemble them in a building berth which includes part of the core-clamping structure. When the core has been built, the remaining part of the clamping structure is fitted and the core is turned upright. If it has been fitted during the stacking stage, the top yoke is then removed, the windings are mounted on the core and the top yoke is then reinterlaced.

Single-phase transformers usually have a three-leg core with high-voltage and low-voltage windings mounted concentrically on the centre leg to reduce leakage flux and minimize the winding impedance. The outer legs form the return path for the magnetizing flux. This type of construction is shown in **Fig. 6.6(a)**. For very high ratings in the region of 500 MVA, it may be economic to use a two-leg construction with two sets of windings connected in parallel and mounted one on each leg, as shown in **Fig. 6.6(b)**. Where a transport height limit applies, a large single-phase transformer may be mounted on a four-leg core where the outer legs are used to return remnant flux, in conjunction with smaller top and bottom yokes. This arrangement is shown in **Fig. 6.6(c)**.

Three-phase transformers are also based usually on a three-leg construction. In this case the high-voltage and low-voltage windings of each phase are mounted concentrically with one phase on each of the three legs. The phase fluxes sum to zero in the top and bottom yokes and no physical return path is necessary. This arrangement is shown in **Fig. 6.6(d)**. Where transport heights are a limitation, a three-phase transformer may be built using a five-leg construction as shown in **Fig. 6.6(e)**. The yoke areas in this case are reduced and the outer legs are included as return flux paths.

Core laminations are compacted and clamped together using yoke clamps connected by flitch plates on the outer faces of the legs, or by tie-bars locking the top and bottom yoke clamps together. **Figure 6.7** shows a large three-phase, three-leg core in which the yokes are clamped by rectangular-section steel frames and insulated straps; flitch plates lock the two clamps together and the core-bands on the legs are temporary, to be removed when the windings are fitted.

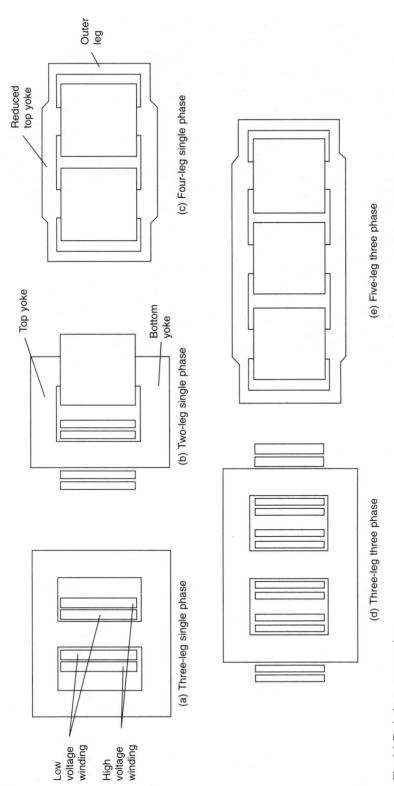

Fig. 6.6 Typical core constructions

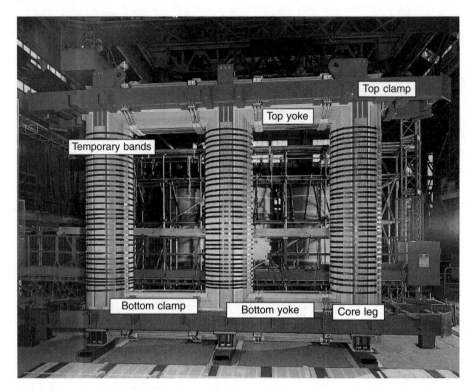

Fig. 6.7 Three-leg core for a three-phase transmission transformer

6.2.2 Windings

Winding conductors may be of copper or aluminium, and they may be in foil or sheet form, or of round or rectangular section. Foil or sheet conductors are insulated from each other by paper or Nomex interleaves, whereas round or rectangular conductors may be coated with enamel or wrapped with paper or some other solid insulation covering.

Low-voltage windings for transformers with ratings up to about 4 MVA may use foils with the full width of the winding, or round conductors. For higher-power transformers a large cross-section may be necessary to carry the current in the low-voltage winding, and it may be economic to use stranded or parallel conductors in order to reduce eddy currents. It may also be necessary to transpose conductors during the winding operation to reduce circulating currents within the winding. For very high current applications, continuously transposed conductor (CTC) may be used, where typically 40 or 50 conductors are machine transposed as shown in **Fig. 6.8**. In a CTC, each strand is enamel insulated and the cable is enclosed in a paper covering. The conductors for low-voltage windings are usually wound in layers, with cooling ducts and interlayer insulation as part of the construction.

High-voltage windings have more turns and carry less current than the low-voltage windings. They are usually formed from round enamelled wire or paper-wrapped rectangular conductors. As with low-voltage windings, if the winding loss is to be minimized, parallel conductors or CTC may be used. The conductors are wound in layers or in discs, as shown in **Figs 6.9(a)** and **6.9(b)** respectively. The labour cost in a layer winding is high, and disc windings are usually considered to be more stable mechanically under the effects of through-current faults.

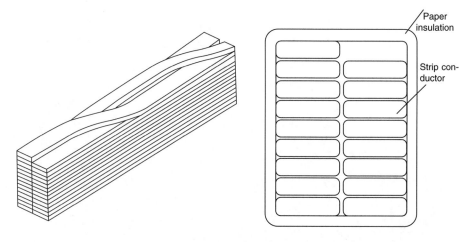

(a) Transposed strip conductor (27 strips in parallel)

(b) Transposed strip conductor in paper envelope

Fig. 6.8 Continuously transposed conductor

Although simple two-winding transformers are widely used, many transformers have three or more windings. These may include regulating windings, tertiary windings to balance harmonic currents or supply auxiliary loads, multiple secondary windings to supply separate load circuits, or multiple primary circuits to connect to power supplies at different voltages or frequencies. The constructional aspects are common between multiple windings, but design aspects are more complicated.

6.2.3 Winding connections

Three-phase transformers are usually operated with the high-voltage and low-voltage windings connected in Y (star), D (delta) or Z (zigzag) connection. The three styles are shown in **Fig. 6.10**.

In star connection, one end of each of the three-phase windings is joined together at a neutral point N and line voltage is applied at the other end; this is shown in **Fig. 6.10(a)**. The advantages of star connection are:

- it is cheaper for a high-voltage winding
- the neutral point is available
- earthing is possible, either directly or through an impedance
- a reduced insulation level (graded insulation) is possible at the neutral
- winding tappings and tapchanger may be located at the neutral end of each phase, with low voltages to earth and between phases
- single-phase loading is possible, with a neutral current flowing

In delta connection, the ends of the three windings are connected across adjacent phases of the supply as shown in **Fig. 6.10(b)**. The advantages of a delta connection are:

- it is cheaper for a high-current low-voltage winding
- in combination with a star winding, it reduces the zero-sequence impedance of that winding

Fig. 6.9(a) Layer-type winding on a horizontal winding machine

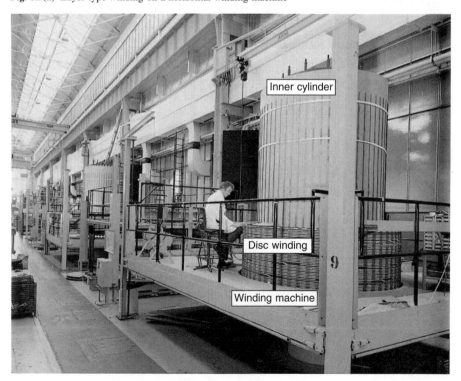

Fig. 6.9(b) Disc-type winding on a vertical winding machine

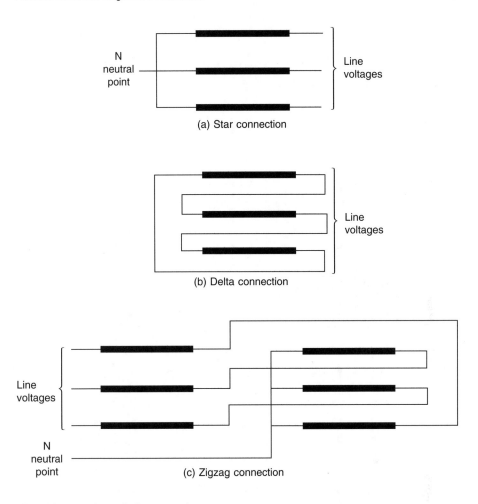

(a) Star connection

(b) Delta connection

(c) Zigzag connection

Fig. 6.10 Three-phase winding connections

The zigzag connection is used for special purposes where two windings are available on each leg and are interconnected between phases as shown in **Fig. 6.10(c)**. The advantages of a zigzag connection are:

- it permits neutral current loading with an inherently low zero-sequence impedance, and it is used in 'earthing transformers' to create an artificial neutral terminal on the system
- voltage imbalance is reduced in systems where the load is not evenly distributed between phases

In order to define the range of possible connections, a designation has been adopted by IEC in which the letters Y, D, Z and N are assigned to the high-voltage windings and y, d, z and n are assigned to the low-voltage windings. Clock-hour designations 1 to 12 are used to signify in 30° steps the phase displacement between primary and secondary windings. Yy0 is the designation for a star–star connection where primary and secondary voltages are in phase. Yd1 is the designation for a star–delta connection with a 30° phase shift between primary and secondary voltages. The common connections for three-phase transformers are shown in **Fig. 6.11**.

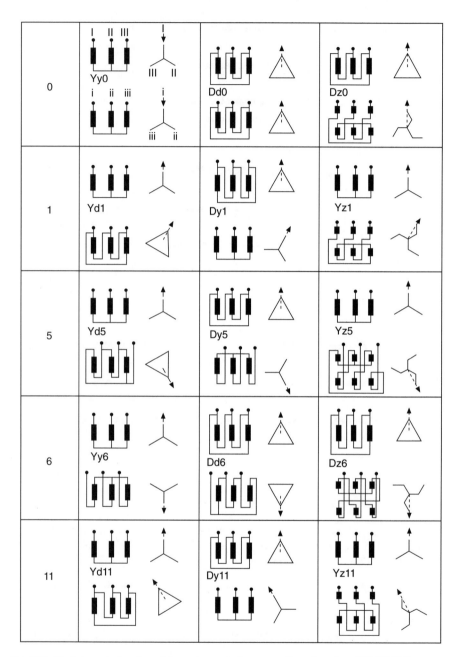

Fig. 6.11 Clock-hour designation of three-phase winding connections (based on IEC 60076-1)

In many transmission transformers, an autotransformer connection is employed. The connection is shown in **Fig. 6.12**, and unlike the two-winding transformer in which primary and secondary are isolated, it involves a direct connection between two electrical systems. This connection can have a cost advantage where the ratio of input and output voltages is less than 5 : 1. Current flowing in the 'series' winding corresponds to that in the higher-voltage system only. The 'common' winding carries

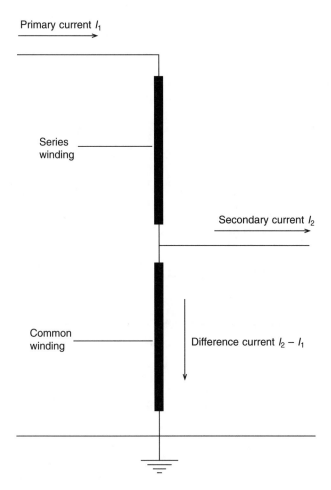

Primary current I_1

Series winding

Secondary current I_2

Common winding

Difference current $I_2 - I_1$

Fig. 6.12 Autotransformer connection

a current which is the difference between the two systems, and being sized for this lower current it can be significantly cheaper. The autotransformer is more susceptible to damage from lightning impulse voltages and it has a lower strength against through-fault currents; both of these weaknesses can be corrected, but at increased cost.

It may occasionally be necessary to make a three-phase to two-phase transformation. This might be in order to supply an existing two-phase system from a new three-phase system, or to supply a two-phase load (such as a furnace) from a three-phase system, or to supply a three-phase load (such as a motor) from a two-phase system. In all these cases the usual method of making the transformation is by using two single-phase transformers connected to each of the systems and to each other by the *Scott connection*.

A Scott connection is shown in **Fig. 6.13**. On one transformer the turns ratio is equal to the transformation ratio, and the mid-point of the winding connected to the three-phase system is brought out for connection to the other single-phase transformer. The second transformer has a turns ratio of 0.866 times the transformation ratio. The primary winding is connected between the third phase of the three-phase supply and the mid-point of the primary winding of the first transformer. The secondary windings

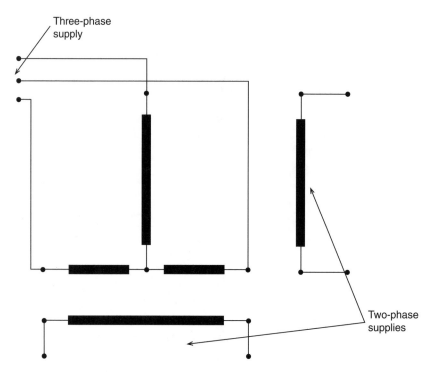

Fig. 6.13 Scott connection of transformers

of the two transformers are connected to the two-phase supply (or load) using a
three- or four-wire connection.

6.2.4 Bushings

Where transformers are enclosed within tanks to contain the insulating and cooling
fluid, it is necessary to link the windings connections to the network using through-
bushings which penetrate the tank.

Low-voltage bushings are generally solid. They are made of porcelain, ceramic or
epoxy insulation with sufficient electrical strength to withstand abnormal voltages
due to lightning activity or switching operations, and to withstand the service voltage
over the lifetime of the transformer. Some low-voltage bushings carry high current
and cooling is necessary. In such cases it is usual to employ a hollow porcelain
construction in which the internal connections are cooled by the oil in the tank.

High-voltage bushings must withstand much higher voltage transients. They are
usually of composite construction with a core of oil-impregnated or resin-bonded
paper in an outer porcelain or epoxy cylinder. This outer cylinder is 'shedded' at the
outside (air) end to increase its electrical strength under wet conditions. A typical
high-voltage bushing with an oil-paper core is shown in **Fig. 6.14(a)**; its internal
construction detail is shown in **Fig. 6.14(b)**.

6.2.5 Tapchangers

It has been explained in **section 6.1** that when a transformer carries load current there
is a variation in output voltage from a transformer which is known as *regulation*. In
order to compensate for this, additional turns are often made available so that the

voltage ratio can be changed using a switch mechanism known as a tapchanger. An off-circuit tapchanger can only be adjusted to switch additional turns in or out of circuit when the transformer is de-energized; it usually has between two and five tapping positions. An on-load tapchanger (OLTC) is designed to increase or decrease the voltage ratio when the load current is flowing, and the OLTC should switch the transformer load current from the tapping in operation to the neighbouring tapping without interruption. The voltage between tapping positions (the step voltage) is normally between 0.8 per cent and 2.5 per cent of the rated voltage of the transformer.

OLTC mechanisms are based either on a slow-motion reactor principle or a high-speed resistor principle. The former is commonly used in North America on the low-voltage winding, and the latter is normally used in Europe on the high-voltage winding.

The usual design of OLTC in Europe employs a selector mechanism to make connection to the winding tapping contacts and a diverter mechanism to control current flows while the tapchanging takes place. The selector and diverter mechanisms

Fig. 6.14(a) High-voltage oil-air bushing

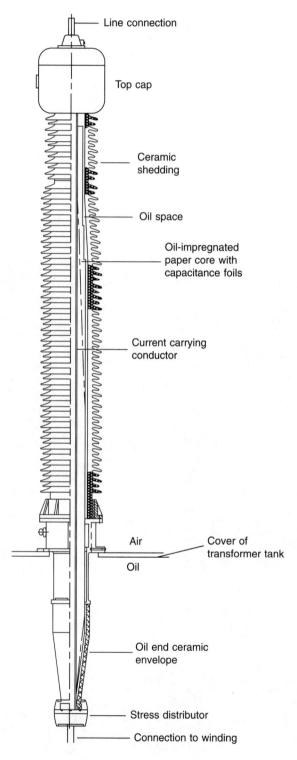

Fig. 6.14(b) Internal construction of high-voltage oil-air bushing

may be combined or separate, depending upon the power rating. In an OLTC which comprises a diverter switch and a tap selector, the tap change occurs in two operations. First, the next tap is selected by the tap switch but does not carry load current, then the diverter switches the load current from the tap in operation to the selected tap. The two operations are shown in seven stages in **Fig. 6.15**.

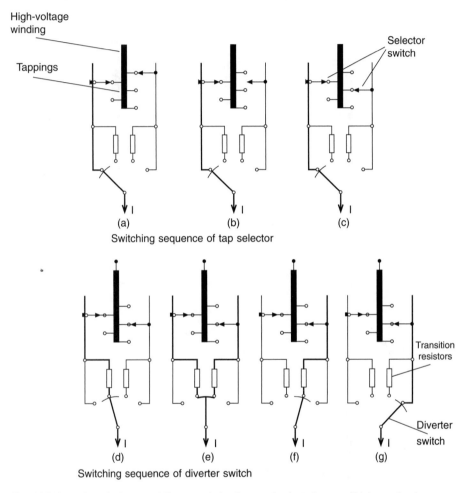

Switching sequence of tap selector

Switching sequence of diverter switch

Fig. 6.15 Operation of selector and diverter switches in an on-load tap changer of high-speed resistor type

The tap selector operates by gearing directly from a motor drive, and at the same time a spring accumulator is tensioned. This spring operates the diverter switch in a very short time (40–60 ms in modern designs), independently of the motion of the motor drive. The gearing ensures that the diverter switch operation always occurs after the tap selection has been completed. During the diverter switch operation shown in **Fig. 6.15(d), (e)** and **(f)**, transition resistors are inserted; these are loaded for 20–30 ms and since they have only a short-time loading the amount of material required is very low.

The basic arrangement of tapping windings is shown in **Fig. 6.16**. The linear arrangement in **Fig. 6.16(a)** is generally used on power transformers with moderate regulating ranges up to 20 per cent. The reversing changeover selector shown in

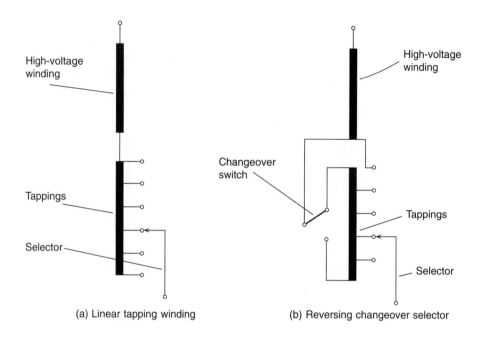

(a) Linear tapping winding   (b) Reversing changeover selector

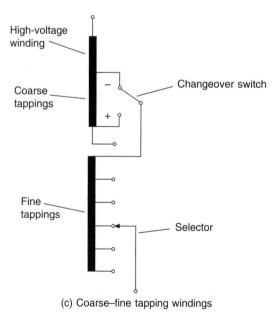

(c) Coarse–fine tapping windings

Fig. 6.16 Basic arrangements of tapping windings

Fig. 6.16(b) enables the voltage of the tapped winding to be added or subtracted from the main winding so that the tapping range may be doubled or the number of taps reduced. The greatest copper losses occur at the position with the minimum number of effective turns. This reversing operation is achieved with a changeover selector which is part of the tap selector of the OLTC. A two-part coarse–fine arrangement shown in **Fig. 6.16(c)** may also be used. In this case the reversing changeover selector for the fine winding can be connected to the 'plus' or 'minus' tapping of the coarse winding, and the copper losses are lowest at the position of the lowest number of effective turns. The coarse changeover switch is part of the OLTC.

Regulation is mostly carried out at the neutral point in star windings, resulting in a simple, low-cost, compact OLTC and tapping windings with low insulation strength to earth. Regulation of delta windings requires a three-phase OLTC, in which the three phases are insulated for the highest system voltage which appears between them; alternatively three single-phase OLTCs may be used.

6.2.6 Cooling equipment

Transformers may be naturally cooled, in which case the cooling medium (oil or air) circulates by thermosyphon forces, or they may be forced cooled, with fans or pumps to circulate the air or oil over the core and through the windings.

In oil-cooled transformers a more economic solution is to use directed oil-flow cooling in which oil is pumped directly into the windings. The oil is cooled by passing (or pumping) it through plate radiators, which may be externally cooled by fans, or by using forced air-cooled tubular construction, or by pumping it through water coolers. Designations for these cooling systems which identify both the cooling fluid and the type of cooling have been assigned by IEC and these are summarized in **Table 6.1**.

6.3   Main classes of transformer

Transformers are used for a wide variety of purposes, with the complete range of voltage and power ratings as well as many special features for particular applications. The following covers the main types.

6.3.1 Transformers for electronics

Transformers for electronic circuits or for low-voltage power supplies are used to match the supply voltage to the operating voltage of components or accessories, or to match the impedance of a load to a supply in order to maximize power throughput.

The core is usually constructed in the low-power transformers from C- and I-laminations or from E- and I-laminations. The windings are usually of round enamelled wire, and the assembly may be varnished or encapsulated in resin for mechanical consolidation and to prevent ingress of moisture.

Increasing numbers of this type operate at high frequencies in the kHz range and use laminations of special steel often containing cobalt to reduce the iron losses.

6.3.2 Small transformers

These are used for stationary, portable or hand-held power supply units, as isolating transformers and for special applications such as burner ignition, shavers, shower heaters, bells and toys. They may be used to supply three-phase power up to 40 kVA at frequencies up to 1 MHz. These transformers are usually air insulated, the smaller units using enamelled windings wires and ring cores and the larger units using C- and I- or E- and I-laminated cores.

Safety is a major concern for these transformers and they are identified as class I, class II or class III. Class I units are insulated and protected by an earth terminal.

Table 6.1 Cooling system designations

For oil immersed transformers a four-letter code is used:

First letter: internal cooling medium in contact with windings:

O	mineral oil or insulating liquid with fire point ≤300°C
K	insulating liquid with fire point >300°C
L	insulating liquid with no measurable fire point

Second letter: circulation mechanism for internal cooling medium:

N	natural thermosyphon flow
F	forced oil circulation, but thermosyphon cooling in windings
D	forced oil circulation, with oil directed into the windings

Third letter: external cooling medium

A	air
W	water

Fourth letter: circulation mechanism for external cooling medium

N	natural convection
F	forced circulation

Examples: ONAN or OFAF

For dry-type transformers a four-letter code is used:

First letter: internal cooling medium in contact with the windings

A	air
G	gas

Second letter: circulation mechanism for internal cooling medium

N	natural convection
F	forced cooling

Third letter: external cooling medium

A	air
G	gas

Fourth letter: circulation mechanism for external cooling medium

N	natural convection
F	forced cooling

Examples: AN or GNAN

Class II transformers have double insulation or reinforced insulation. Class III transformers have outputs at Safety Extra-Low Voltages (SELV) below 50 V ac or 120 V dc.

6.3.3 Distribution transformers

These are used to distribute power to domestic or industrial premises. They may be single phase or three phase, pole mounted or ground mounted, and they have ratings ranging from 16 kVA up to 2500 kVA.

The windings and core are immersed in mineral oil, with natural cooling, and there are two windings per phase. The primary (high-voltage) winding has a highest voltage ranging from 3.6 kV to 36 kV; the secondary (low-voltage) winding voltage does not exceed 1.1 kV. The high-voltage winding is usually provided with off-circuit tappings of ± 2.5 per cent, or +2 × 2.5 per cent, −3 × 2.5 per cent.

The preferred values of rated output are 16, 25, 50, 100, 160, 250, 400, 630, 1000, 1600 and 2500 kVA, and the preferred values of short-circuit impedance are 4 per cent or 6 per cent. Losses are assigned from lists, for instance from BS 7821-1, or by using a loss-capitalization formula.

The core and windings of a typical distribution transformer rated at 800 kVA, 11 000/440 V are shown in **Fig. 6.17**.

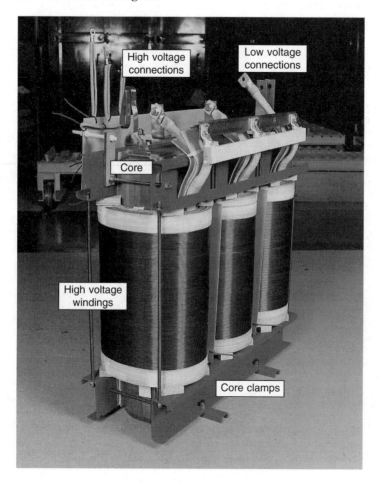

Fig. 6.17 Core and windings of an 800 kVA, 11 000/440 V distribution transformer

6.3.4 Supply transformers

These are used to supply larger industrial premises or distribution substations. Ratings range from 4 MVA to 30 MVA, with primary windings rated up to 66 kV and secondary windings up to 36 kV.

Transformers in this class are fluid cooled. Most supply transformers use mineral oil, but for applications in residential buildings, oil rigs and some factories the coolant may be synthetic esters, silicone fluid or some other fluid with a higher fire point than mineral oil.

6.3.5 Transmission (or intertie) transformers

These are among the largest and highest voltage transformers in use. They are used to transmit power between high-voltage networks. Ratings range from 60 MVA to 1000 MVA and the windings are rated for the networks which they link, such as 33, 66, 132, 275 and 400 kV in the UK, or voltages up to 500 kV or 800 kV in other countries. The impedance of a transmission transformer is usually 18 per cent in the UK, or 8 per cent in continental Europe, but for some system conditions, an impedance of up to 30 per cent is used.

Transmission transformers are oil filled, and are usually fitted with oil pumps and radiator fans to assist cooling of the windings and cores. They are usually fitted with OLTCs, but some networks at 400 kV and 275 kV are linked by transformers without regulating windings. The core and windings of a five-legged transmission transformer rated at 1000 MVA and 400 kV/275 kV/11 kV are shown in **Fig. 6.18.**

Fig. 6.18 Core and windings of a 1000 MVA, 400/275 kV transmission transformer

6.3.6 Generator (or step-up) transformers

Power is usually generated in large power stations at typically 18–20 kV, and generator transformers are used to step up this voltage to the system voltage level. These transformers are usually rated at 400, 500, 630, 800 or 1000 MVA.

Base-load power stations usually operate continuously at full load, and the power is transformed to the highest operating voltage in one step. Combined-cycle gas turbine stations may be embedded in the transmission and distribution system at lower operating voltages such as 132 kV.

Generator transformers are usually fitted with regulating windings and OLTCs.

6.3.7 Phase-shifting transformers

Where power is transmitted along two or more parallel transmission lines, the power flow divides between the lines in inverse proportion to the line impedances. Higher power is therefore transmitted through the line with lowest impedance and this can result in overload on that line, when the parallel line is only partly loaded. Phase-shifting transformers are used to link two parallel lines and to control power flow by injecting a voltage 90° out of phase (*in quadrature*) with the system voltage into one line, at either leading or lagging power factor. Where the transformer controls the phase angle but not the voltage, the unit is known as a *quadrature booster*. Where the voltage is also controlled, the unit is known as a *phase-shifting transformer.* **Fig. 6.19** shows a 2000 MVA, 400 kV quadrature booster transformer on site; the unit is split between two tanks in order to meet construction limitations of size and weight.

Fig. 6.19 2000 MVA 400 kV quadrature booster transformer in two tanks on site

6.3.8 Converter transformers

Where power is transmitted through an HVDC system, a converter station is used to change ac power to dc using multiple rectifier bridges. DC power is converted back to ac using inverter bridges. Converter transformers handle ac power and power at mixed ac/dc voltages by combining the power flow through 12 phases of rectifier/ inverter bridges through dc valve windings.

The insulation structure must withstand all normal and abnormal conditions when ac voltage is mixed with dc voltage of differing polarities over the operating temperature range. The presence of dc currents may also cause dc saturation of the core, leading to abnormal magnetizing currents and variations in sound.

A phase of a three-phase converter transformer bank typically comprises a high-voltage primary winding and two secondary ac/dc valve windings. Three such

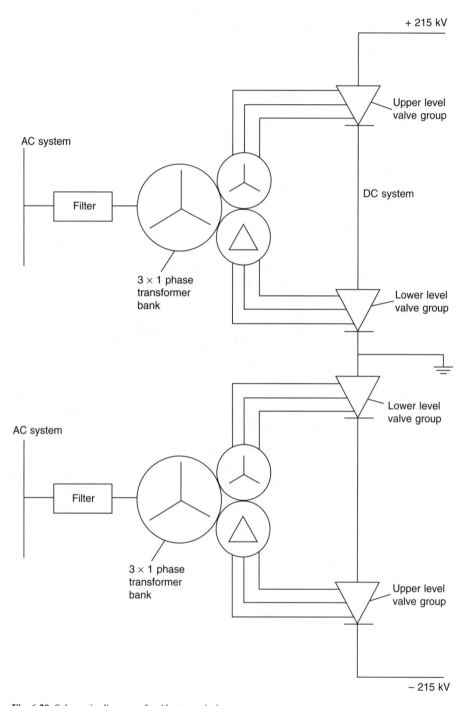

Fig. 6.20 Schematic diagram of ac/dc transmission system

transformers together form the two secondary three-phase systems; one is connected in delta and the other in star. Each secondary system feeds a six-pulse bridge and the two bridges are connected in series to form a 12-pulse arrangement, as shown schematically in **Fig. 6.20**. Two such transformer banks are used with the secondary circuits connected in opposite polarity to form a +/–215 kV dc transmission system.

6.3.9 Railway transformers

Transformers for railway applications may be trackside units to supply power to the track, or on-board transformers in the locomotive or under the coaches, to power the drive motors.

Trackside transformers are subjected to uneven loading depending upon the position of the train in the railway system. On-board transformers are designed for the lowest possible weight, resulting in a high-loss performance. Modern train control systems using thyristors or GTOs subject the transformers to severe harmonic currents that require special design consideration.

6.3.10 Rectifier and furnace transformers

Special consideration is needed for transformers in industrial applications involving arc furnaces or heavy-current dc loads in electrochemical plant. The primary windings in such cases are usually rated at 33 kV or 132 kV in the UK, but the secondary windings carry many thousands of amperes and are rated at less than 1 kV. Current sharing between parallel paths in the transformer becomes important because of the magnetic fields created by the high currents. These strong magnetic fields can cause excess heating in magnetic steels if these are used in the structure of the transformer, because of the flow of proximity currents in the steel. To reduce this excess heating, non-magnetic steel is often used to form part of the tank or the cover.

The OLTCs in furnace transformers are subject to a heavy duty; they may perform hundreds of thousands of operating cycles a year, which is more than a lifetime's duty for many transmission transformers.

6.3.11 Dry-type transformers

A dry-type construction is possible where a higher-temperature class of insulation is required than is offered by cellulose and a class 'O' or class 'K' fluid. Dry-type transformers use non-cellulosic solid insulation and the windings may be varnish dipped to provide a class 'C' capability, or vacuum encapsulated in epoxy resin to form a class 'F' or class 'H' system. Ratings can be up to 20 MVA at voltages up to 36 kV. Overload performance can be provided by fans.

This type of transformer is more expensive than a fluid-filled equivalent, and because of the reduced fire risk they are used in special applications where the public are involved, such as underground tunnels, residential blocks of flats or oil rigs.

A typical cast-resin transformer rated 2500 kVA, 11 000/440 V is shown in **Fig. 6.21**.

6.3.12 Gas-filled transformers

For applications where low flammability is paramount, designs have been developed in which the transformer is insulated and cooled with SF_6 gas. This provides an alternative to dry-type construction, but it has only been used on a limited scale. High-voltage SF_6 transformers are available at high cost, but this may be justifiable

Core clamp

Resin encapsulated
winding

Cooling fan

Fig. 6.21 Dry-type 2500 kVA, 11 000/440 V transformer with cast-resin encapsulation

in cases of high land values, where the overall 'footprint' of the unit can be reduced
by the elimination of fire-fighting equipment.

6.4 Rating principles

6.4.1 Rated power

The rated power of each winding in a transformer refers to a continuous loading and
it is a reference value for guarantees and tests concerning load losses and temperature
rise. Where a transformer winding has different values of apparent power under
different cooling conditions, the highest value is defined as the rated power. A two-
winding transformer is given only one value of rated power, which is identical for
both windings, but a multi-winding transformer may have reduced levels of rated
power on auxiliary windings.

When a transformer has rated voltage applied to the primary winding and rated
current flows through the secondary winding, the transformer is carrying rated power
for that set of windings. This definition of rated power, used by IEC, implies that the
rated power is the value of power input to the transformer, including its own absorption
of active and reactive power. The secondary voltage under full load differs from the
rated voltage by the voltage drop in the transformer.

In North America a different definition is adopted in ANSI/IEEE standard C57.1200.
Here, rated power is based on the output that can be delivered at rated secondary
voltage, and allowance must be made so that the necessary primary voltage can be
applied to the transformer. The difference between basing rated power on loaded
voltages (IEC) or open-circuit voltages (ANSI/IEEE) is significant.

6.4.2   Overloading

Although transformers are rated for continuous operation, it is possible to supply overloads for limited periods. The analysis of overloading profiles is based on the deterioration of cellulose. At temperatures above the rated temperature, cellulose degrades at a faster rate, and the load cycle should therefore include a period of operation at lower temperature to balance effects of these faster ageing rates at higher temperature. In general, the ageing rate doubles for every 8°C rise in temperature.

When the ambient temperature is below the rated ambient temperature it is also permissible to overload the transformer, the limit of the overloading being the operating temperature of the winding. Under severe operational pressures it is sometimes necessary to overload a transformer well beyond its nameplate rating, but such an operation will usually result in shorter life. IEC 354 provides a guide to overloading of oil-filled transformers based on these principles.

A maximum winding temperature of 140°C should not be exceeded even under emergency conditions, because free gas can be produced by cellulose at temperatures above this, and gas bubbles may cause dielectric failure of the windings.

6.4.3   Parallel operation of transformers

For operational reasons it may be necessary to operate transformers connected directly in parallel. For successful parallel operation, the transformers must have:

- the same voltage ratings and ratio (with some tolerance)
- the same phase-angle relationship (clock-hour number)
- the same percentage impedance (with some tolerance)

It is not advisable to parallel operate transformers with widely different power ratings as the natural impedance for optimal design varies with the rating of the transformer. The power divides between parallel-connected transformers in a relationship which is inversely proportional to their impedances; a low-impedance transformer operated in parallel with a higher-impedance unit will pass the greater part of the power and may be overloaded. A mismatch in loading of up to 10 per cent is normally acceptable.

6.5   Test methods

6.5.1   Specification testing

Transformers for power supply units and isolating transformers are subject to production testing, where compliance is generally checked by inspection and electrical tests. The electrical tests establish earthing continuity (for class I), no-load output voltage, dielectric strength between live parts of the input circuit and accessible conducting parts of the transformer, and dielectric strength between input and output circuits.

Tests on larger transformers are carried out as part of the manufacturing process to ensure that transformers meet the characteristics specified by the purchaser. The tests prescribed by IEC are grouped in three categories, these being routine tests, type tests and special tests:

- *routine tests*, to be carried out on all transformers
- *type tests*, to be carried out on new designs or the first unit of a contract
- *special tests*, to be carried out at the specific request of the purchaser

(a) *Routine tests*
These are carried out on all transformers and include:

- measurement of winding resistance using a dc measurement circuit or a bridge
- measurement of voltage ratio and check of phase displacement. The voltage ratio is checked on each tap position and the connection symbol of three-phase transformers is confirmed.
- measurement of short-circuit impedance and load loss. The short-circuit impedance (including reactance and ac resistance in series) is determined by measuring the primary voltage necessary to circulate full-load current with the secondary terminals short-circuited. The load loss (copper loss) is measured using the same circuit.
- measurement of no-load loss and magnetizing current, at rated voltage and frequency. This is measured on one set of windings with the other windings open-circuit. Because the power factor at no-load may be between 0.1 and 0.2, special low power factor wattmeters are necessary to ensure accuracy. A three-wattmeter connection is used for load loss measurement of three-phase transformers, because the previously used two-wattmeter method is inaccurate.
- dielectric tests, consisting of applied voltage and induced voltage tests, sometimes linked with a switching surge test. The applied voltage test verifies the integrity of the winding insulation to earth and between windings, and the induced voltage test verifies the insulation between turns and between windings. The general requirements set down by IEC are shown in **Table 6.2**. For transformers rated up to 300 kV, the induced voltage test is for one minute at an overvoltage of between 2.5 and 3.5 times rated voltage, carried out at a higher frequency to avoid core saturation. For rated voltages over 300 kV, the induced test is for 30 minutes at 1.3 or 1.5 times rated voltage, with an initial short period at a higher voltage to initiate activity in any vulnerable partial discharge sites; this is linked with a partial discharge test. This long-term test will not prove the insulation against switching overvoltages, and so for this it must be linked to a switching impulse test.
- tests on OLTCs to confirm the timing of the tapchanger mechanism and to prove the capability on full-voltage and full-current operation.

(b) *Type tests*
These are carried out on new designs, or on the first unit of a contract. They include:

- a temperature rise test using the normal cooling equipment to verify the temperature rise of the windings under full-load or overload conditions. For smaller transformers it may be possible to carry out the temperature rise test at full voltage by supplying a suitable load. For larger transformers it is more usual to supply the total losses (load loss and no-load loss) under short-circuit conditions and to establish the top oil temperature and the average winding temperature at a lower voltage. Where a number of transformers are available, it is possible to supply them in a 'back-to-back' connection and to make the test at full voltage and full current. The test is carried out over a period of 8 to 24 hours to establish steady-state temperatures, and it may be associated with Dissolved Gas Analysis (DGA) of oil samples taken during the test to identify possible evidence of microdeterioration in the insulation at high local temperatures.

Table 6.2 Dielectric test values for transformers

Rated withstand voltages for windings of transformers <300 kV

Highest voltage for equipment kV (rms)	Rated short duration power frequency voltage test kV (rms)	Rated lightning impulse voltage test kV (peak)
≤1.1	3	–
3.6	10	20 or 40
7.2	20	40 or 60
12	28	60 or 75
24	50	95 or 125
36	70	145 or 170
72.5	140	325
145	230 or 275	550 or 650
245	325 or 360 or 395	750 or 850 or 950

Rated withstand voltages for windings of transformers ≥300 kV, in conjunction with a 30 m test at 1.3× or 1.5× rated voltage

Highest voltage for equipment kV (rms)	Rated switching impulse withstand voltage kV (peak)	Rated lightning impulse withstand voltage kV (peak)
300	750 or 850	850 or 950 or 1050
420	950 or 1050	1050 or 1175 or 1300 or 1425
525	1175	1425 or 1550
765	1425 or 1550	1550 or 1800 or 1950

Notes: 1 In general the switching impulse test level is ≈ 83% of the lightning impulse test level.
2 These test levels are not applicable in North America.

- a lightning impulse voltage test to prove performance under atmospheric lightning conditions. The specified impulse waveform has a 1 μs front time (time-to-crest) and a 50 μs tail time (time from crest to half-crest value). The winding is tested at a level prescribed by IEC which is linked to the rated voltage. This is set out in **Table 6.2**. One test application is made at 75 per cent of the test value (Reduced Full Wave, or RFW) and this is followed by three test applications at full test value (FW). The test is considered satisfactory if there are no discrepancies between the recorded oscillograms of applied voltage and between the recorded oscillograms of neutral current.

(c) *Special tests*

These are carried out at the specific request of the purchaser. The more usual special tests are:

- chopped-impulse wave test. The purchaser may wish to simulate the situation when a lightning surge propagated along a transmission line is chopped to earth by a lightning protection rod gap mounted on the bushing. The Chop Wave (CW) is specified as an FW waveform with a chop to earth occurring between 3 and 7 seconds after the crest. The chop-wave test is combined with the lightning impulse voltage test and the test sequence is 1-RFW, 1-FW, 2-CW and 2-FW test applications, sometimes followed by a further RFW for comparison purposes. A successful test shows no discrepancies between applied voltage oscillograms or between neutral current oscillograms, although if rod gaps are used to perform the chopped-wave test there may be acceptable deviations between neutral current oscillograms after the time of chop. If a triggered gap is used to control time of chop the oscillograms should be identical.

- determination of zero-sequence impedance, measured between the line terminals of a star-connected or zigzag-connected winding connected together, and its neutral terminal. The impedance is expressed in ohms per phase, and it is a measure of the impedance presented by the transformer to a three-phase fault to earth.

- short-circuit withstand test to verify the integrity of the transformer under through-fault conditions. IEC recommendations allow the withstand performance of a winding to be demonstrated by calculation or by test. The resource needed to test fully a large power transformer is substantial, and few test facilities are available in the world.

- determination of sound levels. The sound produced by a transformer is predominantly due to magnetostriction in the core and is related to core flux density. The sound level can be measured in terms of sound power, sound pressure or sound intensity. The last method was developed especially for measurement of low sound levels in the presence of high sound ambients. In transformers which have been designed with low flux density to reduce the magnetostriction effect, the predominant sound may be produced by the windings as a result of movements caused by the load currents. This load-current generated sound should be measured by the sound intensity method.

6.5.2   In-service testing

Two types of in-service testing are used. Surveillance testing involves periodic checks, and condition monitoring offers a continuous check.

(a) *Surveillance testing – oil samples*
When transformers are in operation many users carry out surveillance testing to monitor operation. The most simple tests are carried out on oil samples taken on a regular basis. Measurement of oil properties such as breakdown voltage, water content, acidity, dielectric loss angle, volume resistivity and particle content all give valuable information on the state of the transformer. DGA gives early warning of deterioration due to electrical or thermal causes, particularly sparking, arcing and service overheating. Analysis of the oil by High-Performance Liquid Chromatography (HPLC) may detect the presence of furanes or furfuranes which will provide further information on moderate overheating of the insulation.

(b) *Condition monitoring*
Sensors can be built into the transformer so that parameters can be monitored on a continuous basis. The parameters which are typically monitored are winding temperature, tank temperature, water content, dissolved hydrogen, load current and voltage transients. The data collection system may simply gather and analyse the information, or it may be arranged to operate alarms or actuate disconnections under specified conditions and limits which represent an emergency.

6.6 Commissioning, maintenance and repair

6.6.1 Commissioning

Power transformers can be transported to site complete with oil, bushings, tapchangers and cooling equipment. It is then a relatively simple matter to lift them onto a pole or plinth and connect them into the system.

Large transformers are subject to weight restrictions and size limitations when they are moved by road or rail and it is necessary to remove the oil, bushings, cooling equipment and other accessories to meet these limitations. Very large transformers are usually carried on custom-built transporters, such as the 112-wheel transporter shown in **Fig. 6.22**; this is carrying a 1000 MVA, 400 kV transmission transformer that has been dismantled to meet a 400 tonne road weight restriction. Once a transformer of this size arrives on site, it must be lifted or jacked onto its plinth for re-erection. In some cases with restricted space it may be necessary to use special techniques such as water skates to manoeuvre the transformer into position.

Fig. 6.22 Road transporter carrying a 1000 MVA, 400 kV transmission transformer

When the transformer has been erected and the oil filled and reprocessed, it is necessary to carry out commissioning tests to check that all electrical connections have been correctly made and that no deterioration has occurred in the insulation system. These commissioning tests are selected from the routine tests and usually include winding resistance and ratio, magnetizing current at 440 V, and analysis of oil samples to establish breakdown strength, water content and total gas content.

6.6.2 Maintenance

Transformers require little maintenance in service, apart from regular inspection and servicing of the OLTC mechanism. The diverter contacts experience significant wear due to arcing, and they must be replaced at regular intervals which are determined by the operating regime. For furnace transformers it may be advisable to filter the oil regularly in a diverter compartment in order to remove carbon particles and maintain the electrical strength.

The usual method of protecting the oil breather system in small transformers is to use silicone gel breathers to dry incoming air; in larger transformers refrigerated breathers continuously dry the air in a conservator. Regular maintenance (at least once a month) is necessary to maintain a silica gel breather in efficient working order.

If oil samples indicate a high water content then it may be necessary to dry the oil using a heating-vacuum process. This also indicates a high water content in the paper insulation and it may be necessary to redry the windings by applying a heating and vacuum cycle on site, or to return the transformer to the manufacturer for reprocessing or refurbishment.

6.6.3 Diagnostics and repair

In the event of a failure, the user must first decide whether to repair or replace the transformer. Where small transformers are involved, it is usually more economic to replace the unit. In order to reach a decision, it is usually necessary to carry out diagnostic tests to identify the number of faults and their location. Diagnostic tests may include the surveillance tests referred to in **section 6.5.2**, and it may also be decided to use acoustic location devices to identify a sparking site, low-voltage impulse tests to identify a winding fault and frequency response analysis of a winding to an applied square wave to detect winding conductor displacement.

If the fault is in a winding, it usually requires either replacement of the winding in a repair workshop or rewinding by the manufacturer, but many faults external to the windings such as connection or core faults can be corrected on site.

Where a repair can be undertaken on site it is essential to maintain dry conditions in the transformer by continual purging using dry air. Any material taken into the tank must be fully processed and a careful log should be maintained of all materials taken into and brought out of the tank.

When a repair is completed, the transformer must be redried and reimpregnated, and the necessary tests carried out to verify that the transformer can be returned to service in good condition.

6.7 Standards

The performance requirements of power transformers are covered by a range of international, regional and national standards. At the highest level are the recommendations published by the International Electrotechnical Commission (IEC). These are performance standards, but they are not mandatory unless referred to in a contract.

Regional standards in Europe are Euro-Norms (ENs) or Harmonized Documents (HDs) published by the European Committee for Electrotechnical Standardization (CENELEC). CENELEC standards are part of European law, ENs must be transposed into national standards and no national standard may conflict with an HD. Many ENs and HDs are based on IEC recommendations, but some have been specifically prepared to match European legislation requirements such as EU Directives.

National standards in the UK are published by the British Standards Institution (BSI). BSI standards are generally identical to IEC or CENELEC standards, but some BS standards address issues not covered by IEC or CENELEC.

In North America the main regional standards are published by the American National Standards Institute (ANSI) in conjunction with the Institute of Electrical and Electronic Engineers (IEEE). ANSI/IEEE standards are generally different from IEC recommendations, but the two are becoming closer as a result of international harmonization following GATT treaties on international trade.

The main transformer standards in IEC, CENELEC and BSI are shown in **Table 6.3**, together with their interrelationship. Where appropriate the equivalent ANSI/ IEEE standard is also referenced.

Table 6.3 Comparison of international, regional and national standards for transformers

IEC	EN/HD	BS	Subject of standard	ANSI/IEEE
60076	HD 398	BS 171	Power transformers	C.57.12.00
60076-1	EN 60076-1	BSEN 60076-1	: general	
60076-2	EN 60076-2	BSEN 60076-2	: temperature rise	C.57.12.90
60076-3	HD 398.3	BS 171-3	: dielectric tests	C.57.12.90
60076-4	–	–	: guide for lightning impulse testing	C.57.98
60076-5	HD 398.5	BS 171-5	: ability to withstand short-circuit	C.57.12.90
60076-6	EN 60289	BSEN 60289	: reactors	C.57.21
60076-7	–	BS 7735	: loading guide – oil filled	C.57.92
60076-8	–	BS5953	: application guide	
60076-9	–	–	: terminals and tapping markings	C.57.12.90
60076-10	EN 60551	BSEN 60551	: determination of sound levels	C.57.12.90
60076-11	HD 464	BS 7806	: dry-type power transformers	C.57.12.01
60076-12	–	–	: loading guide – dry type	C.57.96
60137	EN 60137	BSEN 60137	Insulated bushings for ac voltages above 1 kV	C.57.19.00
–	HD 506	BS 7616	Bushings for liquid-filled transformers 1 kV to 36 kV	
60214	EN 60214	BSEN 60214	Tap-changers	
60214-1	EN 60214	BSEN 60214	: for power transformers	C.57.131
60214-2	–	–	: application guide	
61378	–	–	Converter transformers	
61378-1	–	–	: transformers for industrial applications	
61248	–	–	Transformers and inductors for electronics and telecommunications	
61558	–	–	Safety of power transformers, power supply units and similar	C.57.12.22
–	HD 428	BS 7281	Oil-filled distribution transformers	
–	HD 428-1	BS 7281-1	: general to 24 kV	
–	HD 428-2	BS 7281-2	: transformers with cable boxes	
–	HD 428-3	BS 7281-3	: transformers for 36 kV	
–	HD 428-4	BS 7281-4	: harmonic currents	
–	HD 538	BS 7844	Dry-type distribution transformers	C.57.12.51
–	HD 538.1	BS 7844-1	: general to 24 kV	
–	HD 538.2	BS 7844-2	: transformers for 36 kV	
–	–	BS 3535	Safety requirements for transformers that may be stationary or portable, single phase or polyphase	

Chapter 7

Switchgear

Mr N.P. Allen
GEC Alsthom Low Voltage Equipment Ltd

Mr T. Harris
ALSTOM T&D Distribution Switchgear

Mr C.J. Jones
VA TECH Reyrolle Ltd

7.1 Introduction

Switchgear is used to connect and disconnect electric power supplies and systems. It is a general term which covers the switching device and its combination with associated control, measuring, protective and regulating equipment, together with accessories, enclosures and supporting structures.

Switchgear is applied in electrical circuits and systems from low voltage, such as domestic 220/240 V applications, right up to transmission networks up to 1100 kV. To meet this range of voltages and powers, a wide range of technology is needed, and this chapter has been split into the main three subdivisions of low voltage, medium voltage (distribution) and high voltage (transmission) in order to deal with this wide range.

The main classes of equipment are:

- disconnectors, or isolators
- switches
- fuse–switch combinations
- circuit breakers
- earthing switches

A *disconnector* is a mechanical switching device which in the open position provides a safe working gap in the electrical system, to withstand normal working system voltage and any overvoltages which may occur. It is able to open or close a circuit if a negligible current is switched, or if no significant change occurs in the voltage between the terminals of the poles. Currents can be carried for specified times in normal operation and under abnormal conditions.

A *switch* is a mechanical device which is able not only to make, carry and interrupt current occurring under normal conditions in a system, but also to close a circuit safely, even if a fault is present. It must therefore be able to close satisfactorily carrying a peak current corresponding to the short-circuit fault level, and it must be able to carry this fault current for a specified period, usually one or three seconds.

A fuse and a switch can be used in combination with ratings chosen so that the fuse operates at currents in excess of the rated interrupting or breaking capacity of the switch. Such a device is known as a *'fuse switch'* if the fuseholder is also used as part of the main moving contact assembly, or a *'switch fuse'* if the fuse is a separate and static part of an assembly which includes the switch connected in series.

Figure 7.1 shows schematically the various combinations of fuse, switch and disconnector that are available.

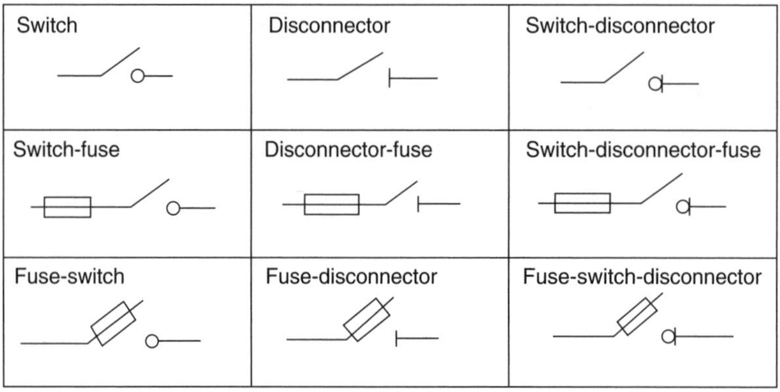

Fig. 7.1 Summary of equipment definitions

A *circuit breaker* is a mechanical switching device which is not only able to make, carry and interrupt currents occurring in the system under normal conditions, but also to carry for a specified time and to make and interrupt currents arising in the system under defined abnormal conditions, such as short-circuits. It experiences the most onerous of all the switching duties and is a key device in many switching and protection systems.

An *earthing switch* is a mechanical device for the earthing and short-circuiting of circuits. It is able to withstand currents for a specified time under abnormal conditions, but it is not required to carry normal service current. An earthing switch may also have a short-circuit making capacity.

7.2 Principles of operation

Since they perform the most arduous duty, circuit breakers are used here for the following brief description of switchgear operating principles.

The heart of a circuit breaker is the contact system which comprises a set of moving contacts, a set of fixed contacts, their current carrying conductors or leads and an opening mechanism which is often spring loaded.

The fixed and moving contacts are normally of copper, with sufficient size and cross-section to carry the rated current continuously. Attached or plated onto the copper are the contact tips or faces which are of silver or an alloy. The contact tip or face is the point at which the fixed and moving contacts touch, enabling the current to flow. It is therefore a requirement not only that a low contact junction resistance is maintained, but also that the contacts are not welded, destroyed or unduly eroded by the high thermal and dynamic stresses of a short-circuit.

As the device closes, the contact faces are forced together by spring or other pressure generated from the mechanism. This pressure is required in order to reduce the contact junction resistance and therefore the ohmic heating at the contacts; it also assists with the destruction or compaction of foreign material such as oxides which may contaminate the contact faces. The closing process is further assisted by arranging the geometry to give a wiping action as the contacts come together; the wiping helps to ensure that points of purely metallic contact are formed.

The breaking or interruption action is made particularly difficult by the formation of an arc as the contacts separate. The arc is normally extinguished as the current reaches a natural zero in the ac cycle; this mechanism is assisted by drawing the arc out to maximum length, therefore increasing its resistance and limiting the arc current. Various techniques are adopted to extend the arc; these differ according to size, rating and application and this is covered in more detail in the following sections.

The interruption of a resistive load current is not usually a problem. In this case the power factor is close to unity. When the circuit breaker contacts open, draw an arc and interrupt the current, the voltage rises slowly from its zero to its peak following its natural 50 Hz or 60 Hz shape. The build-up of voltage across the opening contacts is therefore relatively gentle, and it can be sustained as the contact gap increases to its fully open point.

However, in many circuits the inductive component of current is much higher than the resistive component. If a short-circuit occurs near the circuit breaker, not only is the fault impedance very low, and the fault current at its highest, but also the power factor may be very low, often below 0.1, and the current and voltage are nearly 90° out of phase. As the contacts open and the current is extinguished at its zero point, the voltage tends instantaneously to rise to its peak value. This is illustrated in **Fig. 7.2**. This results in a high rate of rise of voltage across the contacts, aiming for a peak transient recovery voltage which is considerably higher than the normal peak system voltage. There is a risk under these circumstances that the arc will restrike even though the contacts are separating, and the design of the circuit breaker must take this into account.

7.3   Low-voltage switchgear

7.3.1   Switches, disconnectors, switch disconnectors and fuse combination units

(a) *Construction and operation*
These switches are categorized by the ability to make and break current to particular duties which are listed for ac applications in **Table 7.1**. The category depends upon a multiple of the normal operating current and an associated power factor.

In all these devices, the contacts are usually of silver-plated copper; it is unusual to find the alloy contact tips used in circuit breakers. The contact system, connecting copperwork and terminals are usually housed within a housing of thermoset plastic or of a thermoplastic with a high temperature capability. The mechanism of the switch may be located in a number of positions, but it is usual to mount it at the side of or below the contact system.

The switching devices may be fitted with auxiliaries which normally take the form of small switches attached to the main device. These auxiliary switches provide means of indication and monitoring the position of the switch.

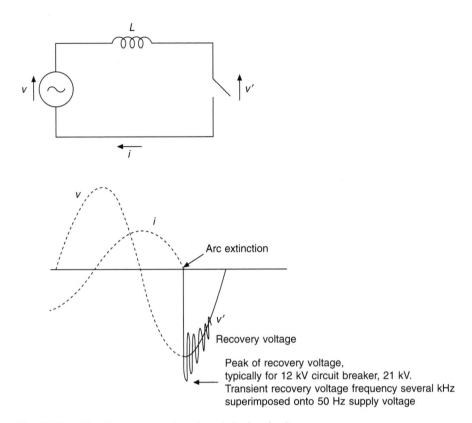

Fig. 7.2 Circuit breaker recovery voltage in an inductive circuit

Table 7.1 Categories of duty and overload current ratings for switches, disconnectors, switch disconnectors and fuse combination units

Category of duty	% of full-load current rating		Application
	making	breaking	
AC21	150%	150%	resistive loads
AC22	300%	300%	resistive and inductive loads
AC23	1000%	800%	highly inductive loads, stalled motor conditions

There are switching devices that are motor assisted to open and close the contacts, and some have tripping devices incorporated into the mechanism to open the switch under circuit conditions outside the protective characteristics of the fuse.

Many types of fuselink can be accommodated on a switch fuse or a fuse switch. In the UK, the bolted type is probably the most common, followed by the 'clip-in' type. There are many designs available to hold the clip-in fuselink within the switch moulding. In continental Europe the DIN type fuselink is the most common; here the fuse tags are formed as knife blades to fit into spring-tensioned clips.

The construction of a *fuse switch* is illustrated generally in **Fig. 7.3**. It consists of a moving carriage containing the fuselinks, which is switched into and out of a fixed system containing the contacts and mechanism. The moving carriage is shown in **Fig. 7.4**; it consists of a plastic and steel framework with copper contacts and terminals

Fig. 7.3 Main features of a fuse switch

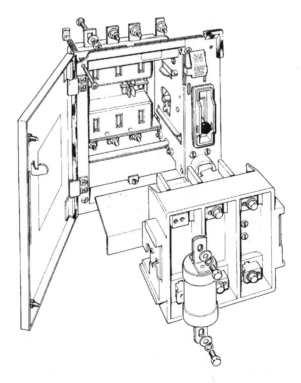

Fig. 7.4 Bill Sovereign fuse switch disconnector (courtesy of Bill Switchgear)

onto which are mounted the fuselinks. This is normally a three-pole or four-pole construction. It is usual to be able to extract the moving carriage to facilitate the changing of fuselinks outside the switch. Removal of the carriage also ensures total isolation of the load from the supply. The moving carriage is moved in and out by a spring-charged mechanism that ensures a quick make and break of the contacts.

In a *switch–fuse* the fuselinks remain static; the simplest form is a knife blade contact in series with a fuselink. In the modern form of switch fuse the fuselinks are mounted on a plastic enclosure which contains the contacts and the mechanism to operate these contacts. The contacts are normally arranged on both load side and supply side of the fuselink in order to ensure that the fuse is isolated when the switch is in the off position. The contact system is driven by a spring-assisted mechanism which is usually designed to provide quick make and break operation independent of the speed at which the operator moves the on–off handle.

In its simplest form, the switch is a knife blade contact, but the modern switch would follow the form of the fuse switch without the fuselinks. Many manufacturers use a fuse switch or switch fuse with solid copper links in place of the fuselinks.

The *disconnector* also follows the construction of the switch, but it must be capable of providing complete isolation when the contacts are open. This usually requires the provision of adequate clearance and creepage distances between live parts.

The *switch disconnector* combines the properties of a switch and a disconnector; this is in most cases what a manufacturer would build for sale.

(b) *Standards and testing*

Switches, disconnectors, switch disconnectors and fuse combination units are designed to comply with the requirements of IEC 947-3 or EN 60947-3. This requires a combination of type tests and routine tests.

A sequence of type tests is required to prove compliance with the standard. This sequence includes the following key tests:

- the making and breaking switching capacity test, which shows that the device is suitable for extreme overload conditions of resistive, inductive or highly inductive loads. The categories of duty and the current overload ratings have been shown in **Table 7.1**. The test is carried out by operating the device a number of times at the assigned rating.
- temperature-rise verification. This is carried out at the highest current rating of the device to prove that under full-load conditions in service the device will not damage cables, terminals and insulating materials or put operators at risk through contact with hot accessible parts. The limits of acceptable temperature rise are stipulated in the standard.
- the operational performance test. This is conducted to prove the mechanical and electrical durability of the device. A number of on-load and off-load switching operations are made, depending upon the make/break duty assigned by the manufacturer.
- dielectric verification to prove that the device has completed the sequence without damage to its insulation system.

A fuse switch or switch fuse is also proven by a fuse-protected short-circuit withstand (breaking) test. In this test, the fuselinks used have to be of the maximum rating, and with a breaking capacity assigned by the manufacturer. No damage such as welding of the contacts must occur to the switch as a result of this test. A fuse-protected making test is also carried out in which the switch is closed onto the declared rated short-circuit current.

A disconnector has to provide isolation properties and additional type tests are performed to prove these. In particular, a leakage current test is conducted after the main test sequence; maximum levels of acceptable leakage current are specified. In addition, the isolator handle is subjected to a force of three times the normal operation force necessary to switch the device off (within given minimum and maximum limits), the contacts of one phase being artificially locked in the off position and the test force being applied to open the switch. The on–off indicators of the disconnector must not give a false indication during and after this test.

Routine tests to be applied to all switch devices include an operational check in which each device is operated five times to check mechanical integrity. Also a dielectric test is carried out at a voltage which depends upon the rated voltage of the device. For a rated operating voltage of 380/415 V the dielectric test voltage is 2500 V.

7.3.2   Air circuit breakers and moulded case circuit breakers

(a) *Construction and operation*
Both the air circuit breaker and the moulded case circuit breaker (mccb) comprise the following features:

- a contact system with arc-quenching and current-limiting means
- a mechanism to open and close the contacts
- auxiliaries which provide additional means of protection and indication of the switch positions

The modern air circuit breaker is generally used as an incoming device on the supply side of a low voltage switchboard, and it represents the first line of protection on the load side of the transformer. An example is shown in **Fig. 7.5**. In addition to the above features, it also includes:

- a tripping and protection system to open the circuit breaker under fault conditions (if required)
- a means of isolating the device from the busbars
- usually an open construction, or the contact system housed in a plastic moulding
- current ratings from 400 A to 6300 A

The mccb may be used as an incoming device, but it is more generally used as an outgoing device on the load side of a switchboard. It is normally mounted into a low-voltage switchboard or a purpose-designed panel board. In addition to the three features listed at the start of this section, it also includes:

- an electronic or thermal/electromagnetic trip sensing system to operate through the tripping mechanism and open the circuit breaker under overload or fault conditions
- all parts housed within a plastic moulded housing made in two halves
- current ratings usually from 10 A to 1600 A

The basis of the main contact system has been explained in **section 7.3.1(a)**. The fixed contacts are usually mounted on a back panel or within a plastic moulding, and the moving contacts are usually supported on an insulated bar or within an insulated carriage.

Fig. 7.5   Air circuit breaker (courtesy of Terasaki (Europe) Ltd

In addition to these main contacts, arcing contacts are provided. The arcing contacts generally comprise a silver–tungsten alloy which provides high electrical conductivity with excellent wear against arc erosion and good anti-welding properties. The arc contacts are positioned in such a way that they make first and break last during opening and closing of the contacts. Their purpose is to ensure that any arcing takes place on the arcing contacts before the main contacts touch; the arcing contacts can then break once the main contacts have been forced together. During opening, the sequence is reversed. This action protects the main contacts from damage by arcing.

Extinguishing of the arc is an important feature in all low-voltage switchgear. All air circuit breakers have devices to extinguish the arc as quickly as possible. It is well known that an arc, once formed, will tend to move away from its point of origin; by forming the contact system carefully, the magnetic field generated by the current flow is used to move the arc into an arcing chamber where its extinction is aided.

Within the arcing chamber are number of metal arc-splitter plates which normally have a slot or 'V' shape cut into them in order to encourage the arc to run into the arcing chamber. A typical arrangement is shown in **Fig. 7.6**. The plates are used to

Fig. 7.6 Contact system and arc chute in an air circuit breaker

split the arc into a number of smaller arcs, having the effect of lengthening and cooling the arc. The arc is then extinguished when the voltage drop along it is equal to the voltage across the open contacts.

If the arc can be extinguished very quickly before the full prospective fault current is reached, the maximum current passed through the circuit breaker can be limited. This current limiting is achieved if the arc is extinguished within 10 ms of the start of the fault, before the peak of the sinusoidal current waveform is reached. Many designs have been tried in an attempt to achieve current limiting. Many of these designs are based on forming the current-carrying conductors within the circuit breaker in such a way that the electromagnetic forces created by the currents force the arc into the arcing chamber very quickly. A particular development of this is to form 'blow-out' coils from the current-carrying conductors.

The opening and closing mechanism of an air circuit breaker generally uses the energy developed in releasing a charged spring. The spring may be charged manually or with the aid of a winding mechanism attached to a motor. The charge is usually held until a release mechanism is activated, and once the spring energy is released to the main mechanism, linkages are moved into an over-toggle mode to close the moving contacts onto the fixed contacts. The force of the main spring is also used to transfer energy into compressing the moving contact springs, and it is these springs which provide the energy to open the contacts when required.

Attached to and interfacing with the main mechanism is the tripping mechanism which is used to open the contacts. The means of tripping an air circuit breaker is a coil, which may be a shunt trip coil, an undervoltage coil or a polarized trip coil. The shunt trip coil is current operated and consists of a winding on a bobbin with a moving core at its centre. When a current flows, the magnetic flux causes the core to move and this core operates the trip mechanism. An undervoltage coil is similar in construction, but the moving core is held in place by the magnetic flux against a spring; if the voltage across the winding falls this allows the spring to release the core and operate the trip mechanism. In the polarizing trip coil, the output of the coil is used to nullify the magnetic field from a permanent magnet which is set into the coil; when the coil is energized, the permanent magnetic field is overcome and using a spring energy a moving core is released to operate the trip mechanism.

The release coil receives its signal from the overcurrent detection device, which is usually an electronic tripping unit. This tripping unit has the capability to detect overloads and short-circuits, and it may be able to detect earth fault currents. The overload protection device detects an overload current and relates this to the time for which the current has flowed before initiating a trip signal; the time–current characteristic which is followed here is determined by the manufacturer within the guidelines of the standard. The general principles of time–current protection are explained more fully in **Chapter 8**, but a typical time–current characteristic is shown for convenience in **Fig. 7.7**. The short-circuit protection enables a high fault current to be detected and the circuit breaker to be opened in less than 20 ms. The short-circuit protection characteristics are usually adjustable in multiples of the full-load rated current, and they may incorporate a time interval which would usually be less than a second.

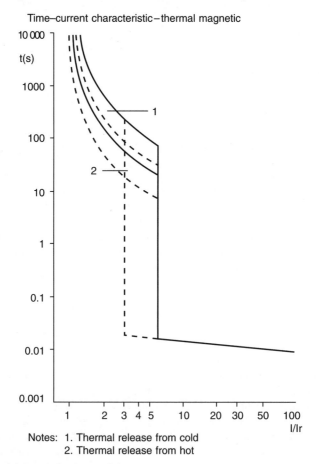

Fig. 7.7 Typical time–current characteristic curves

The mccb also incorporates an electronic trip device similar to that described above, but this may not include all the sophistication of the air circuit breaker system. Some mccbs will include an electromechanical thermal trip and electromagnetic trip in place of the electronic trip unit. These trips are shown diagrammatically in **Fig 7.8**. The electromechanical thermal trip consists of a heater and bimetal, the heater being part of the main current-carrying circuit. In overload conditions the heater causes deflection of the bimetal which is time related depending upon the current flow through the heater. The bimetal deflection results in operation of the trip mechanism and hence a time–current characteristic can be drawn for the circuit breaker. An electromagnetic trip usually comprises a U-shaped coil around the current-carrying circuit of the mccb, together with a moving pole piece. When a fault current passes, the magnetic flux within the U-shaped coil pulls in the pole piece and this movement is used to trip the circuit breaker.

(b) *Testing and standards*
Low-voltage air circuit breakers and mccbs are designed and tested according to the requirements of IEC 947-2 or EN 60947-2.

A series of type tests is carried out to prove compliance with the standard. This series of tests includes:

- a short-circuit capability test. This determines the maximum short-circuit current the device can withstand, while after the test retaining the ability to pass full-load current without overheating and allow the circuit breaker limited use. This test is carried out in a short-circuit test laboratory able to deliver exactly the current specified in the standard; large generators and transformers are necessary for this work, making it expensive.

Fig. 7.8 Electromechanical and electromagnetic trips used in a mccb

- a short-time withstand test. This establishes that the circuit breaker can carry a short-circuit current for a period of time; in many cases this would be 50 kA for 1 s. A circuit breaker with this capability can be used as an incoming device to a distribution system. Based on the principles of co-ordination explained in **Chapter 8**, this allows other devices within the distribution system to open before the incoming circuit breaker, thereby not necessarily shutting off all of that distribution system.
- an endurance test. This determines the number of open–close operations the circuit breaker can withstand before some failure occurs which means that the device will no longer carry its rated current safely. The tests are carried out with and without current flowing through the contacts.
- opening during overload or fault conditions. This test verifies that the device opens according to the parameters set by the standard and those given by the manufacturer. It determines the time–current characteristic of the circuit breaker, an example of which has been shown in **Fig 7.7**.
- a temperature rise test. This is carried out after all the above type tests. The standard specifies the maximum temperature rise permissible at the terminals and ensures that the device will be safe to use after short-circuits have been cleared and after a long period in service.

7.3.3 Miniature circuit breakers

Like the mccb, the miniature circuit breaker (mcb) has a contact system and means of arc quenching, a mechanism and tripping and protection system to open the circuit breaker under fault conditions.

The mcb has advanced considerably in the past 25 years. Early devices were generally of the 'zero-cutting' type, and during a short-circuit the current had to pass through a zero before the arc was extinguished; this provided a short-circuit breaking capacity of about 3 kA. Most of these early mcbs were housed in a bakelite moulding. The modern mcb is a much smaller and more sophisticated device. All the recent developments associated with moulded case circuit breakers have been incorporated into mcbs to improve their performance, and with breaking capacities of 10 kA to 16 kA now available, mcbs are used in all areas of commerce and industry as a reliable means of protection.

Most mcbs are of single-pole construction for use in single-phase circuits. The complete working system is housed within a plastic moulding, the typical external appearance of which being shown in **Fig. 7.9**. A schematic showing the principal parts of the mcb is shown in **Fig. 7.10**. The contact system comprises a fixed and a moving contact, and attached to each is a contact tip which provides a low-resilience contact junction to resist welding.

Modern mcbs are fitted with arc chutes consisting of metal plates which are held in position by insulating material. The arc chute does not necessarily surround the contact; in some designs arc runners are provided to pull the arc into the arc chute.

The tripping mechanism usually consists of a thermal-magnetic arrangement. The thermal action is provided by a bimetal with, in some cases, a heater. For ratings in the range 6–63 A the bimetal forms part of the current path, the heat generated within the bimetal itself being sufficient to cause deflection. The deflection is then used to activate the tripping mechanism. The characteristics of the bimetal are chosen to provide particular delays under certain overload or fault currents according to the required time–current characteristic. A high-resistance bimetal is used for low-current

Fig. 7.9 Miniature circuit breaker, external view

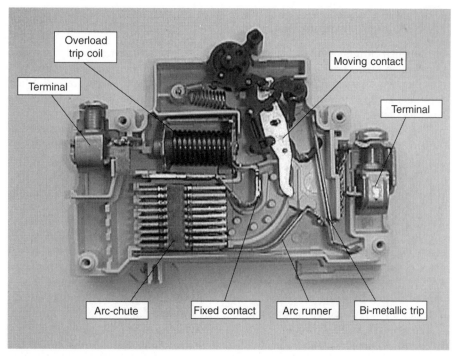

Fig. 7.10 Principal parts of an mcb

devices and a lower-resistance bimetal for high-current devices. In very low-current mcbs a heater may be incorporated around the bimetal in order to generate sufficient heat to deflect it.

The magnetic tripping element usually consists of a coil which is wrapped around a tube, there being a spring-loaded slug within the tube. Movement of the slug operates the tripping mechanism to open the mcb. It can also be used to assist in opening the contacts by locating the coil near to the moving contact. When a fault current flows, the high magnetic field generated by the coil overcomes the spring force holding the slug in position; the slug then moves to actuate the tripping mechanism and forces the contacts apart by striking the moving contact arm. For low mcb ratings the coil is formed from thin wire with many turns; for higher ratings the wire is thicker, with fewer turns. The magnetic trip is set by the manufacturer according to the required characteristics. These characteristics are defined in the standard and form 'types' which are shown in **Table 7.2**.

MCBs are designed and tested according to the requirements of IEC 898.

Table 7.2 Magnetic trip settings for mcbs

MCB type	Minimum trip current	Maximum trip current	Application
B	$3I_n$	$5I_n$	Domestic: resistive or small inductive loads such as lighting and socket outlets
C	$5I_n$	$10I_n$	Light industrial: inductive loads such as fluorescent lighting and motors
D	$10I_n$	$20I_n$	Very inductive loads such as welding machines

7.3.4   Residual current devices

The residual current device (rcd) is used to detect earth fault currents and to interrupt supply if an earth current flows. The main application is to prevent electrocution but rcds can also be used to protect equipment, especially against fire. The earth fault currents that operate an rcd can range from 5 mA up to many amperes. For typical domestic applications the typical trip current would be 30 mA.

The rcd can be opened and closed manually to switch normal load currents, and it opens automatically when an earth fault current flows which is about 50 per cent or more of the rated tripping current.

The main features of an rcd are shown in **Fig 7.11**. The key component is a toroidal transformer, upon which the load current (live) and return current (neutral) conductors are wound in opposite directions. The toroid also carries a detecting winding. If no earth fault current is flowing then the load and return currents are equal. In this case the mmfs generated by the load and return current windings are equal; there is no resultant flux in the toroid and the detecting winding does not generate any current. When a fault current flows there is a difference between the load and return currents which generates a resultant flux in the toroid and induces a current in the detecting winding. The current generated in the detecting winding operates a relay which opens the main contacts of the rcd.

From a very small output the detecting winding has to produce sufficient power to operate the tripping mechanism. Two alternative methods are used. In the first method, the output signal from the detecting coil is electronically amplified and the

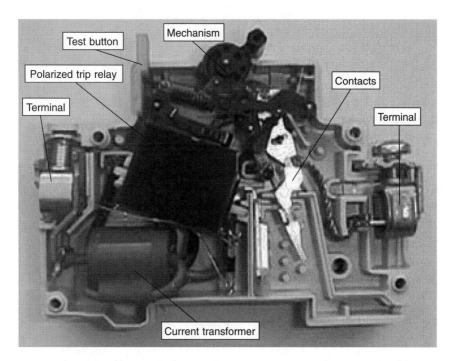

Fig. 7.11 Residual current device

second method uses a polarized relay operating on a sensitive mechanical trip mechanism. The operation of a polarized trip relay has already been described for circuit breakers in **section 7.3.2(a)**; it is based on the magnetic output of a small coil nullifying the field from a permanent magnet, causing the release of an armature. The basic operation is illustrated in **Fig. 7.12**.

The operation of an rcd has here been described for single-phase operation, but it may also be applied in a three-phase application where typically it might be used in a light industrial system for protection against fire. There are two arrangements of a three-phase rcd. Either the three phases are wound around a current transformer, or the three phases and the neutral are wound onto a balancing transformer.

The rcd has only limited breaking capacity and it is not a replacement for overcurrent protection devices such as the mcb. The residual current breaker with overcurrent (rcbo) is now available; this is an rcd with an overcurrent tripping mechanism and enhanced contacts to cope with interruption of fault conditions.

RCDs are designed and tested according to the requirements of IEC 1008 and IEC 1009.

7.3.5 Standards

The leading international standards adopted for low-voltage switchgear are shown in **Table 7.3**.

7.4 Medium-voltage (distribution) switchgear

Distribution switchgear, also commonly referred to as medium-voltage switchgear, is generally acknowledged to cover the range 1 kV to 36 kV. 72.5 kV and even 132 kV

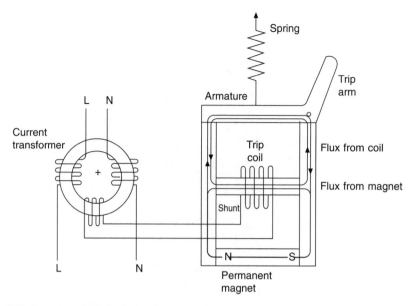

Fig. 7.12 Operation of polarized trip relay

Table 7.3 International standards relating to low-voltage switchgear

IEC	EN/HD	Subject of standard
IEC 898	EN 60898	Miniature circuit breakers (mcbs)
IEC 347-2	EN 60947-2	Air circuit breakers (acbs) and moulded case circuit breakers (mccbs)
IEC 947-3	EN 60947-3	Switches – fuse combination switches
IEC 1008	EN 61008	Residual current devices (rcds)

can be considered as distribution voltages rather than transmission voltages, and the equipment overlaps with high-voltage or transmission switchgear in this range.

Switchgear in the distribution-voltage range differs considerably from low-voltage switchgear. There are similarities with aspects of transmission switchgear, but many functions and practices are different.

7.4.1 Types of circuit breaker

In the past, oil- and air-break devices have predominated, but now several alternative methods of arc interruption are used in distribution-voltage circuit breakers.

Early *oil designs* featured plain-break contacts in a tank of oil capable of withstanding the considerable pressure built up from large quantities of gas generated by long arcs. During the 1920s and 1930s various designs of arc control device were introduced to improve performance. These were designed such that the arc created between the contacts produces enough energy to break down the oil molecules, generating gases and vapours which by the cooling and de-ionizing of the arc resulted in successful clearance at current zero. During interruption, the arc control device encloses the contacts, the arc, a gas bubble and oil. Carefully designed vents allow the gas to escape as the arc is lengthened and cooled. **Figure 7.13** shows an axial-blast arc control device. The use of oil switchgear is reducing significantly in most areas of the world because of the need for regular maintenance and the risk of fire in the event of failure.

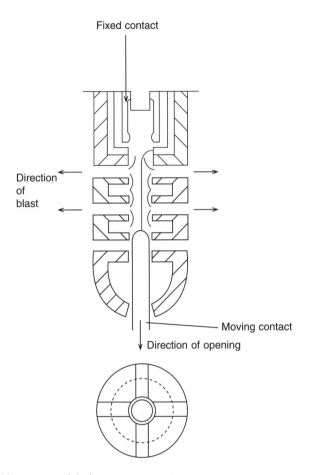

Fig. 7.13 Axial-blast arc control device

Sulphur hexafluoride (SF₆) is the most effective gas for the provision of insulation and arc interruption. It was initially introduced and used predominantly in transmission switchgear, but it became popular at distribution voltages from the late 1970s.

The main early types of SF_6 circuit breaker are known as 'puffer' types. These use a piston and cylinder arrangement, driven by an operating mechanism. On opening, the cylinder containing SF_6 gas is compressed against the piston, increasing the gas pressure; this forces the gas through an annular nozzle in the contact giving a 'puff' of gas in the arc area. The gas helps to cool and de-ionize the arc, resulting in arc interruption at current zero. The dielectric properties of SF_6 are quickly re-established in the heat of the arc, giving a rapid increase in dielectric strength as the voltage across the contacts begins to rise. The energy required to drive the puffer type of circuit breaker is relatively high, resulting in the need for a powerful mechanism.

Other designs of SF_6 circuit breaker started to be introduced in the 1980s. These make more use of the arc energy to aid interruption, allowing the use of a lighter and cheaper operating mechanism. Whilst in the puffer design the compressed SF_6 is made to flow over an arc which is basically stationary, these newer designs move the arc through the SF_6 to aid interruption. In the rotating-arc SF_6 circuit breaker a coil is wound around the fixed contact. As the moving contact withdraws from the fixed contact the current is transferred to an arcing contact and passes through the coil. The current in the coil sets up an axial magnetic field with the arc path at right angles to the magnetic flux, and rotation of the arc is induced according to Fleming's left-hand rule (see **Chapter 2, eqn 2.16**). The arc rotation is proportional to the magnitude of the current. Careful design is required in order to ensure an adequate phase shift between the flux and the current to maintain adequate movement of the arc as the current falls towards its zero crossing. The operation of a rotating-arc interrupter is illustrated in **Figs 7.14** and **7.15**.

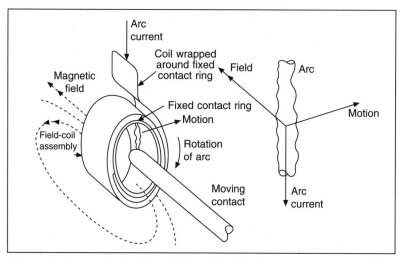

The arc has been extended towards the coil axis by transverse movement of the contact bar. It is shown after transfer to the arcing tube and consequent production of the magnetic field. It is thus in the best plane to commence rotation

Fig. 7.14 Operation of a rotating-arc SF$_6$ interrupter

Other types of self-extinguishing circuit breaker use an insulating nozzle which is similar to that in a puffer breaker, but with a coil to produce a magnetic field for moving the arc. The thermal energy in the arc is used to increase the local pressure. This results in arc movement which again is proportional to the current, and successful interruption follows. In some designs a small puffer is also used to ensure there is enough gas flow to interrupt low levels of current, where the thermal energy in the arc is relatively low.

The various types of SF$_6$ high-voltage switchgear are referred to in **section 7.5.2**, and **Fig. 7.24** shows the main configurations.

The main alternative to the SF$_6$ circuit breaker in the medium-voltage range is the *vacuum interrupter*. This is a simple device, comprising only a fixed and moving contact located in a vacuum vessel, but it has proved by far the most difficult to develop. Although early work started in the 1920s, it was not until the 1960s that the first vacuum interrupters capable of breaking large currents were developed, and commercial circuit breakers followed about a decade later.

The principle of operation of a vacuum interrupter is that the arc is not supported by an ionized gas, but is a metallic vapour caused by vaporization of some of the contact metal. At current zero the collapse of ionization and vapour condensation is very fast, and the extremely high rate of recovery of dielectric strength in the vacuum ensures a very effective interrupting performance. The features of a vacuum interrupter

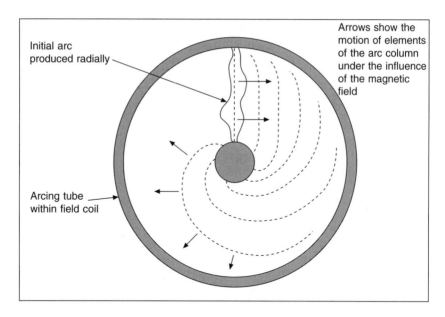

Initial arc
produced radially

Arcing tube
within field coil

Arrows show the
motion of elements
of the arc column
under the influence
of the magnetic
field

Development of a spiral arc. The arc is motivated to move sideways but this
tendency is limited by the shape of the electrodes and it develops quickly into
a spiral with each element tending to move sideways as shown by the arrows

Fig. 7.15 Operation of a rotating-arc SF$_6$ interrupter

which are key to its performance are the contact material, the contact geometry and
ensuring that the envelope (a glass or ceramic tube with welded steel ends) remains
vacuum-tight throughout a working life in excess of 20 years. Even today, specialized
manufacturing techniques are necessary to achieve this. A typical vacuum interrupter
is shown in **Fig. 7.16**.

The circuit breaker is located within a switchgear housing, some types of which
are described later. The main insulation in the housing is usually air, although some
designs now have totally sealed units filled with SF$_6$ gas. Structural isolation is
required to support current-carrying conductors; this is normally some type of cast
resin. Thermoplastic materials which can be injection moulded are often used for
smaller components, but larger items such as bushings which are insulation covered
are usually made from thermosetting materials such as polyurethane or epoxy resin
mixed with filler to improve its mechanical and dielectric properties.

7.4.2 Main classes of equipment

(a) Primary switchgear
Primary substations up to 36 kV are usually indoor, although they can be outdoor
where space is not a problem. Indoor equipment is usually metal enclosed, and it can
be subdivided into metalclad, compartmented and cubicle types. In the metalclad
type, the main switching device, the busbar section and cable terminations are segregated
by metal partitions. In compartmented types the components are housed in separate
compartments but the partitions are non-metallic. The third category covers any form
of metal-enclosed switchgear other than the two described.

The vertically isolated, horizontally withdrawn circuit breaker is the traditional

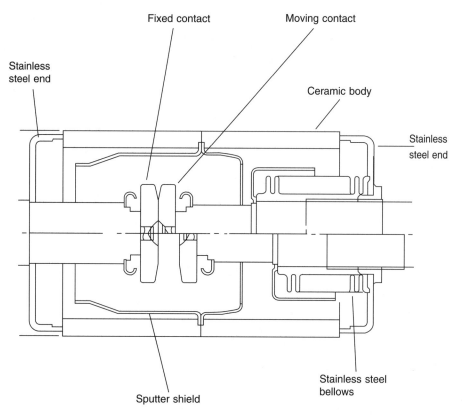

Fig. 7.16 Typical vacuum interrupter (courtesy Meidensha Corporation, Japan)

arrangement in the UK or UK-influenced areas. This originates from the bulk-oil circuit breaker, for which there was a need for easy removal for frequent maintenance, but manufacturers have utilized the already available housings to offer vacuum or SF_6 circuit breakers as replacements for oil circuit breakers, or even for new types to this design. The design also provides convenient earthing of the circuit or busbars by alternative positions of the circuit breaker in the housing; this is achieved by disconnecting the circuit breaker, lowering it from its service position, moving it horizontally and then raising and reconnecting the connections between the busbar or cable and earth. A typical section is shown in **Fig. 7.17**.

Horizontally isolated, horizontally withdrawn equipment has been used extensively in mainland Europe, with air-break, small oil volume and latterly with SF_6 and vacuum circuit breaker truck designs. A section of this type of switchgear is shown in **Fig. 7.18**. Since the circuit breaker cannot be used as the earthing device in this design, integral earthing switches or portable earthing devices are required.

Fixed-position circuit breakers were introduced in the 1970s; these depart from the withdrawable arrangements and are based on a 'sealed-for-life' concept. Due to the early problems with vacuum interrupters, some users were reluctant to accept fixed-position vacuum switchgear, but now that excellent long-term reliability of vacuum interrupters has been established, newer designs have been introduced again recently. Several arrangements in which the housing is completely sealed and filled with SF_6 as a dielectric medium are now available and an example is shown in **Fig. 7.19**. Because of the compact dimensions that SF_6 permits these are gaining

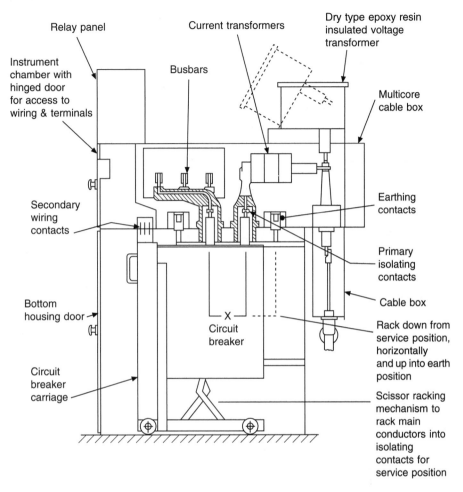

Fig. 7.17 Section of vertically isolated, horizontally withdrawn switchgear

popularity, particularly at 36 kV, where there can be a cost advantage not only in the price, but also in the reduced size of the equipment and the substation required.

(b) *Secondary switchgear*
Secondary distribution switchgear is that which is connected directly to the electrical utility transformers which provide low-voltage supplies to customers. The systems distributing power from the primary substations can be conveniently divided into overhead circuits and underground cable networks.

Most faults on overhead systems are caused by lightning strikes, branches brushing the conductors, clashing of conductors in high wind, or large birds bridging the lines; they are usually transient in nature. In most cases the fault duration is short and the circuit breakers used in these circuits, known as *reclosers*, are programmed to close again a very short time after they have opened. This allows the fault to be cleared with a minimum of disruption to consumers. It is normal to programme the recloser to open then close up to four times in order to allow time for the fault to disappear. If after this time the fault is still present, it is assumed that the fault is not transient; the recloser then locks out and the faulty section of line is isolated. A recloser

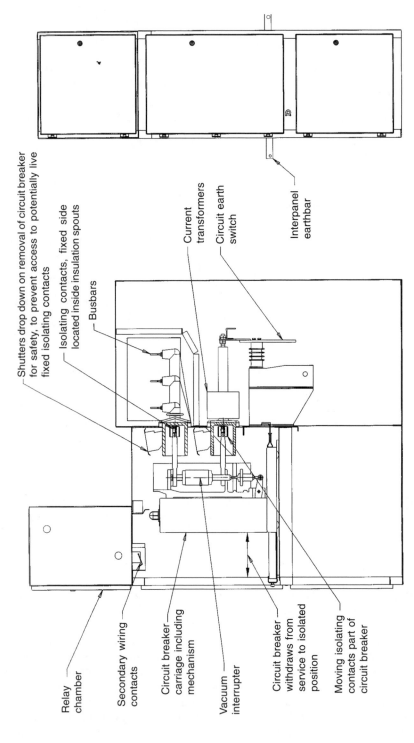

Shutters drop down on removal of circuit breaker for safety, to prevent access to potentially live fixed isolating contacts

Isolating contacts, fixed side located inside insulation spouts

Busbars

Current transformers

Circuit earth switch

Interpanel earthbar

Relay chamber

Secondary wiring contacts

Circuit breaker carriage including mechanism

Vacuum interrupter

Circuit breaker withdraws from service to isolated position

Moving isolating contacts part of circuit breaker

Fig. 7.18 Section of horizontally isolated, horizontally withdrawn switchgear

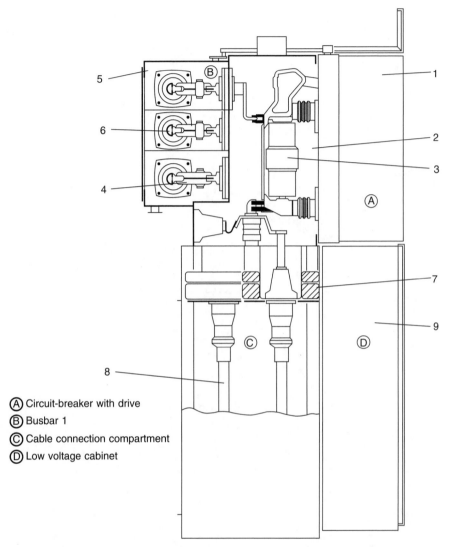

A Circuit-breaker with drive
B Busbar 1
C Cable connection compartment
D Low voltage cabinet

1 Drive for circuit-breaker and disconnector/ earthing switch with control panel	5 Busbar compartment
2 Circuit-breaker compartment	6 Busbar
3 Vacuum interrupter	7 Current transformer
4 Disconnector/earthing switch blades (in position "Disconnector MADE")	8 Connecting cable
	9 Cabinet for control and auxiliary devices (low-voltage cabinet)

Fig. 7.19 Section of 36 kV SF_6 gas-insulated switchgear (courtesy ALSTOM T&D)

normally controls several circuits. An 11 kV pole-mounted auto-recloser is shown in **Fig. 7.20**. Reclosers are available with SF_6 and vacuum interrupters; the equipment used in conjunction with reclosers are switches and sectionalizers, which can be air-break switches, vacuum or SF_6 devices, expulsion fuses and off-load switching devices.

Most urban areas in the UK and Europe are supplied by underground cable. In this case the step-down transformer (for example, 11 000/415 V) which supplies a consumer

Fig. 7.20 11 kV pole-mounted auto-recloser (courtesy FKI plc – Engineering Group)

circuit is connected to a protective device such as a fuse switch or circuit breaker in a ring. A simplified distribution network with the main components is shown in **Fig. 7.21**. The 'ring main unit' usually has two load-break switches with the control protective device combined into a single unit. In the UK this type of unit is ruggedly designed for outdoor use, but in continental Europe indoor designs are more normal. The traditional oil-filled unit has been replaced largely by SF_6 equipment over recent years. The ring main unit has normally been manually operated with a spring-operated mechanism to ensure that its operation is not dependent on the force applied by the operator. However, as utilities press to improve the reliability of supplies to consumers and to reduce the time lost through outages caused by faults, the use of automatically operated motor-driven spring mechanisms is now on the increase.

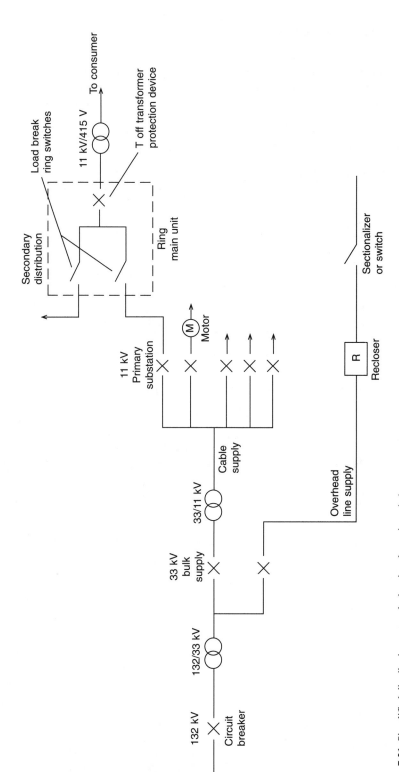

Fig. 7.21 Simplified distribution network showing the main switchgear components

7.4.3 Rating principles

The main consideration in the choice of switchgear for a particular system voltage is the current it is capable of carrying continuously without overheating (the rated normal current) and the maximum current it can withstand, interrupt and make onto under fault conditions (the rated short-circuit current). Studies and calculations based on the circuits of interest will quantify these parameters. The switchgear rating can then be chosen from tables of preferred ratings in the appropriate standard, such as IEC 694 and IEC 56. **Table 7.4** shows figures taken from IEC 56 showing the co-

Table 7.4 Co-ordination of rated values for circuit breakers, taken from IEC56

Rated voltage U (kV)	Rated short-circuit breaking current I_{sc} (kA)	Rated normal current I_n (A)							
3.6	10	400							
	16		630		1250				
	25				1250	1600		2500	
	40				1250	1600		2500	4000
7.2	8	400							
	12.5	400	630		1250				
	16		630		1250	1600			
	25		630		1250	1600		2500	
	40				1250	1600		2500	4000
12	8	400							
	12.5	400	630		1250				
	16		630		1250	1600			
	25		630		1250	1600		2500	
	40				1250	1600		2500	4000
	50				1250	1600		2500	4000
17.5	8	400	630		1250				
	12.5		630		1250				
	16		630		1250				
	25				1250				
	40				1250	1600		2500	
24	8	400	630		1250				
	12.5		630		1250				
	16		630		1250				
	25				1250	1600		2500	
	40					1600		2500	4000
36	8		630						
	12.5		630		1250				
	16		630		1250	1600			
	25				1250	1600		2500	
	40					1600		2500	4000
52	8			800					
	12.5				1250				
	20				1250	1600	2000		
72.5	12.5			800	1250				
	16			800	1250				
	20				1250	1600	2000		
	31.5				1250	1600	2000		

ordination of rated values for circuit breakers. Ratings vary commonly from less than 400 A in a secondary system to 2000 A or more in a primary substation, with fault currents of 6 kA in some overhead line systems up to 40 kA and above in some primary circuits.

The standards also specify overvoltages and impulse voltages which the switchgear must withstand. Normal overvoltages which occur during switching or which may be transferred due to lightning striking exposed circuits must be withstood. The switchgear also needs to provide sufficient isolating distances to allow personnel to work safely on a part of the system that has been disconnected. The lightning impulse withstand voltage and the normal power frequency withstand voltage are specified relative to the normal system voltage in the standards.

7.4.4 Test methods

Type tests are performed on a single unit of the particular type and rating in question. They are performed to demonstrate that the equipment is capable of performing the rated switching and withstand duties without damage, and that it will provide a satisfactory service life within the limits of specified maintenance. The main type tests are:

- dielectric tests. Lightning impulse is simulated using a standard waveshape having a very fast rise time of 1.2 μs and a fall to half value of 50 μs. Power frequency tests are performed at 50 Hz or 60 Hz. The method of performing the tests is detailed in standards, for instance IEC 60, 'Guide to high voltage testing techniques'. This requires the switchgear to be arranged as it would be in service. If outdoor rated, the tests need to be carried out in dry and wet conditions, the latter using water of specified conductivity which is sprayed onto the equipment at a controlled rate during application of the high voltage.

- temperature rise tests. Again the switchgear is arranged in its service condition, and the normal continuous current is applied until the temperature of the main components has stabilized. The final temperature, measured by thermocouples, must not exceed values stated in the specifications; these values have been chosen to ensure that no deterioration of metal or insulation is caused by continuous operation.

- short-circuit and switching tests. These are conducted over a range of currents from 10 per cent to 100 per cent of the short-circuit rating. The tests are at low power factor, which represents the most onerous switching conditions seen in service. The different levels represent short-circuits at various points in the system, from close to the circuit breaker terminals to a long distance along the cable or line. Whilst the current to interrupt varies with these positions, another major factor is the rate of rise of the transient recovery voltage appearing across the circuit breaker contacts; this can have a significant influence on the circuit breaker performance and it plays a key part in the design. The ability to interrupt the short-circuit current with a decaying dc component superimposed on the power frequency must also be demonstrated. The decrement or rate of decay of the dc voltage is specified in the standards, but the dc component which is present when the contacts open depends upon the opening time of the circuit breaker (it being reduced if the opening time is long). **Figure 7.22** shows the determination of short-circuit making and breaking currents and of the dc component. Circuit breakers must also be tested when closing onto a fault condition, and then carrying the full short-circuit current for a specified period, usually one or three seconds.

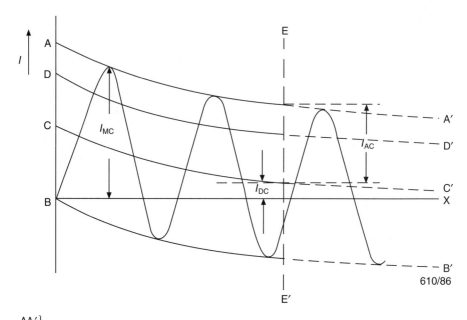

AA'		
BB'	}	= envelope of current-wave
BX		= normal zero line
CC'		= displacement of current-wave zero-line at any instant, the dc component
DD'		= rms value of the ac component of current at any instant, measured from CC'
EE'		= instant of contact separation (initiation of the arc), with dc component I_{DC} indicated
I_{MC}		= making current
I_{AC}		= peak value of ac component of current at instant EE'
$\dfrac{I_{AC}}{\sqrt{2}}$		= rms value of the ac component of current at instant EE'
I_{DC}		= dc component of current at contact separation point
$\dfrac{I_{DC} \times 100}{I_{AC}}$		= percentage value of the dc component

Fig. 7.22 Determination of short-circuit making and breaking currents and of dc component

- mechanical endurance test. To demonstrate that the equipment has a satisfactory life, this test is performed over a specified number of operations which depends upon the type of equipment and duty; the number of open–close operations can vary from 1000 to 10 000. The tests are carried out without electrical load, but at the rated output of the mechanism of the device.

Routine tests are performed by the manufacturer on every unit of switchgear to ensure that the construction is satisfactory and that the operating parameters are similar to the unit which was subjected to the type test. These tests are specified in the standards, and the minimum performed will include the following:

- power frequency voltage withstand dry tests on the main circuit
- voltage tests on the control and auxiliary circuits
- measurement of the resistance of the main circuit
- mechanical operating tests

7.4.5 Commissioning and maintenance

Switchgear is usually transported in sections which need to be carefully erected and connected together at the substation.

Commissioning tests are performed after installation; these are necessary to ensure that the connections are sound and that the equipment is functioning satisfactorily. These tests will normally include checks of current and voltage transformers, checking of the protection scheme, circuit breaker operation and high-voltage withstand tests.

With the demise of oil-filled switchgear the requirement for maintenance has been significantly reduced. SF_6 and vacuum circuit breakers which are 'sealed for life' should require no servicing of the interrupting device. Manufacturers provide instructions on simple inspection and servicing of other parts of the circuit breaker; this is usually limited to cleaning, adjustment and lubrication of the mechanism.

7.4.6 Standards

Some of the commonly used standards and specifications for distribution switchgear are given in **Table 7.5**.

Table 7.5 Standards and specifications for distribution switchgear

IEC	BS	Electricity Association	Subject	N. American
56	5311		High voltage ac circuit breakers	
129	EN60129		AC disconnectors and earth switches	
265	EN60265		High voltage switches	
298	EN60298		AC metal enclosed switchgear and controlgear for rated voltages above 1 kV up to and including 52 kV	
420	EN60420		High voltage ac switch fuse combinations	
694	EN60694		Common specifications for high voltage switchgear and controlgear standards	
		41-26	Distribution switchgear for service up to 36 kV (cable connected)	
		41-27	Distribution switchgear for service up to 36 kV (overhead line connected)	
			Overhead, pad mounted, dry vault, and submersible automatic circuit reclosers and fault interrupters for ac systems	ANSI/IEEE C37.60
			Overhead, pad mounted, dry vault, and submersible automatic line sectionalizers for ac systems	ANSI/IEEE C37.63

7.5 High-voltage (transmission) switchgear

7.5.1 System considerations

The transmission of high powers over long distances necessitates the use of High (HV), Extra High (EHV) or Ultra High (UHV) voltages. Historically these voltages were classed as 72.5 kV and above. The lower voltages were introduced first, and as the technologies have developed these have increased so that now the highest transmission voltages being used are 1100 kV. For long transmission distances it is also possible to use dc systems; these have particular advantages when two systems are to be connected, but they are beyond the scope of this section, which considers only ac transmission switchgear.

In the UK the 132 kV system was first established in 1932, and subsequent growth has required expansion of the system up to 275 kV (1955) and 420 kV (1965). Other countries have developed similarly, so a multiple voltage transmission system exists in virtually every major electricity-using nation.

In recent years there has been some harmonization of the rated voltages of transmission systems, the agreed levels set down in IEC 694 being 72.5, 100, 123, 145, 170, 245, 300, 362, 420, 550 and 800 kV. Systems will be operated at a variety of normal voltage levels such as 66, 132, 275 and 400 kV, but the maximum system operating voltage may be higher than these nominal ratings, so the next highest standard value will be selected as the rated voltage of the system.

In order to connect, control and protect these transmission systems it is necessary to use switchgear of various types and ratings.

A switchgear installation which allows the connection and disconnection of the interconnecting parts of a transmission system is referred to as a substation. A substation will include not only the switchgear, but also transformers and the connections to overhead lines or cable circuits. The main functional elements of the switchgear installation in a substation are circuit breakers, switches, disconnectors and earthing switches. Some of the most common configurations for substations are illustrated in **Fig. 7.23**.

Switching conditions on transmission systems are similar in principle to those on distribution systems, with a requirement to interrupt normal load currents, fault currents and to undertake off-load operations. Because of the different circuit parameters in transmission and distribution networks, the switching duties impose particular requirements which have to be taken into account during the design and type testing of switchgear equipment. A particular requirement for transmission is the ability to deal with the onerous transient recovery voltages which are generated during interruption of faults on overhead lines when the position of the fault is close to the circuit breaker terminals; this is defined in IEC 56 as the short line fault condition. A further requirement is the energization of long overhead lines, which can generate large transient overvoltages on the system unless countermeasures are taken.

7.5.2 Types of circuit breaker

The first designs of transmission circuit breaker used minimum-volume oil interrupter technology similar to that which had been developed for distribution switchgear.

The 1940s saw the development and introduction throughout the world of air-blast circuit breakers. These devices used compressed air at pressures up to 1200 psi, not only to separate the contacts but also to cool and de-ionize the arc drawn between the contacts. Early air-blast designs used interrupters that were not permanently pressurized, and reclosed when the blast was shut off. Isolation was achieved separately by a series-connected switch which interrupted any residual resistive and/or capacitive grading current, and had the full rated fault-making capacity.

The increase of system voltages to 400 kV in the late 1950s coincided with the discovery, investigation and specification of the kilometric or short line fault condition. Air-blast interrupters are particularly susceptible to high rates of rise of recovery voltage (RRRV), and parallel-connected resistors are needed to assist interruption. At the increased short-circuit current levels of up to 60 kA which were now required,

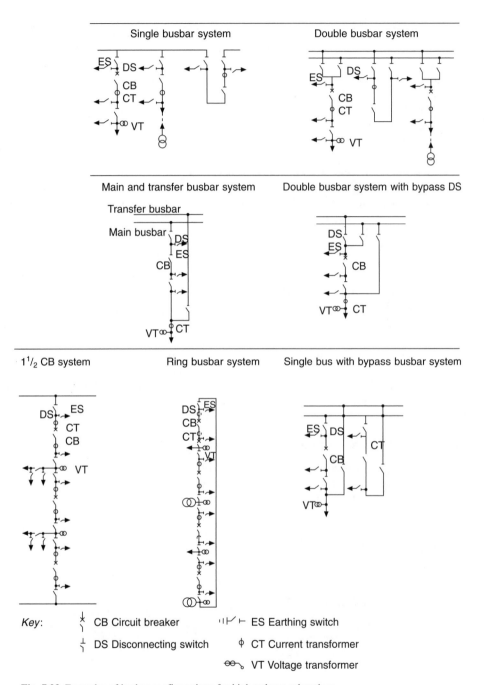

Fig. 7.23 Examples of busbar configurations for high-voltage substations

parallel resistors of up to a few hundred ohms per phase were required, and the duty of breaking the resistor current could no longer be left to the series-connected air switch. To deal with this, permanently pressurized interrupters incorporating parallel-resistor switching contacts were used. The original pressurized-head circuit breakers used multiple interrupters per phase in order to achieve the required fault-switching capacity at high transmission voltages; 10 or 12 interrupters were used for 400 kV. These were proved by direct unit testing at short-circuit testing stations, each having a direct output of 3 GVA. It was in fact the ability to test which determined the number of interrupters used in the design of a circuit breaker.

The introduction of synthetic testing methods increased the effective capacity of short-circuit testing stations by an order of magnitude, and this allowed the proving of circuit breakers having fewer interrupters. 400 kV air-blast circuit breakers with only four or six interrupters per phase were being supplied and installed up to the late 1970s.

The next step-change in technology occurred with the introduction of SF_6 circuit breakers. The merits of electronegative gases such as SF_6 had long been recognized, and freon had been used in the late 1930s to provide the primary insulating medium in 33 kV metalclad switchgear installed in two UK power stations. The reliability of seals was a problem, and it was only during the nuclear reactor programme that sufficiently reliable gas seals were developed to enable gas-insulated equipment to be reconsidered.

The first practical application of SF_6 in switchgear was as the insulating medium for instrument transformers in transmission substations in the late 1950s and early 1960s. These were quickly followed by the use of SF_6 not only as an insulating medium but also as the means of extinguishing arcs in circuit breakers. The first generation of SF_6 circuit breakers were double-pressure devices which were based on air-blast designs, but using the advantages of SF_6. These double-pressure designs were quickly followed by puffer circuit breakers which have already been described in **section 7.4.1**. Puffer circuit breakers, now in the form of second or third generation of development, form the basis of most present-day transmission circuit breaker designs. **Figure 7.24** illustrates the various forms of SF_6 interrupter.

7.5.3 Main classes of equipment

Transmission substations may be built indoors, or outdoors, but because of the space required it is usually difficult to justify an indoor substation. The strongest justification for an indoor substation is the effect of harsh environments on the design, operation and maintenance of the equipment.

The two main types of transmission switchgear currently available are conventional or Air-Insulated Switchgear (AIS) and gas-insulated metal-enclosed switchgear.

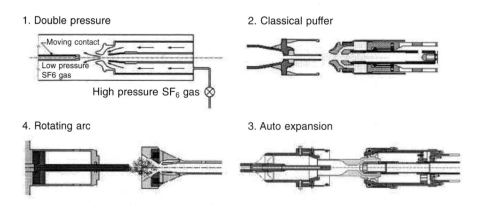

1. Double pressure   2. Classical puffer

Moving contact

Low pressure SF6 gas

High pressure SF_6 gas

4. Rotating arc   3. Auto expansion

Fig. 7.24 Various forms of SF_6 interrupter

(a) *Conventional or Air-Insulated Switchgear (AIS)*

Here the primary insulating medium is atmospheric air, and in the majority of AIS the insulating elements are of either solid or hollow porcelain.

In circuit elements such as disconnectors and earthing switches, where the switching is done in air, the insulators are solid, and provide not only physical support and insulation, but also the mechanical drive for the elements. Examples of disconnectors and earth switch arrangements are shown in **Fig. 7.25**. The choice of disconnector design is dependent on the substation layout and performance requirements.

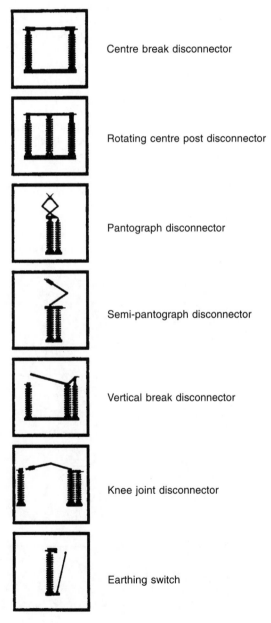

Centre break disconnector

Rotating centre post disconnector

Pantograph disconnector

Semi-pantograph disconnector

Vertical break disconnector

Knee joint disconnector

Earthing switch

Fig. 7.25 Examples of disconnector and earth switch arrangements

Equipment such as circuit breakers which need an additional insulating or interrupting medium use hollow porcelain insulators that contain the active circuit elements in the appropriate insulating medium. As previously outlined, initial AIS installations used oil as the insulating and interrupting medium, which in many cases was replaced by the use of compressed air, and modern designs almost always use SF_6 gas.

The layout of a substation is governed by the need to ensure that air clearances meet the dielectric design requirements, including conditions when access is required to the substation for maintenance. Rules for the minimum clearances and the requirements for creepage distances on the external porcelain insulators are contained in the standards. For locations with atmospheric pollution it is necessary to provide extended creepage distances on insulators. Guidelines for the selection of creepage distance are given in IEC 815.

A recent development in AIS technology is the use of composite insulators for functional elements in place of the traditional porcelain. There is a risk of explosion with porcelain insulators when they are filled with insulating gas if a mechanical fault occurs with the equipment. A composite insulator consists of silicone rubber sheds on a filament glass-fibre tube; it is not brittle and will not explode if the insulator is ruptured. Composite insulators are now being used on instrument transformers and circuit breakers for AIS, and their use is likely to become more common.

An example of an outdoor AIS substation is shown in **Fig. 7.26**.

(b) *Gas-insulated metal-enclosed switchgear*
In metal-enclosed switchgear there is an external metal enclosure which is intended to be earthed; it is complete except for external connections. The earliest transmission

Fig. 7.26 Example of an AIS substation

metal-enclosed equipment used oil as the insulating medium, but since the introduction of SF_6, all transmission metal-enclosed switchgear is based on SF_6 as the insulating medium.

The main significance of the Gas-Insulated Substation (GIS) principle is the dramatic reduction of insulating clearances made possible by SF_6, resulting in much more compact substation layouts. GIS are easier and more economic to accommodate where land use is critical, such as in many urban areas. It also offers considerable advantages in applications where environmental pollution is severe, such as in coastal, desert or heavily industrialized areas. Environmental conditions can also influence the need for maintenance of the switchgear and the implementation of that maintenance; GIS is so compact that transmission substations can be indoor, reducing the maintenance problems considerably. The greater reliability of GIS is another factor which is encouraging its increased application.

GIS switchgear is generally of modular design. All individual elements such as busbars, disconnectors, instrument transformers and circuit breakers are contained in interconnecting earthed enclosures filled with SF_6 at up to 7 bar. At the highest voltage levels, the three phases are housed in separate phase-segregated enclosures. At the lower voltages it is common for all three phases to be housed in the same enclosure and with some designs this philosophy is extended up to 300 kV. Because of the possibility of heating effects from induced currents it is normal for phase-segregated designs to use non-magnetic materials (typically aluminium) for the enclosures, but for phase-integrated designs it is possible to use magnetic steels for some ratings without overheating constraints.

An example of a GIS substation is shown in **Fig. 7.27**.

Fig. 7.27 Example of a GIS substation

7.5.4 Rating principles

As for distribution switchgear, the main consideration in the choice of rating for a particular system is the rated normal current and the rated short-circuit current. Power system studies using load flow and transient analysis techniques will establish the values of the parameters for individual systems. It is also necessary to select lightning impulse rated values from the range of standard values specified in standards (as for distribution switchgear) and to specify for transmission systems with a rated voltage of 300 kV and above a rated switching impulse withstand voltage. The rated switching surge voltage level is defined as the peak voltage of a unipolar standard 250/2500 μs waveform; this represents the transient voltages which are generated by the operation of circuit breakers when switching overhead lines, shunt reactors and other circuits.

An important consideration in the choice of equipment ratings for transmission systems is insulation co-ordination. Insulation co-ordination is the selection of the electric strength of equipment and its application in relation to the voltages which can appear on the system taking into account the characteristics of any protective devices, so as to reduce to an economically and operationally acceptable level the probability that the resulting voltage stresses imposed on the equipment will cause damage to the equipment or affect continuity of service. For system voltages up to 300 kV, experience has demonstrated that the most important factor in determining system design is the stress due to lightning; for voltages over 300 kV the switching overvoltage increases in importance. It is now common to use metal oxide surge arrestors (MOA) to limit lightning overvoltage levels on transmission systems. In addition, the MOA can in many cases provide adequate limitation of switching overvoltages, but for longer overhead lines it may be required to use a circuit breaker with controlled (or point-on-wave) switching in order to keep overvoltages at acceptable levels. The traditional technique of using circuit breakers with parallel pre-insertion resistors is now being replaced by the use of the MOA and controlled switching.

7.5.5 Test methods

It is a requirement of international standards that switchgear must be so designed and manufactured that it satisfies the test specifications with regard to its insulating capacity, switching performance, protection against contact, current-carrying capacity and mechanical function. Evidence of this is obtained by type testing a prototype or sample of the switchgear. In addition, routine tests are performed on each individual item of switchgear manufactured, either on completed or subassembled units in the factory or on site.

Type tests are performed on transmission switchgear in a similar manner to those applied to distribution switchgear (see **section 7.4.4**), but because of the difference in rated values the necessary test values, the test equipment and techniques differ in detail. Some tests which are specific to transmission switchgear are:

- *dielectric tests.* In addition to lightning impulse and power frequency tests it is necessary to perform switching impulse tests for equipment with a rated voltage above 300 kV.
- *short-circuit and switching tests.* The test parameters for transmission equipment involve much higher energy requirements than for distribution equipment, so

special techniques have been developed over the years. To provide the high voltage and high currents needed for transmission switchgear, it is usual to provide a 'synthetic' test source where the high voltage and high currents are provided in the same test from different sources.

Routine tests are performed on all switchgear units manufactured, or on subassemblies, to ensure that the performance of the factory-built unit will match that of the type-tested unit. AIS is normally fully assembled in the factory and routine tested as a complete functioning unit. Routine tests will include mechanical operations to check operating characteristics, resistance checks of current-carrying paths, tests on control and auxiliary circuits and power-frequency withstand tests to verify insulation quality. For GIS, particularly at higher voltage levels, it is not practical to assemble fully a complete installation in the factory. Therefore subassemblies of GIS or transportable assemblies are individually tested prior to shipment; for GIS at 145 kV or 245 kV this might involve a complete 'bay' or circuit of equipment, but at 420 kV or 550 kV the size of the equipment dictates that only subassemblies can be transported to site as a unit. Tests on transportable assemblies include power-frequency withstand with partial discharge detection, verification of gas tightness, mechanical operations and main circuit resistance.

7.5.6   Commissioning and maintenance

Due to the physical size of transmission switchgear it is necessary to transport the equipment to site in subassemblies and then to reassemble to complete switchgear on site. This procedure is more extensive than for distribution switchgear and it results in more elaborate commissioning tests.

After routine testing at the factory, AIS is dismantled into subassemblies for transport to site and then reassembled at the substation. A short series of operational and functional checks are performed before the equipment is put into service.

Site commissioning procedures for GIS are more elaborate than for AIS, since GIS can be particularly sensitive to assembly defects or to particulate contamination; because of the small electrical clearances made possible by SF_6, it is essential that all particulate contamination down to a size of about 1 mm is excluded from the equipment. For these reasons extensive power-frequency withstand tests are performed on site on the completed installation. The requirements for commissioning of GIS are covered in IEC 517.

Modern designs of transmission switchgear are becoming simpler and inherently more reliable, with reduced maintenance requirements. The traditional preventive maintenance approach, with maintenance performed at prescribed intervals irrespective of the condition of the switchgear, is being replaced by predictive maintenance in which maintenance is needed only when the condition of the equipment warrants intervention or where operational duties are severe. Diagnostic techniques and condition monitoring to support this are now becoming available; in particular, it is very difficult to apply conventional partial discharge tests to a complete GIS installation and special diagnostic techniques have been developed to monitor its structural integrity.

Many items of transmission switchgear are still in service after 30, 40, or even 50 years of satisfactory service. A typical design life for modern transmission switchgear is 40 years, but this is conservative in comparison with the figures being achieved on the older equipment still in service today.

7.6 Standards

Some of the commonly used standards and specifications for transmission switchgear are given in **Table 7.6**.

Table 7.6 Standards and specifications for transmission switchgear

IEC	EN/HD	BS	Subject	N. American – ANSI
60056	HD348	5311	High voltage ac circuit breakers	C37.04 C37.06
60071-1	EN60071-1		Insulation co-ordination. Pt 1 – Definitions, principles and rules	C92.1
60129	EN60129	EN60129	AC disconnectors and earthing switches	C37.32
60265-2	EN60265-2	EN60265-2	High-voltage switches for rated voltages of 52 kV and above	
60427	EN60427	EN60427	Synthetic testing of high-voltage ac circuit breakers	C37.081
60517	EN60517	EN60517 5524	Gas-insulated metal-enclosed switchgear rated 72.5 kV and above	C37.55
60694	EN60694	EN60694	Specifications for high-voltage switchgear and controlgear standards	C37.09
61166	EN61166	EN61166	Seismic qualification of high-voltage ac circuit breakers	C37.81
61259	EN61259	EN61259	Gas-insulated metal-enclosed switchgear rated 72.5 kV and above – switching of bus-charging currents by disconnectors	

References

7A.   Ryan, H.M. (ed.), *High Voltage Engineering and Testing*, Peter Peregrinus, 1994.
7B.   Ryan, H.M. and Jones, G.R. *SF₆ Switchgear*, Peter Peregrinus, 1989.
7C.   Flurscheim, C.H. (ed.), *Power Circuit Breaker Theory and Design*, Peter Peregrinus, 1985.

Chapter 8

Fuses and protection relays

Dr D.J.A. Williams

8.1 Protection and co-ordination

Fuse and protection relays are specialized devices for ensuring the safety of personnel working with electrical systems and for preventing damage due to various types of faults. Common applications include protection against overcurrents, short-circuits, overvoltage and undervoltage.

The main hazard arising from sustained overcurrent is damage to conductors, equipment or the source of supply by overheating, possible leading to fire. A short-circuit may melt a conductor, resulting in arcing and the possibility of fire; the high electromechanical forces associated with a short-circuit also cause mechanical stresses which can result in severe damage. A heavy short-circuit may also cause an explosion. Rapid disconnection of overcurrents and short-circuits is therefore vital. An important parameter in the design and selection of protective devices is the *prospective current*; this is the current which would flow at a particular point in an electrical system if a short-circuit of negligible impedance were applied. The prospective current can be determined by calculation if the system impedance or *fault capacity* at that point is known.

In addition, personnel working with electrical equipment and systems must be protected from electric shock. A shock hazard exists when a dangerous voltage difference is sustained between two exposed conducting surfaces which could be touched simultaneously by different parts of the body. This voltage normally arises between earth and metalwork which is unexpectedly made live and if contact is made between the two through the body, a current to earth is caused. Fuses can provide protection where there is a low-resistance path to earth, because a high current flows and blows the fuse rapidly. To detect a wide range of currents flowing to earth it is necessary to use current transformers and core-balance systems; these operate a protective device, for example a circuit breaker such as the rcd described in **section 7.3.4** for low-voltage systems.

When designing an electrical protection system it is also necessary to consider co-ordination so that when a fault occurs, the minimum section of the system around the fault is disconnected. This is particularly important where disconnection has safety implications, for instance in a hospital. An illustration of co-ordination is shown in **Fig. 8.1**.

Protective devices are described by a *time–current characteristic*. In order to achieve co-ordination between protective devices, their time–current characteristics must be sufficiently separated, as shown in **Fig. 8.2**, so that a fault downstream of both of them operates only the device nearest to the fault. A variety of shapes of

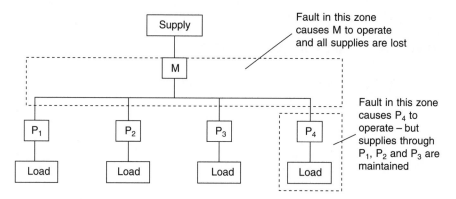

Fault in this zone causes M to operate and all supplies are lost

Fault in this zone causes P₄ to operate – but supplies through P₁, P₂ and P₃ are maintained

Fig. 8.1 Example of co-ordination

time–current characteristic for both fuses and protection relays are available for different applications.

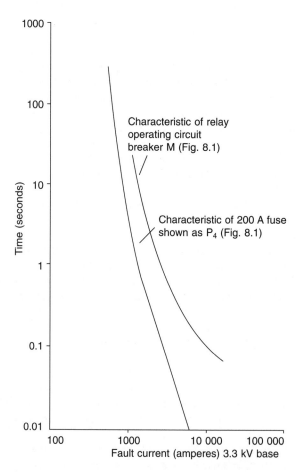

Fig. 8.2 Time–current characteristics for a fuse and a protection relay controlling a circuit breaker (reproduced with permission from GEC Measurements)

Another consideration when designing electrical protection is back-up. In some circuits it is desirable to have a device such as an mcb (see **section 7.3.3**) which disconnects lower overcurrents and can be reset. Higher overcurrents, which the mcb cannot disconnect without being damaged, are disconnected by an upstream fuse in series with the mcb. This arrangement has the advantage that reclosure onto a severe fault is less likely, because replacement of the fuse would be necessary in this case as well as reclosure of the mcb. It is also possible for a protective device to fail, for example because of mishandling. The effects of a failure can be minimized or avoided by a back-up protective device which operates under such conditions. In **Fig. 8.1**, protective device M backs up each of the protective devices P. However, if M operates as back-up device, co-ordination is lost because all four branches of the circuit lose supply, not just the faulty branch.

8.2 Fuses

8.2.1 Principles of design and operation

A fuse consists of a replaceable part (the fuselink) and a fuse holder. Examples of fuse holders are shown in **Fig. 8.3**.

The simplest fuselink is a length of wire. It is mounted by screw connections in a holder which partly encloses it. When an overcurrent or short-circuit current flows, the wire starts to melt and arcing commences at various positions along it. The arc voltage causes the current to fall and once it has fallen to zero, the arcs are extinguished. The larger the wire cross-section, the larger is the current that the fuselink will carry without operating. In the UK, fuses of this type are specified for use at voltages up to 250 V and currents up to 100 A. They are known as *semi-enclosed* or *rewireable* fuses.

The most common fuselink is the *cartridge* type. This consists of a barrel (usually of ceramic) containing one or more elements which are connected at each end to caps fitted over the ends of the barrel. The arrangement is shown in **Figs 8.4** and **8.5**. If a high current breaking capacity is required the cartridge is filled with sand of high chemical purity and controlled grain size. The entire fuselink is replaced after the fuse has operated and a fault has been disconnected. Cartridge fuses are used for a much wider range of voltages and currents than semi-enclosed fuses.

Fuselinks can be divided into *current-limiting* and *non-current-limiting* types. A sand-filled cartridge fuselink is of the current-limiting type; when it operates it limits the peak current to a value which is substantially lower than the prospective current. A non-current-limiting fuse, such as a semi-enclosed fuse, does not limit the current significantly.

The element shown in **Fig. 8.4** is a notched tape. Melting occurs first at the notches when an overcurrent flows and this results in a number of controlled arcs in series. The voltage across each arc contributes to the total voltage across the fuse, and this total voltage results in the current falling to zero. Because the number of arcs is limited, the fuselink voltage should not be high enough to cause damage elsewhere in the circuit. The characteristic development of current and voltage during the operation of a fuse is shown in **Fig. 8.6**.

The function of the sand is to absorb energy from the arcs and to assist in quenching them; when a high current is disconnected, the sand around the arcs is melted.

The element is usually of silver because of its resistance to oxidation. Oxidation of the element in service would affect the current that could be carried without

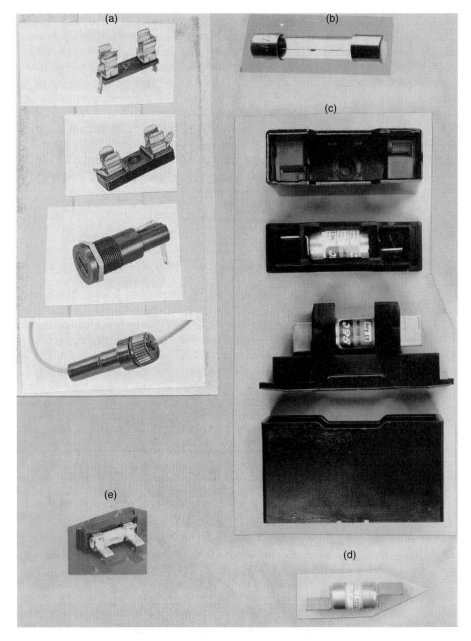

Fig. 8.3 Fuse holders for miniature and compact low-voltage fuses: (a) fuse holders for miniature fuse-links; (b) Miniature fuselink (reproduced with permission from Littelfuse); (c) Fuse holders for compact low-voltage fuselinks; (d) Compact low-voltage fuselink; (e) Rewirable fuse (not to the same scale)

melting, because the effective cross-section of the element is changed. Silver-plated copper elements are also used.

Many elements include an *m-effect blob*, which can be deposited on wire (**Fig. 8.3(b)**) or notched tape. The blob is of solder-type alloy which has a much lower melting point than the element. If a current flows which is large enough to melt

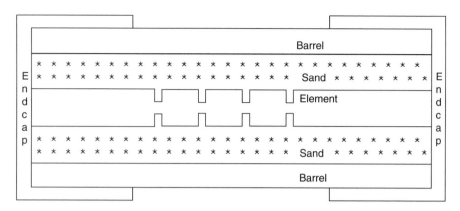

Fig. 8.4 Cross-section through a low-voltage cartridge fuselink

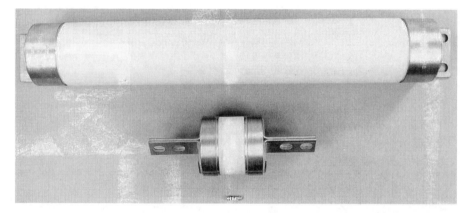

Fig. 8.5 High-voltage and low-voltage cartridge fuselinks (high-voltage fuse (top) and two low-voltage fuses showing a range of fuse sizes) (reproduced with permission from ERA Technology Ltd)

only the m-effect blob, the solder diffuses into the silver. This creates a higher local resistance in the element and the fuse operates at a lower current than it would have done in the absence of the blob.

Other types include the *expulsion fuse* which is used at high voltage, and the *Universal Modular Fuse (UMF)* which is used on Printed Circuit Boards (PCBs).

Fuses offer long life without deterioration in their characteristics or performance, and cartridge fuses have the particular advantage that they contain the arc products completely.

8.2.2 Rating principles and properties

(a) *Current ratings (IEC)*
The rated current of a fuse is the maximum current that a fuselink will carry indefinitely without deterioration. In the case of ac ratings, an rms symmetrical value is given.

The current rating printed on a fuselink applies only at temperatures below a particular value. Derating may be necessary at high ambient temperatures and where fuses are mounted in hot locations such as an enclosure with other heat-generating equipment.

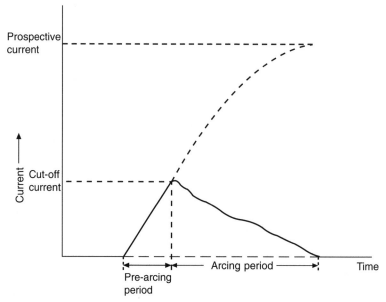

Fig. 8.6 Current and voltage during the operation of a fuse (reproduced with permission from ERA Technology Ltd)

(b) *Voltage ratings (IEC)*

The rated voltage of a fuse is the nominal voltage for which it was designed. Fuselinks will perform satisfactorily at lower voltages, but at much lower voltages, the reduction in current caused by the resistance of the fuselink should be considered. In the case of ac ratings, the rms symmetrical value is given, and for dc ratings the mean value, including ripple, is given.

IEC recommendations are moving towards harmonized low-voltage ac supplies of 230, 400 and 690 V, but although the nominal voltage is being changed in many countries it will be possible for the voltage to remain at its previous non-harmonized level for several years. In Europe, the nominal voltage is 230 V, and the permitted variations will allow supplies to remain at 240 V and 220 V. Fuselinks marked 230 V may have been designed originally for use with higher or lower voltages, and problems may therefore arise when replacing fuselinks because a device manufactured for use at 220 V would not be safe to use on a 240 V system. A fuselink designed for 240 V could safely be used at 220 V. Similar considerations apply where the voltage is changed from 415 V to 400 V, or 660 V to 690 V.

(c) *Variations in rating principles*

The IEC rating principles are used worldwide, except in North America, where UL (Underwriters Laboratory) standards apply.

The rated current to a UL standard is the minimum current required to operate the fuse after many hours, and the current that it will carry indefinitely (the IEC rated current) is approximately 80 per cent of this rating.

The voltage rating marked on a UL fuselink is the *maximum* voltage at which it can be used, whereas that marked on an IEC fuselink is the *nominal* voltage.

These differences must be considered when replacing fuselinks, particularly in the case of miniature cartridge fuselinks which are interchangeable. In general it is preferable to replace a fuselink with one of the same rating from the same manufacturer; this ensures that its characteristics are as similar as possible to those of the previous fuselink.

The IEC and UL ratings of fuse holders also differ. The IEC rating is the highest rated current of a fuselink with which it is intended to be used. A higher rating may be given in North America, this being related to the maximum current that does not cause overheating when a link of negligible resistance is used.

(d) Frequency ratings

Fuses are most commonly used in ac circuits with frequencies of 50 Hz or 60 Hz and a fuse designed for one of these frequencies will generally operate satisfactorily at the other. If the arc extinguishes at current zero, then the maximum arcing time on a symmetrical fault will be 10 ms at 50 Hz and 8 ms at 60 Hz.

Fuse manufacturers should be consulted about the suitability of fuses for other frequencies, which may include 17.67 Hz for some railway supplies, 400 Hz for aircraft and higher frequencies for some electronic circuits.

In dc circuits there is no current zero in the normal waveform and fuselinks designed for ac may not operate satisfactorily. Separate current and voltage ratings are given for fuselinks tested for use in dc circuits. DC circuits can be more inductive for a given current than ac systems, and since the energy in the inductance is dissipated in the fuse it is necessary for the dc voltage rating to be reduced as the time constant (L/R) of a circuit increases.

(e) Breaking capacity

The breaking capacity of a fuse is the current which can be interrupted at the rated voltage. The required breaking capacity will depend upon the position of the fuse in the supply system. For instance, 6 kA may be suitable for domestic and commercial applications, but 80 kA is necessary at the secondary of a distribution transformer.

The power factor of a short-circuit affects the breaking capacity, and appropriate values are used when testing fuses.

(f) Time–current characteristics

The time–current characteristic of a fuse is a graph showing the dependence upon current of the time before arcing starts (the pre-arcing time); an example has been shown in **Fig. 8.2**. The total operating time of a fuse consists of the pre-arcing time and the arcing time. When pre-arcing times are longer than 100 ms and the arc is then extinguished at its first current zero (that is an arcing time of less than 10 ms on a 50 Hz supply) then the time–current characteristic can be taken to represent the total operating time.

The *conventional time*, the *conventional fusing current* and the *conventional non-fusing current* are often shown on time–current characteristics. These values are defined in the standards. All fuses must operate within the conventional time when carrying the conventional fusing current; when carrying the conventional non-fusing current they must not operate within the conventional time.

(g) I^2t

I^2t is defined as the integral of the square of the current let through by a fuse over a period of time. Values are given by manufacturers for pre-arcing I^2t and total let-through I^2t.

Table 8.1 shows typical values of I^2t for low-voltage cartridge fuses of selected current ratings; values differ between manufacturers.

Table 8.1 Example of the variation of I^2t with current rating

Current rating (A)	Pre-arcing I^2t (kA$^2\cdot$s)	Total I^2t (kA$^2\cdot$s)
16	0.3	0.8
40	3.0	8.0
100	30	80
250	300	800
630	3000	8000

The heat generated in a circuit during a short-circuit or fault condition before the fuse disconnects is given by the product of I^2t and the circuit resistance. As the let-through I^2t becomes constant above a particular level of fault current, the heat generated does not increase for prospective currents above this value, unless the breaking capacity is exceeded.

(h) *Power dissipation*
The resistance of a fuse will result in dissipation of power in the protected circuit when normal currents are flowing. This should be considered when designing the layout of a protection system.

(i) *Cut-off current*
A current-limiting fuse prevents a fault current from rising above a level known as the *cut-off current*. This is illustrated in **Fig. 8.6**. The cut-off current is approximately proportional to the cube root of the prospective current, and the maximum current is therefore very much lower than it would be if a non-current-limiting protection device were used.

8.2.3 Main classes of equipment

Fuses are produced in many shapes and sizes, and various types are illustrated in **Figs 8.5, 8.7**, **8.8** and **8.9**. The main three categories are:

- miniature   (up to 250 V)
- low voltage   (up to 1000 V ac or 1500 V dc)
- high voltage (greater than 1000 V ac)

All three categories include current-limiting and non-current-limiting types.

(a) *Miniature fuses*
Cartridge fuses have in the past been the most common form of miniature fuse, but the UMF (see **Fig. 8.7**) is becoming increasingly used on PCBs. A UMF is much smaller than a cartridge fuse, and it is mounted directly on the PCB, whereas a cartridge fuse is mounted in a holder. Subminiature fuses have pins for mounting on PCBs.

Miniature cartridge fuses and subminiature fuses are rated for use at 125 V or 250 V. UMFs have additional voltage ratings of 32 V and 63 V which make them more suitable for many types of electronic circuit. Miniature fuses are available with current ratings from 2 mA to 10 A. The maximum sustained power dissipation which is permitted in cartridge fuses ranges from 1.6 W to 4 W.

Miniature fuses may have a low, intermediate or high breaking capacity. All three ranges are available for UMFs, and these are shown in **Table 8.2**.

Fig. 8.7 Examples of UMFs (photograph printed with the permission of Littelfuse)

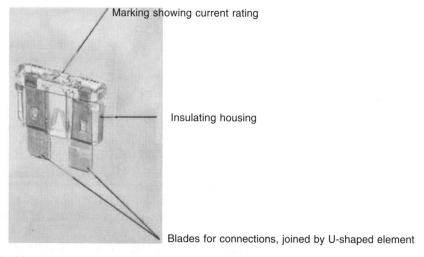

Marking showing current rating

Insulating housing

Blades for connections, joined by U-shaped element

Fig. 8.8 Blade-type automotive fuselink (reproduced with permission from Littelfuse)

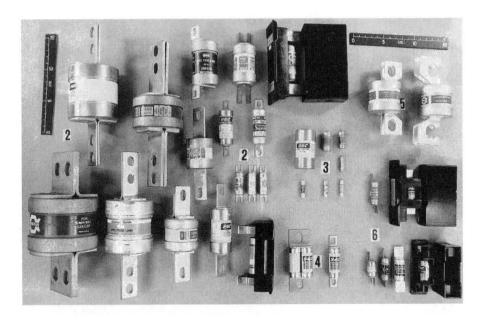

Fig. 8.9 A range of low-voltage fuselink types: (2) general purpose industrial fuselinks, (3) fuselinks for domestic purposes, (4) fuselinks for protecting semiconductors, (5) fuselinks for use in UK electricity supply networks, and (6) compact fuselinks for industrial applications (reproduced with permission from ERA Technology Ltd)

Table 8.2 Breaking capacity of UMFs

Voltage rating (V)	Breaking capacity (A)		Maximum overvoltage (kV)
32	low	35*	0.33
63	low	35*	0.5
125	low	35*	0.8
250	low	100	1.5
250	intermediate	500	2.5
250	high	1500	4.0

*or 10 × rated current, whichever is the greater.

Cartridge fuses are available with low or high breaking capacity. Low breaking capacity types have glass barrels without sand filler and a visual check can therefore be made on whether or not the fuse has operated. High breaking capacity cartridge fuselinks are generally sand filled and have ceramic barrels; they can interrupt currents of up to 1500 A.

A range of speeds of operation is available. Time-lag (surge-proof) fuses are required in circuits where there is an inrush current pulse, for instance when capacitors are charged or when motors or transformers are magnetized. The fuse must not be operated by these normal-operation surges, which must not cause deterioration of the fuse. The five categories of time-lag are medium time-lag (M), time-lag (T), long time-lag (TT), very quick-acting (FF) and quick-acting (F). They are available as cartridge fuses and the last two are used in the protection of electronic circuits. The letters shown in brackets are marked on the end caps.

UMFs are available in similar categories, which are super quick-acting (R), quick-acting (F), time-lag (T) and super time-lag (S).

The time–current characteristics of miniature fuses of the same type but with different ratings are similar in shape. Time can therefore be plotted against multiples of rated current and it is unnecessary to show separate characteristics for each current rating. Examples of time–current characteristics are shown in **Fig. 8.10**.

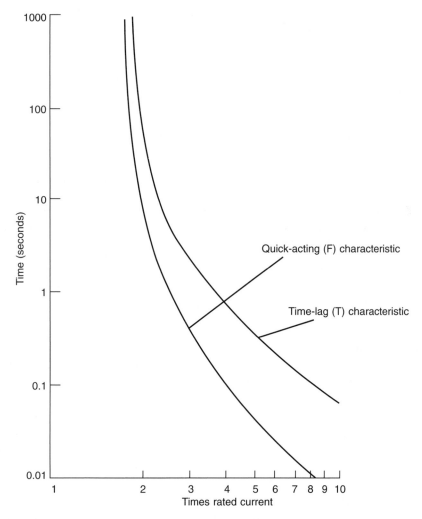

Fig. 8.10 Time–current characteristics of time-lag and quick-acting miniature fuses (reproduced with permission from ERA Technology Ltd)

Cartridge fuses have various types of elements. Fast-acting types have a straight wire, and time-delay types use wire with an m-effect blob (**Fig. 8.3(b)**), helical elements on a heat-absorbing former or short elements with springs connecting them to the end caps.

In addition to the most common types which have been described, miniature fuses are produced in a wide range of shapes and sizes. As an example of this, a blade-type automotive fuselink is shown in **Fig. 8.8**; the element in this fuse is visible through the plastic casing.

(b) *Low-voltage fuses*

A wide range of low-voltage fuses is available for industrial and domestic applications. These fuses have ratings appropriate for national or international single-phase or three-phase supplies, for example 220, 230, 240, 400, 415, 660 and 690 V.

Widely differing systems for domestic protection are used in different countries and these cannot be described separately here. As an example, in the UK current-limiting cartridge fuses are used in plugs which supply appliances, the consumer unit supplying an entire property may have current-limiting cartridge fuses, semi-enclosed fuses or miniature circuit breakers and another fuse is installed by the supply authority on the incoming supply.

Industrial fuses may have general-purpose (type 'g') fuselinks which will operate correctly at any current between 1.6 times the rated current (the conventional fusing current) and the breaking capacity. Such fuses must not be replaced by type 'a' back-up fuselinks, which have a higher minimum breaking current and do not necessarily operate safely below this current; this type 'a' back-up fuselink is used to save space.

Type 'gG' fuses are used for general application. These have a full breaking capacity range and provide protection for cables and transformers and back-up protection for circuit breakers. Specialized fuses are available for the protection of motors, semiconductors, street lighting, pole-mounted transformers and other purposes. **Reference 8A** provides further detail. Common applications are motor protection and semiconductor protection and these are described briefly below.

Fuselinks for motor-starter protection must be able to withstand starting pulses without deterioration. 'gM' fuselinks are designed for this purpose and they have a dual rating. A designation 100M160, for example, means that the fuselink has a continuous rating of 100 A and the general-purpose characteristics of a fuselink rated at 160 A.

Fuselinks for semiconductor protection are designed to operate with an arc voltage which does not damage the semiconductor device; this voltage is therefore lower than for other types of fuselink. Arc voltages at several supply voltages are shown in **Table 8.3** for typical semiconductor fuselinks.

Table 8.3 Peak arc voltages for semiconductor fuselinks

Supply voltage (V)	Peak arc voltage (V)
230	700
400	900
690	1400

Semiconductor fuselinks also have lower let-through I^2t and cut-off current because semiconductors are susceptible to damage by heat and overcurrents. These fuses operate at higher temperatures than normal to achieve the necessary protection, and forced air cooling may be used to increase their current rating.

(c) *High-voltage fuses*

High-voltage fuses can be of current-limiting or non-current-limiting type. The latter are expulsion fuses which do not contain the arc products when they operate; they can be very noisy and are therefore normally used outdoors.

Current-limiting high-voltage fuses are enclosed (as already shown in **Fig. 8.5**) and they may be used for the protection of motors, transformers and shunt power capacitors. The rated current of the fuselink is normally higher than the expected current. These fuses are normally used in three-phase systems and are tested at 87 per cent of their rated voltage. In a three-phase earthed neutral system the voltage rating should be at least 100 per cent of the line-to-line voltage, and in a single-phase system it should be at least 115 per cent of the circuit voltage.

Further information can be found in **references 8A**, **8B** and **8C**.

8.2.4   Test methods

(a) *Type tests*
Before production of a type of fuselink commences, type tests are performed to ensure that preproduction fuselink samples comply with relevant national or international standards. Measurements of power dissipation, time–current characteristic, overload withstand capability, breaking capacity and resistance are included in these type tests.

(b) *Production tests*
Routine testing of many important fuse characteristics is not possible because tests such as breaking capacity are destructive. Extensive testing in production would also be very costly. Fuse manufacturers therefore make production fuselinks as identical as possible to the samples used for type testing.

The quality of fuselinks depends upon the quality of the components supplied to the fuse manufacturer. Key items such as barrels, filling material, element material and end caps are therefore regularly inspected and tested when received.

The dimensions and straightness of barrels are checked and their ability to withstand mechanical and thermal shock and internal pressure is tested. End cap dimensions are checked to ensure that they fit closely over the barrel. The moisture content, chemical composition and grain size of the filler are measured. The diameter or thickness of the element wire or tape is checked and its resistance per metre is measured. Where elements are produced from tape and notched, the dimensions and pitch of the notches are tightly controlled.

During assembly checks are made to ensure that the fuselink is completely filled with sand and that the element resistance is correct. After assembly the overall dimensions are checked and the resistance is once more measured. A visual check including the markings is then made.

Other tests are made in the case of specialized fuselinks. For example, the condition of the elements in a high-voltage fuselink is examined using X-ray photography.

In addition to these routine tests, manufacturers may also occasionally take sample fuselinks from production and subject them to some or all of the type tests.

(c) *Site checks*
Before use, every fuselink should be checked visually for cracks and tightness of end caps and the resistance should be checked. It should also be checked that the ratings, especially current, voltage, breaking capacity and time–current characteristic, are correct for the application. In the case of semi-enclosed, rewireable fuses care should be taken to use the appropriate diameter of fuse wire. Fuse holders should be checked to ensure that the clips or means of connection are secure and correctly aligned.

If a fuselink has been dropped onto a hard surface or subjected to other mechanical stress it should not be used; damage may not be visible but it could cause the fuse to malfunction with potentially serious results.

If a fault occurs and the fuselink is overloaded, it should be replaced even if it has not operated. This situation arises especially in three-phase systems where one or two of the three fuses may operate to clear the fault.

Fuselinks (as opposed to rewireable fuses) cannot be safely repaired; they must always be replaced.

8.2.5 Standards

Many national and international standards exist because of the number of different fuse types. **Tables 8.4**, **8.5** and **8.6** summarize the position for miniature, low-voltage and high-voltage fuses respectively. IEC recommendations are listed, together with related EN and BS standards and North American standards covering the same field.

Table 8.4 Comparison of international, regional and national standards for miniature fuses

IEC	EN	BS	Subject	N. American
127			Miniature fuses	
127-1	60127-1	EN60127-1	: definitions and general requirements	
127-2	60127-2	EN60127-2	: cartridge fuselinks	
127-3	60127-3	EN60127-3	: sub-miniature fuselinks	
127-4			: universal modular fuses (UMFs)	
127-5	60127-5	EN60127-5	: quality assurance	
127-6	60127-6	EN60127-6	: fuse holders for miniature cartridge fuselinks	
127-9			: test holders and test circuits	
127-10			: user guide	
			Blade-type electric fuses	SAE J 1284
			Miniature blade-type electric fuses	SAE J 2077
			Fuses for supplementary overcurrent protection	UL 198G
			Automotive glass tube fuse (32 V)	UL 275

In order to comply with the EMC directive, the following statement has been added to most of the UK fuse standards:

> Fuses within the scope of this standard are not sensitive to normal electromagnetic disturbances, and therefore no immunity tests are required. Significant electromagnetic disturbance generated by a fuse is limited to the instant of its operation. Provided that the maximum arc voltages during operation in the type test comply with the requirements of the clause in the standard specifying maximum arc voltage, the requirements for electromagnetic compatibility are deemed to be satisfied.

8.3 Protection relays

8.3.1 Principles of design and operation

Systems incorporating protection relays can disconnect high currents in high-voltage circuits which are beyond the scope of fuse systems.

In general, relays operate in the event of a fault by closing a set of contacts or by triggering a thyristor. This results in the closure of a trip-coil circuit in the circuit breaker which then disconnects the fault. The presence of the fault is detected by current transformers, voltage transformers or bimetal strips.

Table 8.5 Comparison of international, regional and national standards for low-voltage fuses

IEC	EN	BS	Subject	N. American
241			Fuses for domestic and similar purposes (report)	
269			Low voltage fuses	
269-1	EN60269	88-1 (related to 1361 and 1362)	: general applications	
269-2		88 secn 2.1	: supplementary requirements for fuses for use by authorized persons (fuses mainly for industrial applications)	
269-2-1		related to 88 secn 2.2 and 88-6	: examples of standardized fuses for industrial applications	
269-3			: supplementary requirements for fuses for use by unskilled persons (fuses mainly for household and similar applications)	
269-3A			: examples of standardized fuses mainly for household and similar applications	
269-3-1		related to 1361 and 1362	: supplementary requirements for fuses for use by unskilled persons (fuses mainly for household and similar applications)	
269-4		88-4	: supplementary requirements for fuselinks for the protection of semiconductor devices	
			Class H fuses	UL 198B
			High-interrupting capacity fuses, current-limiting types	UL 198C
			Class K fuses	UL 198D
			Class R fuses	UL 198E
			Plug fuses	UL 198F
			Class T fuses	UL 198H
			DC fuses for industrial use	UL 198L
			Fuseholders	UL 512
			Low-voltage cartridge fuses	NEMA FU
			Fuses (125, 250, 600 V)	CSA C22-2 No 59
			Specification for low-voltage cartridge fuses (600 V or less)	ANSI/IEEE C97.1
			Application guide for low-voltage ac non-integrally fused power circuit breakers (using separately mounted current-limiting fuses)	ANSI/IEEE C37.27

Electromechanical and solid-state relays are both widely used, but the latter are becoming more widespread because of their bounce-free operation, long life, high switching speed and additional facilities that can be incorporated into the relay. Additional facilities can, for instance, include measurement of circuit conditions and transmission of the data to a central control system by a microprocessor relay. This type of relay can also monitor its own function and diagnose any problems that are found. Solid-state relays can perform any of the functions of an electromechanical relay whilst occupying less space, but electromechanical relays are less susceptible to interference and transients. Electromechanical relays also have the advantage of providing complete isolation and they are generally cheaper than solid-state devices.

Table 8.6 Comparison of international, regional and national standards for high-voltage fuses

IEC	EN	BS	Subject	N. American
282			High-voltage fuses	
282-1	60282-1	EN60282-1 2692-1	: current-limiting fuses	
282-2		2692-2	: expulsion and similar fuses	
282-3		related to 2692-3	: determination of short-circuit power factor for testing current-limiting fuses, expulsion fuses and similar fuses	
549		related to 5564	High-voltage fuses for the protection of shunt capacitors	
644	60644-1	EN60644-1 5907	Specification for high-voltage fuselinks for motor circuit applications	
787		6553	Guide for the selection of fuselinks of high-voltage fuses for transformer circuit applications	
			Service conditions and definitions for high-voltage fuses, distribution enclosed single-pole air switches, fuse disconnecting switches, and accessories	ANSI/IEEE C37.40
			Design tests for high-voltage fuses, distribution enclosed single-pole air switches, and accessories	ANSI/IEEE C37.41
			Specifications for distribution cut-outs and fuselinks	ANSI/IEEE C37.42
			Specifications for distribution oil cut-outs and fuselinks	ANSI/IEEE C37.44
			Specifications for power fuses and fuse disconnecting switches	ANSI/IEEE C37.46
			Specifications for distribution fuse disconnecting switches, fuse supports and current-limiting fuses	ANSI/IEEE C37.47
			Guide for application, operation and maintenance of high-voltage distribution cut-outs and fuselinks, secondary fuses, distribution enclosed single-pole air switches, power fuses, fuse disconnecting switches and accessories	ANSI/IEEE C37.48
			High-voltage fuses	NEMA SG2

Contacts in electromechanical relays may have to close in the event of a fault after years of inactivity and twin sets of contacts can be used to improve reliability. The contact material must be chosen to withstand corrosive effects of a local environment because a film of corrosion would prevent effective contact being made. Dust in the atmosphere can also increase the contact resistance and result in failure. Both corrosion and dust contamination can be avoided by complete enclosure and sealing of the relay.

The contacts must also withstand arcing during bounce on closure and when opening, but this is usually less important than resisting corrosion. Because of the need for high reliability the contacts are usually made of or plated with gold, platinum, rhodium, palladium, silver or various alloys of these metals.

Solid-state relays are not affected directly by corrosion or dust, but temperature and humidity may effect them if conditions are severe enough.

Electromechanical relays operate by induction, attraction or thermally where a bimetallic strip is used to detect overcurrent. The first two types are most common and their principles are described, along with those of solid-state relays, in the following sections. Further information on protection relays can be found in **references 8C, 8D** and **8E**.

(a) *Induction relays*

An induction relay has two electromagnets, labelled E_1 and E_2 in **Fig. 8.11**. Winding A of electromagnet E_1 is fed by a current transformer which detects the current in the protected circuit. Winding B in electromagnet E_1 is a secondary, and it supplies the winding on E_2. The phases of the currents supplied to E_1 and E_2 differ and therefore the magnetic fluxes produced by the two electromagnets have different phases. This results in a torque on the disc mounted between the electromagnets, but the disc can only move when a certain torque level is reached because it is restricted by a hair spring or a stop. Normal currents in the protected circuit do not therefore cause movement of the disc.

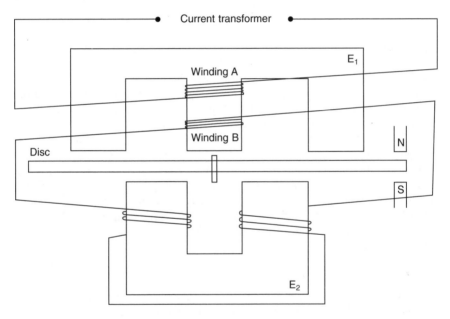

Fig. 8.11 Principle of operation of an induction relay

When the disc does turn, its speed depends upon the current supplied by the current transformer and the eddy current braking effected by a permanent magnet located near the edge of the disc. When the disc rotates through a certain angle, the relay contacts close and the time for this to occur can be adjusted by the position of the stop or the angle through which the disc has to rotate. This adjustment allows protection co-ordination to be achieved by means of 'time grading'. For example, in the radial feeder shown in **Fig. 8.12**, the minimum time taken for the protection relay to operate would be set higher at points closer to the supply. A fault at point X, a considerable distance from the supply, would cause operation of the relay set at a minimum time of 0.4 s and the fault would be disconnected before it caused operation of relays nearer the supply, thus preventing the unnecessary tripping of healthy circuits.

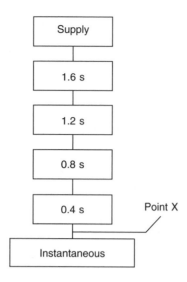

Fig. 8.12 Time grading

Induction relays have Inverse Definite Minimum Time (IDMT) time–current characteristics in which the time varies inversely with current at lower fault currents, but attains a constant minimum value at higher currents. This constant minimum value depends upon the adjustments previously described.

Further adjustment is possible by means of tappings on the relay winding A in **Fig. 8.11**. For example, if a current transformer has a secondary winding rated at 1 A, tappings could be provided in the range 50 per cent to 200 per cent in 25 per cent steps, corresponding to currents of 0.5 A to 2 A in 0.25 A steps. If the circuit is uprated, it may then be possible to adjust the relay rather than replace it. For example, if the 100 per cent setting is used when the maximum current expected in the protected circuit is 400 A, the 150 per cent setting could be used if the maximum current is increased to 600 A.

(b) *Attracted-armature relays*

The basis of operation of an attracted-armature relay is shown in **Fig. 8.13**. The electromagnet pulls in the armature when the coil current exceeds a certain value and the armature is linked to the contacts and when it moves it opens normally closed contacts and closes normally open contacts. The time required for operation is only a few seconds, and it depends upon the size of the current flowing in the coil.

These devices are called instantaneous relays. They have a range of current settings which are provided by changing the tapping of the coil or by varying the air gap between the electromagnet and the armature.

(c) *Solid-state relays*

The first solid-state relays were based on transistors and performed straightforward switching, but now they can often perform much more complicated functions by means of digital logic circuits, microprocessors and memories.

Currents and voltages are measured by sampling incoming analogue signals, and the results are stored in digital form. Logic operations such as comparison are then performed on the data to determine whether the relay should operate to give an alarm

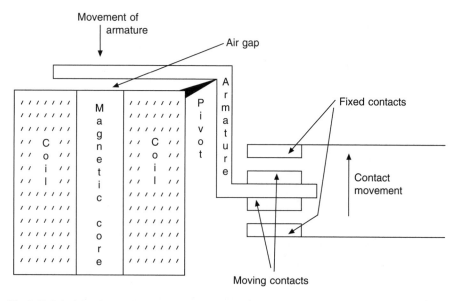

Fig. 8.13 Principle of operation of an attracted-armature relay

or trip a circuit breaker. Digital data can also be stored on computer for subsequent analysis, for instance after a fault has occurred.

Solid-state relays may incorporate a variety of additional circuits. **Figure 8.14** shows, for example, a circuit which imposes a time delay on the output signal from a relay. Such circuits may also be used to shape the time–current characteristic of a protection relay in various ways. A comparison of typical time–current characteristics from electromechanical and solid-state relays is shown in **Fig. 8.15**. There are many other possible functions such as power supervision to minimize power use, an arc sensor to override time delays and measurement of true rms values in the presence of harmonics.

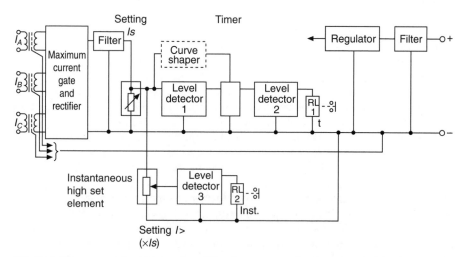

Fig. 8.14 Components of a solid-state time-delayed overcurrent relay (reproduced with permission from GEC)

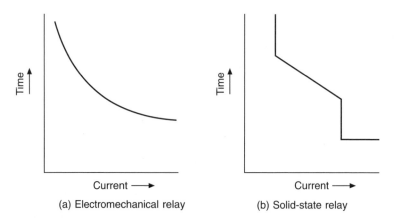

(a) Electromechanical relay   (b) Solid-state relay

Fig. 8.15 Comparison of time–current characteristics of electromechanical and solid-state relays (reproduced with permission from ERA Technology Ltd)

Solid-state relays may require shielding against electromagnetic interference arising from electrostatic discharges or high-voltage switching. Optical transmission of signals is sometimes used to reduce the effects of this interference.

The electronics in solid-state relays can be damaged by moisture, and the relays are usually encapsulated to prevent this.

8.3.2   Rating principles and properties

All components of the protection system including current transformers, relays and circuit breakers must have the correct current, voltage and frequency rating, and the I^2t let-through and interrupting capacity of the entire system depends upon the circuit breaker. A protection relay must have a minimum operating current which is greater than the rated current of the protected circuit, and other properties of the relay must be chosen correctly in order for it to operate the circuit breaker as required by the application.

Manufacturers publish information regarding the selection, installation and use of relays. The following points in particular will need to be considered.

Heat is generated within a relay in use and if several relays are grouped together in an enclosed space provision should be made to ensure that temperature rises are not excessive.

Protection levels, time delays and other characteristics of both electromechanical and solid-state relays can be changed on site. For example, overvoltage protection may be set to operate at levels between 110 per cent and 130 per cent, and **Fig. 8.16** shows the effect of adjusting the operating time of an overcurrent relay.

In general, electromechanical relays can be adjusted continuously and solid-state relays are adjusted in steps. Many other adjustments can be made. For example, DIP switches in solid-state relays can be used to set the rated voltage and frequency and to enable phase-sequence supervision.

The adjustment of operating times is used to provide time grading and co-ordination, which have already been explained with reference to **Fig. 8.12**. An alternative system uses protection relays which operate only when a fault is in a clearly defined zone; this is illustrated in **Fig. 8.17**. To achieve this, a relay compares quantities at the boundaries of a zone. Protection based on this principle can be quicker than with time-graded systems because no time delays are required.

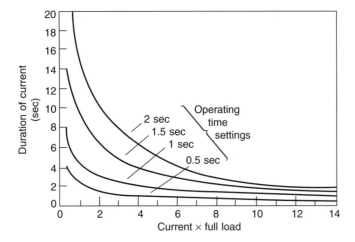

Fig. 8.16 Adjustment of time–current characteristics for an overcurrent relay

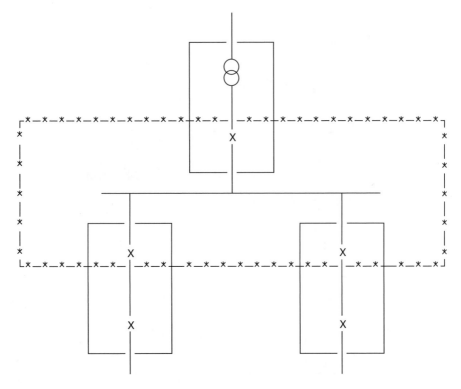

Fig. 8.17 Overlapping zones of protection. The protection zone marked –x–x–x overlaps the zones marked —. X marks the positions of circuit breakers with their associated protection relays

8.3.3 Main classes of relay

Protection relays may be 'all-or-nothing' types, such as overcurrent tripping relays, or they may be measuring types which compare one quantity with another. An example of the latter is in synchronization, when connecting together two sources of power.

A protection relay may be classified according to its function. Various functions are noted below, and details for a wide range of functions are give in **reference 8F**. Common applications are:

- undervoltage and overvoltage detection
- overcurrent detection
- overfrequency and underfrequency detection

These functions may be combined in a single relay. For example, the relay shown in **Fig. 8.18** is used in power stations and provides overvoltage and undervoltage, overfrequency and underfrequency protection, and it rapidly disconnects the generator in the case of a failure in the connected power system.

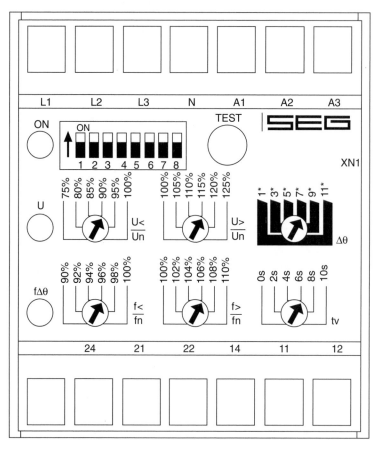

Fig. 8.18 Mains decoupling relay showing adjustments for voltage and frequency protection (reproduced with permission from SEG)

Another application in power systems is the protection of transmission lines by distance relays. Current and voltage inputs to a distance relay allow detection of a fault within a predetermined distance from the relay and within a defined zone. The fault impedance is measured and if it is less than a particular value, then the fault is within a particular distance. This is illustrated in the R–X diagrams shown in case (a) of **Fig. 8.19**; if the R and X values derived from the measured fault impedance $R + iX$ result in a point within the circle, then the relay operates. If directional fault detection is required, the area of operation is moved, as shown in case (b).

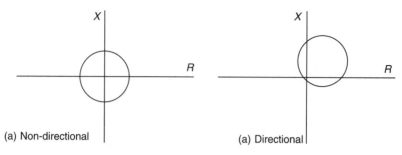

(a) Non-directional   (a) Directional

Fig. 8.19 Operating areas for distance protection relays

Another form of protection for lengths of conductor is the pilot wire system. Current transformers are placed at each end of a conductor and are connected by pilot wires. Relays determine whether the currents at the two ends of the conductor are the same, and they operate if there is an excessive difference. In the balanced voltage system shown in **Fig. 8.20(a)**, no current flows in the pilot wire unless there is a fault. In a balanced current system, current does flow in the pilot wire in normal conditions, and faults are detected by differences in voltage at the relays which are connected between the pilot wires; this is shown in **Fig. 8.20(b)**.

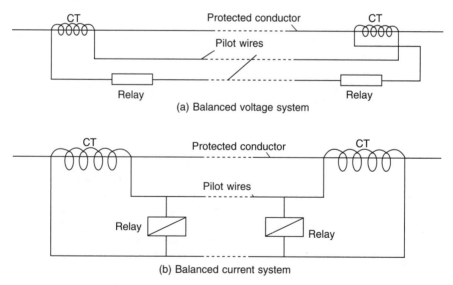

Fig. 8.20 Pilot wire protection systems

Other applications involving relays include:

- the protection of motor starters against overload
- checking phase balance
- the protection of generators from loss of field
- the supervision of electrical conditions in circuits

8.3.4 Test methods

(a) *Production tests*

Manufacturers of electromechanical relays often produce their own components and inspect them before assembly. Components for solid-state relays are generally bought in from specialist manufacturers and there is an incoming check to eliminate those that would be subject to early failure or excessive drift. Other characteristics such as memory are checked as appropriate.

After assembly, manufacturers test all protection relays to ensure that they comply with relevant national and international standards. The calibration of adjustable settings is checked.

Fault conditions that the relays are designed to protect against can be simulated, and typical inputs to relays from current or voltage transformers can be duplicated by test sets; these can be used to check, for example, the correct functioning of relays for phase comparison and the proper disconnection of overcurrent and earth faults of various impedances. Such tests can be performed by setting up a test circuit or by means of a computer-based power system simulator which controls the inputs to the relays. The latter method allows the effect of high-frequency transients and generator faults to be investigated, and it is independent of an actual power supply; it may, for example, be used for the testing of selectable protection schemes in distance relays.

Other tests include:

- *environmental tests* are performed to check, for example, the effects on performance of temperature and humidity
- *impact, vibration and seismic tests* are performed on both solid-state and electromechanical relays, although the latter are more prone to damage from such effects
- *voltage transients* are potentially damaging to solid-state relays, and the relays are tested to ensure they can withstand a peak voltage of 5 kV with a rise time of 1.2 μs and a decay time of 50 μs.

(b) *Site tests*

Primary injection tests are applied during initial commissioning and these should show up any malfunction associated with protection relays. Secondary injection tests should be performed if there is a maloperation which may be related to the protection relay. Details of these injections tests are given in **reference 8E**.

Periodic inspection and testing is necessary throughout the lifetime of a protection relay and computerized equipment is available for this purpose. Some relatively complicated protection schemes incorporate automatic checking systems which send test signals at regular intervals; the test signals can also be sent manually. Digital relays may include continuous self-checking facilities.

8.4 Standards

There are many standards covering various types of relay and aspects of their use. Some standards which apply to relays in general are also relevant to protection relays and some apply specifically to protection relays. The key IEC recommendations together with equivalent BS and EN standards and related North American standards are summarized in **Table 8.7**.

Table 8.7 Comparison of international, regional and national standards for protection relays

IEC	EN/CECC	BS	Subject	N. American
255			Electrical relays	
255-0-20		5992-1	: contact performance	
255-1-00		5992-2	: all-or-nothing electrical relays	
255-3		related to 142 secns 3.1 and 3.2	: single input energizing quantity measuring relays with dependent or independent time	
255-4			: single input energizing quantity measuring relays with dependent specified time	
255-5		5992-3	: insulation tests	
255-6	EN60255-6	EN60255-6 related to 142 secn 2.2	: measuring relays and protection equipment	
255-7	related to CECC 16000, EN116000-2	related to CECC 16000, EN116000-2	: test and measurement procedures for all-or-nothing relays	
255-8		142 secn 2.3	: thermal electrical relays	
255-9			: dry reed make contact units	
255-10			: application of the IEC QA system for electronic components to all-or-nothing relays	
255-11			: interruptions to and alternating component (ripple) in dc auxiliary energizing quantity of measuring relays	
255-12		related to 142 secn 4.2	: directional relays and power relays with two input energizing quantities	
255-13		related to 142 secn 4.1	: biased (percentage) differential relays	
255-14		5992-4	: endurance test for electrical relay contacts – preferred values for contact loads	
255-15		5992-5	: endurance test for electrical relay contacts – characteristics of test equipment	
255-16			: impedance measuring relays	
255-17			: thermal electrical relays for motor protection	
255-18		5992-6	: dimensions for general purpose all-or-nothing relays	
255-19	CECC 160100	QC 160100	: electromechanical relays of assessed quality	
255-20			: protection (protective) systems	
255-21		142 secns 1.4 and 1.5	: vibration, shock, bump and seismic tests on measuring relays and protection equipment	
255-22		142 secn 1.4	: electrical disturbance tests	
			Relays and relay systems associated with electric power systems	ANSI/IEEE C37.90
			Surge withstand capability tests for protective relays and relay systems	ANSI/IEEE C37.90.1
			Guide for protective relay applications to power	ANSI/IEEE C37.91
			Guide for power system protective relay applications of audio tones over telephone channels	ANSI/IEEE C37.93

(contd)

Table 8.7 (*contd*)

IEC	EN/CECC	BS	Subject	N. American
			Guide for protective relaying of utility–consumer interconnections	ANSI/IEEE C37.95
			Guide for protective relay applications to power system buses	ANSI/IEEE C37.97
			Seismic testing of relays	ANSI/IEEE C37.98
			Guide for differential and polarizing relay circuit testing	ANSI/IEEE C37.103
			Qualifying class 1E protective relays and auxiliaries for nuclear power generating stations	ANSI/IEEE C37.105
			Testing procedures for relays and electrical and electronic equipment	ANSI/EIA 407-A
			Solid-state relays	ANSI/EIA 407-A

References

8A.   Williams, D.J.A., Turner, H.W. and Turner, C., *User's Guide to Fuses*, 2nd edn, ERA Technology Ltd, UK, 1993.

8B.   Wright, A. and Newbery, P.G., *Electric Fuses*, Peter Peregrinus Ltd, London & New York, 1982.

8C.   Wright, A. and Christopoulos, C., *Electric Power System Protection*, Chapman & Hall, London, 1993.

8D.   GEC Measurements, 'Protective relays – Applications guide', 3rd edn, The General Electric Co., UK, 1987.

8E.   Jones, G.R., Laughton, M.A. and Say, M.G., *Electrical Engineer's Reference Book*, Butterworth-Heinemann, Oxford, UK, 1993.

8F.   Recommended practice for protection and co-ordination of industrial and commercial power systems, *IEEE*, Wiley Interscience, 1986.

Chapter 9

Wires and cables

Mr V.A.A. Banks and Mr P.H. Fraser
BICCGeneral

Mr A.J. Willis
Brand-Rex Ltd

9.1 Scope

Thousands of cable types are used throughout the world. They are found in applications ranging from fibre-optic links for data and telecommunication purposes through to EHV underground power transmission at 275 kV or higher. The scope here is limited to cover those types of cable which fit within the general subject matter of the pocket book.

This chapter therefore covers cables rated between 300/500 V and 19/33 kV for use in the public supply network, in general industrial systems and in domestic and commercial wiring. Optical communication cables are included in a special section. Overhead wires and cables, submarine cables and flexible appliance cords are not included.

Even within this relatively limited scope, it has been necessary to restrict the coverage of the major metallic cable and wire types to those used in the UK in order to give a cursory appreciation.

9.2 Principles of power cable design

9.2.1 Terminology

The voltage designation used by the cable industry does not always align with that adopted by users and other equipment manufacturers, so clarification may be helpful.

A cable is given a voltage rating which indicates the maximum circuit voltage for which it is designed, not necessarily the voltage at which it will be used. For example, a cable designated 0.6/1 kV is suitable for a circuit operating at 600 V phase-to-earth and 1000 V phase-to-phase. However, it would be normal to use such a cable on distribution and industrial circuits operating at 230/400 V in order to provide improved safety and increased service life. For light industrial circuits operating at 230/400 V it would be normal to use cables rated at 450/750 V, and for domestic circuits operating at 230/400 V, cable rated at 300/500 V would often be used. Guidance on the cables that are suitable for use in different locations is given in BS 7540.

The terms *LV (Low Voltage), MV (Medium Voltage)* and *HV (High Voltage)* have different meanings in different sectors of the electrical industry. In the power cable industry the following bands are generally accepted and these are used in this chapter:

LV – cable rated from 300/500 V to 1.9/3.3 kV
MV – cable rated from 3.8/6.6 kV to 19/33 kV
HV – cable rated at greater than 19/33 kV

Multicore cable is used in this chapter to describe power cable having two to five cores. Control cable having seven to 48 cores is referred to as *multicore control cable*.

Cable insulation and sheaths are variously described as *thermoplastic, thermosetting, vulcanized, cross-linked, polymeric* or *elastomeric*. All extruded plastic materials applied to cable are *polymeric*. Those which would remelt if the temperature during use is sufficiently high are termed *thermoplastic*. Those which are modified chemically to prevent them from remelting are termed thermosetting, *cross-linked* or *vulcanized*. Although these materials will not remelt, they will soften and deform at elevated temperatures if subjected to excessive pressure. The main materials within the two groups are as follows:

- *thermoplastic*
  - polyethylene   (PE)
  - medium-density polyethylene   (MDPE)
  - polyvinyl chloride   (PVC)
- *thermosetting, cross-linked* or *vulcanized*
  - cross-linked polyethylene   (XLPE)
  - ethylene-propylene rubber   (EPR)

Elastomeric materials are polymeric. They are rubbery in nature, giving a flexible and resilient extrusion. Elastomers are normally cross-linked, such as EPR.

9.2.2 General considerations

Certain design principles are common to power cables, whether they be used in the industrial sector or by the electricity supply industry.

For many cable types the conductors may be of copper or aluminium. The initial decision made by a purchaser will be based on price, weight, cable diameter, availability, the expertise of the jointers available, cable flexibility and the risk of theft. Once a decision has been made, however, that type of conductor will generally then be retained by that user, without being influenced by the regular changes in relative price which arise from the volatile metals market.

For most power cables, the form of conductor will be solid aluminium, stranded aluminium or stranded copper, although the choice may be limited in certain cable standards. Solid conductors provide for easier fitting of connectors and setting of the cores at joints and terminations. Cables with stranded conductors are easier to install because of their greater flexibility, and for some industrial applications a highly flexible conductor is necessary.

Where cable route lengths are relatively short a multicore cable is generally cheaper and more convenient to install than single-core cable. Single-core cables are sometimes used in circuits where high load currents require the use of large conductor sizes, between 500 mm^2 and 1200 mm^2. In these circumstances the parallel connection of two or more multicore cables would be necessary in order to achieve the required rating and this presents installation difficulties, especially at termination boxes. Single-core cable might also be preferred where duct sizes are small, where longer cable runs are needed between joint bays or where jointing and termination requirements dictate their use. It is sometimes preferable to use 3-core cable in the main part of the route length, and to use single-core cable to enter the restricted space of a termination box. In this case, a transition from one cable type to the other is achieved using trifurcating joints which are positioned several metres from the termination box.

Armoured cables are available for applications where the rigours of installation are severe and where a high degree of external protection against impact during service is required. *Steel Wire Armour (SWA)* or *Steel Tape Armour (STA)* cables are available. Generally SWA is preferred because it enables the cable to be drawn into an installation using a pulling stocking which grips the outside of the oversheath and transfers all the pulling tension to the SWA. This cannot normally be done with STA cables because of the risk of dislocating the armour tapes during the pull. Glanding arrangements for SWA are simpler and they allow full usage of its excellent earth fault capability. In STA the earth fault capability is much reduced and the retention of this capability at glands is more difficult. The protection offered against a range of real-life impacts is similar for the two types.

9.2.3 Paper-insulated cables

Until the mid-1960s, paper-insulated cables were used worldwide for MV power circuits. There were at that time very few alternatives apart from the occasional trial installation or special application using PE or PVC insulation.

The position is now quite different. There is a worldwide trend towards XLPE cable and, in the UK, the industrial sector has adopted XLPE-insulated or EPR-insulated cable for the majority of MV applications, paper-insulated cable now being restricted to minor uses such as extensions to older circuits or in special industrial locations. The use of paper-insulated cables for LV has been superseded completely by polymeric cables in all sectors throughout the world.

The success of polymeric-insulated cables has been due to the much easier, cleaner and more reliable jointing and termination methods that they allow. However, because of the large amount of paper-insulated cable still in service and its continued specification in some sectors for MV circuits, its coverage here is still appropriate.

Paper-insulated cables comprise copper or aluminium phase conductors which are insulated with lapped paper tapes, impregnated with insulating compound and sheathed with lead, lead alloy or corrugated aluminium. For mechanical protection, lead or lead alloy sheathed cables are finished off with an armouring of steel tapes or steel wire and a covering of either bitumenized hessian tapes or an extruded PVC oversheath. Cables that are sheathed with corrugated aluminium need no further metallic protection, but they are finished off with a coating of bitumen and an extruded PVC oversheath; the purpose of the bitumen in this case is to provide additional corrosion protection should water penetrate the PVC sheath at joints, in damaged areas or by long-term permeation.

There are, therefore, several basic types of paper-insulated cable, and these are specified according to existing custom and practice as much as to meet specific needs and budgets. Particular features of paper-insulated cables used in the electricity supply industry and in industrial applications are given in **sections 9.3.1.1** and **9.3.2.1** respectively.

The common element is the paper insulation itself. This is made up of many layers of paper tape, each applied with a slight gap between the turns. The purity and grade of the paper is selected for best electrical properties and the thickness of the tape is chosen to provide the required electrical strength.

In order to achieve acceptable dielectric strength all moisture and air is removed from the insulation and replaced by *Mineral Insulating Non-Draining (MIND)* compound. Its waxy nature prevents any significant migration of the compound during the lifetime of the cable, even at full operating temperature. This is in contrast to oil-filled HV cables, which must be pressurized throughout their service life to keep the insulation fully impregnated. Precautions are taken at joints and terminations to ensure that there is no local displacement of MIND compound which might cause premature failure at these locations. The paper insulation is impregnated with MIND compound during the manufacture of the cable, immediately before the lead or aluminium sheath is applied.

3-core construction is preferred in most MV paper-insulated cables. The three cores are used for the three phases of the supply and no neutral conductor is included in the design. The parallel combination of lead or aluminium sheath and armour can be used as an earth continuity conductor, provided that circuit calculations prove its adequacy for this purpose. Conductors of 95 mm^2 cross-section and greater are sector shaped so that when insulated they can be laid up in a compact cable construction. Sector-shaped conductors are also used in lower cross-sections, down to 35, 50 and 70 mm^2 for cables rated at 6, 10 and 15 kV respectively.

3-core 6.6 kV cables and most 3-core 11 kV cables are of the *belted* design. The cores are insulated and laid up such that the insulation between conductors is adequate for the full line-to-line voltage (6.6 kV or 11 kV). The laid-up cores then have an additional layer of insulating paper, known as the *belt layer,* applied and the assembly is then lead sheathed. The combination of core insulation and belt insulation is sufficient for phase-to-earth voltage between core and sheath (3.8 kV or 6.35 kV).

15, 22 and 33 kV 3-core cables and some 11 kV 3-core cables are of *screened* design. Here each core has a metallic screening tape and the core insulation is adequate for the full phase-to-earth voltage. The screened cores are laid up and the lead or aluminium sheath is then applied so that the screens make contact with each other and with the sheath.

The bitumenized hessian serving or PVC oversheath is primarily to protect the armour from corrosion in service and from dislocation during installation. The PVC oversheath is now preferred because of the facility to mark cable details, and its clean surface gives a better appearance when installed. It also provides a smooth firm surface for glanding and for sealing at joints.

9.2.4 Polymeric cables

PVC and PE cables were being used for LV circuits in the 1950s and they started to gain wider acceptance in the 1960s because they were cleaner, lighter, smaller and easier to install than paper-insulated types. During the 1970s the particular benefits of XLPE and HEPR insulations were being recognized for LV circuits and today it is these cross-linked insulations, mainly XLPE, which dominate the LV market with PVC usage in decline. LV XLPE cables are more standardized than MV polymeric types, but even so there is a choice of copper or aluminium conductor, single core or multicore, SWA or unarmoured, and PVC or Low Smoke and Fume (LSF) sheathed. A further option is available for LV in which the neutral and/or earth conductor is a layer of wires applied concentrically around the laid-up cores rather than as an insulated core within the cable. In this case the concentric earth conductor can replace the armour layer as the protective metal layer for the cable.

For MV cables, polyethylene and PVC were shown to be unacceptable for use as general cable insulation in the years following the 1960s because their thermoplastic nature resulted in significant temperature limitations. XLPE and EPR were required in order to give the required properties. They allowed higher operating and short-circuit temperatures within the cable as well as the advantages of easier jointing and terminating than for paper-insulated cables. This meant that in some applications a smaller conductor size could be considered than had previously been possible in the paper-insulated case.

MV polymeric cables comprise copper or aluminium conductors insulated with XLPE or EPR and covered with a thermoplastic sheath of MDPE, PVC or LSF material. Within this general construction there are options of single-core or 3-core types, individual or collective screens of different sizes and armoured or unarmoured construction. Single-core polymeric cables are more widely used than single-core paper-insulated cables, particularly for electricity supply industry circuits. Unlike paper-insulated cables, polymeric 3-core cables normally have circular-section cores. This is mainly because the increases in price and cable diameter are usually outweighed in the polymeric case by simplicity and flexibility of jointing and termination methods using circular cores.

Screening of the cores in MV polymeric cables is necessary for a number of reasons, which combine to result in a two-level screening arrangement. This comprises extruded semiconducting layers immediately under and outside the individual XLPE or EPR insulation layer, and a metallic layer in contact with the outer semiconducting layer. The semiconducting layers are polymeric materials containing a high proportion of carbon, giving them an electrical conductivity well below that of a metallic conductor, but well above that required for an insulating material. The term should not be confused with the semiconductor materials used in electronic components.

The two semiconducting layers must be in intimate contact in order to avoid partial discharge activity at the interfaces, where any minute air cavity in the insulation would cause a pulse of charge to transfer to and from the surface of the insulation in each half-cycle of applied voltage. These charge transfers result in erosion of the insulation surface and premature breakdown. In order to achieve intimate contact, the insulation and screens are extruded during manufacture as an integral triple layer and this is applied to the individual conductor in the same operation. The inner layer is known as the conductor screen and the outer layer is known as the core screen or dielectric screen.

When the cable is energized, the insulation acts as a capacitor and the core screen has to transfer the associated charging current to the insulation on every half-cycle of the voltage. It is therefore necessary to provide a metallic element in contact with the core screen so that this charging current can be delivered from the supply. Without this metallic element, the core screen at the supply end of the cable would have to carry a substantial longitudinal current to charge the capacitance which is distributed along the complete cable length, and the screen at the supply end would rapidly overheat as a result of excessive current density. However, the core screen is able to carry the current densities relating to the charging of a cable length of say 200 mm, and this allows the use of a metallic element having an intermittent contact with the core screen, or applied as a collective element over three laid-up cores. A 0.08 mm thick copper tape is adequate for this purpose.

The normal form of *armouring* is a single layer of wire laid over an inner sheath of PVC or LSF material. The wire is of galvanized steel for 3-core cables and aluminium for single-core cables. Aluminium wire is necessary for single-core cables to avoid magnetizing or eddy-current losses within the armour layer. In unarmoured cable the screen is required to carry the earth fault current resulting from the failure of any equipment being supplied by the cable or from failure of the cable itself. In this case, the copper tape referred to previously is replaced by a screen of copper wires of cross-section between say 6 mm^2 and 95 mm^2, depending on the earth fault capacity of the system.

9.2.5   Low Smoke and Fume (LSF) cables

Following a number of recent fire disasters, there has been a strong demand for cables which behave more safely in a fire. Cables have been developed to provide the following key areas of improvement:

- improved resistance to ignition
- reduced fire propagation
- reduced smoke emission
- reduced acid gas emission

An optimized combination of these properties is achieved in LSF cables.

The original concept of LSF cables was identified through the requirements of underground railways in the 1970s. At that time, the main concern was to maintain sufficient visibility that orderly evacuation of passengers through a tunnel could be managed if the power to their train were interrupted by a fire involving the supply cables. This led to the development of a smoke test known as the '3-metre cube', this being based on the cross-section of a London Underground tunnel.

The demand for LSF performance has since spread to a wide range of products and applications, and the term LSF now represents a generic family of cables. Each LSF cable will meet the 3-metre cube test, but ignition resistance, acid gas emission and fire propagation performance is specified as appropriate to a particular product and application. For instance, a power cable used in large arrays in a power station has very severe fire propagation requirements, while a cable used in individual short links to equipment would have only modest propagation requirements.

9.3   Main classes of cable

9.3.1   Cables for the electricity supply industry

9.3.1.1   Paper-insulated cables

Until the late 1970s, the large quantities of paper-insulated cables used in the UK electricity supply industry for MV distribution circuits were manufactured according to BS 6480. These cables incorporated lead sheaths, steel wire armour and bitumenized textile beddings or servings. An example is illustrated in **Fig. 9.1**. The lead sheath provided an impermeable barrier to moisture and a return path for earth fault currents, and the layer of steel wire armour gave mechanical protection and an improved earth fault capacity.

Following successful trials and extensive installation in the early 1970s, a new standard (ESI 09-12) was issued in 1979 for Paper-Insulated Corrugated Aluminium

Fig. 9.1 Lead-sheathed, paper-insulated MV cable for the electricity supply industry (courtesy of BICC General)

Sheathed (PICAS) cable. This enabled the electricity supply industry to replace expensive lead sheath and steel-wire armour with a corrugated aluminium sheath which offered a high degree of mechanical protection and earth fault capability, while retaining the proven reliability of paper insulation. The standard was limited to three conductor cross-sections (95, 185 and 300 mm^2), using stranded aluminium conductors with belted paper insulation. An example of PICAS cable is shown in **Fig. 9.2**. PICAS cable was easier and lighter to install than its predecessor and it found almost universal acceptance in the UK electricity supply sector.

9.3.1.2 MV polymeric cables

High-quality XLPE cable has been manufactured for over 25 years. IEC 502 (revised in 1998 as IEC 60502) covered this type of cable and was first issued in 1975. A comparable UK standard BS 6622 was issued in 1986 and revised in 1999.

The following features are now available in MV XLPE cables, and these are accepted by the majority of users in the electrical utility sector:

- copper or aluminium conductors
- semiconducting conductor screen and core screen
- individual copper tape or copper wire screens
- PVC or MDPE bedding
- copper wire collective screens
- steel wire or aluminium armour
- PVC or MDPE oversheaths

Early experience in North America in the 1960s resulted in a large number of premature failures, mainly because of poor cable construction and insufficient care in avoiding

Fig. 9.2 Paper-Insulated Corrugated Aluminium Sheathed (PICAS) cable for the electricity supply industry (courtesy of BICCGeneral)

contamination of the insulation. The failures were due to 'water treeing', which is illustrated in **Fig. 9.3**. In the presence of water, ionic contaminants and oxidation products, electric stress gives rise to the formation of tree-like channels in the XLPE insulation. These channels start either from defects in the bulk insulation (forming 'bow-tie' trees) or at the interfaces between the semiconducting screens and the insulation (causing 'vented' trees). Both forms of tree cause a reduction in electrical strength of the insulation and can eventually lead to breakdown. Water treeing has largely been overcome by better materials in the semiconducting screen and by improvements in the quality of the insulating materials and manufacturing techniques, and reliable service performance has now been established.

The UK electricity supply industry is to adopt XLPE-insulated or EPR-insulated cable for MV distribution circuits. Each distribution company is actively assessing the best construction for its particular needs. An example of the variation between companies is the difference in practice between solidly bonded systems and the use of earthing resistors to limit earth fault currents. In the former case the requirement might typically be to withstand an earth fault current of 13 kA for 3 seconds. In the latter case only 1 kA for 1 second might be specified, and the use of copper-wire collective screens in place of steel-wire armour is viable. The specific designs being used by the UK electricity supply industry are now incorporated into BS 7870-4. Examples of XLPE cable designs being considered by UK distribution companies are shown in **Figs 9.4**, **9.5** and **9.6**.

Fig. 9.3 Example of water treeing in a polymeric cable (courtesy of Dr J.C. Fothergill, University of Leicester)

9.3.1.3 LV polymeric cables

Protective Multiple Earthed (PME) systems which use *Combined Neutral and Earth (CNE)* cables have become the preferred choice in the UK public supply network, both for new installations and for extensions to existing circuits. This is primarily because of the elimination of one conductor by the use of a common concentric neutral and earth, together with the introduction of new designs which use aluminium for all phase conductors.

Before CNE types became established, 4-core paper-insulated sheathed and armoured cable was commonly used. The four conductors were the three phases and neutral, and the lead sheath provided the path to the substation earth. The incentive for PME was the need to retain good earthing for the protection of consumers. With the paper cables, while the lead sheath itself could adequately carry prospective earth fault currents back to the supply transformer, the integrity of the circuit was often jeopardized by poor and vulnerable connections in joints and at terminations. By using the

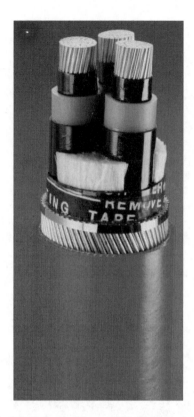

Fig. 9.4 Example of lead-sheathed XLPE-insulated cable for use in the UK electricity supply industry (courtesy of BICCGeneral)

Fig. 9.5 Example of 3-core SWA XLPE-insulated cable for use in the UK electricity supply industry (courtesy of BICCGeneral)

neutral conductor of the supply cable for this purpose, the need for a separate earth conductor was avoided.

The adoption of 0.6/1 kV cables with extruded insulation for underground public supply in the UK awaited the development of cross-linked insulation systems with a performance similar to paper-insulated systems in overload conditions. An example of the cables which have been developed is the *'Waveform'* CNE type which is XLPE-insulated and has the neutral/earth conductor applied concentrically in a sinusoidal form. Insulated solid aluminium phase conductors are laid up to form a three-phase cable, and the CNE conductor consists of a concentric layer of either aluminium or copper wires.

If the wires in the CNE conductor are of aluminium, they are sandwiched between layers of unvulcanized synthetic rubber compound to give maximum protection against corrosion. This construction is known as *'Waveconal'* and is illustrated in **Fig. 9.7**. Where the CNE conductors are of copper, they are partially embedded in the rubber compound without a rubber layer over the wires. This is termed *'Wavecon'* and is illustrated in **Fig. 9.8**. Some electricity companies have adopted Wavecon types because of concern over excessive corrosion in the aluminium CNE conductor. Waveform cables are manufactured in accordance with BS 7870-3.

Both waveform types are compact, with cost benefits. The aluminium conductors and synthetic insulation result in a cable that is light and easy to handle. In addition,

Fig. 9.6 Example of single-core copper-wire screen XLPE-insulated cable for use in the UK electricity supply industry (courtesy of BICCGeneral)

Fig. 9.7 Construction of a 'Waveform' XLPE-insulated CNE cable; 'Waveconal' (courtesy of BICCGeneral)

the waveform lay of the CNE conductors enables service joints to be readily made without cutting the neutral wires, as they can be formed into a bunch on each side of the phase conductors.

9.3.2 Industrial cables

'Industrial cables' are defined as those power circuit cables which are installed on the customer's side of the electricity supply point, but which do not fall into the category of 'wiring cables'.

Generally these cables are rated 0.6/1 kV or above. They are robust in construction and are available in a wide range of sizes. They can be used for distribution of power around a large industrial site, or for final radial feeders to individual items of plant. Feeder cables might be fixed, or in cases such as coal-face cutting machines and mobile cranes they may be flexible trailing or reeling cables.

Many industrial cables are supplied to customers' individual specifications and since these are not of general interest they are not described here. The following sections focus on those types of cable which are manufactured to national standards and which are supplied through cable distributors and wholesalers for general use.

Fig. 9.8 Construction of a 'Waveform' XLPE-insulated CNE cable; 'Wavecon' (courtesy of BICCGeneral)

9.3.2.1   Paper-insulated cables

For ratings between 0.6/1 kV and 19/33 kV, paper-insulated cables for fixed installations were supplied in the UK to BS 480, and then to BS 6480 following metrication in 1969. These cables comprise copper or aluminium phase conductors insulated with lapped paper tapes, impregnated with MIND compound and sheathed with lead or lead alloy. For mechanical protection they were finished with an armouring of steel tapes or steel wire and a covering of bitumenized hessian tapes or an extruded PVC oversheath.

3-core cables of this type with steel-wire armour have been preferred for most applications and these have become known as *Paper-Insulated Lead-Covered Steel Wire Armoured (PILCSWA)* cables. Single-core cables to BS 6480 do not have armouring; this is partly because the special installation conditions leading to the selection of single core do not demand such protection, and partly because a non-magnetic armouring such as aluminium would be needed to avoid eddy current losses in the armour. These single-core cables are known as *Paper-Insulated Lead Covered (PILC)*.

It has already been observed that paper-insulated cables are now seldom specified for industrial use, but BS 6480 remains an active standard.

9.3.2.2 Polymeric cables for fixed installations

XLPE-insulated cables manufactured to BS 5467 are generally specified for 230/400 V and 1.9/3.3 kV LV industrial distribution circuits. These cables have superseded the equivalent PVC-insulated cables to BS 6346 because of their higher current rating, higher short-circuit rating and better availability.

For MV applications in the range 3.8/6.6 kV to 19/33 kV, XLPE-insulated wire-armoured cables to BS 6622 are usually specified.

Multicore LV and MV cables are normally steel-wire armoured. This armouring not only provides protection against impact damage for these generally bulky and exposed cables, but it is also capable of carrying very large earth fault currents and provides a very effective earth connection.

Single-core cables are generally unarmoured, although aluminium wire-armoured versions are available. Single-core cables are usually installed where high currents are present (for instance in power stations) and where special precautions will be taken to avoid impact damage. For LV circuits of this type, the most economic approach is to use unarmoured cable together with a separate earth conductor, rather than to connect in parallel the aluminium wire armour of several single-core cables. For MV applications each unarmoured cable has a screen of copper wires which would, together, provide an effective earth connection.

Even in the harsh environment of coal mines, XLPE-insulated types are now offered as an alternative to the traditional PVC- and EPR-insulated cables used at LV and MV respectively. In this application the cables are always multicore types having a single or double layer of steel-wire armour. The armour has to have a specified minimum conductance because of the special safety requirements associated with earth faults and this demands the substitution of some steel wires by copper wires for certain cable sizes.

Where LSF fire performance is needed, LV wire-armoured cables to BS 6724 are the established choice. These cables are identical in construction and properties to those made to BS 5467 except for the LSF grade of sheathing material and the associated fire performance. Cables meeting all the requirements of BS 6724 and, in addition, having a measure of fire resistance such that they continue to function in a fire are standardized in BS 7846, further details of which are given in **section 9.3.3**. Similarly, BS 7835 for MV wire-armoured cable, which is identical to BS 6622 apart from the LSF sheath and fire performance, was issued in 1996 and is currently under revision.

The only other type of standardized cable used for fixed industrial circuits is multicore control cable, often referred to as auxiliary cable. Such cable is used to control industrial plant, including equipment in power stations. It is generally wire armoured and rated 0.6/1 kV. Cables of this type are available with between 5 and 48 cores. The constructions are similar to 0.6/1 kV power cables and they are manufactured and supplied to the same standards (BS 5467, BS 6346 and BS 6724, as appropriate).

9.3.2.3 Polymeric cables for flexible connections

Flexible connections for both multicore power cables and multicore control cables are often required in industrial locations. The flexing duty varies substantially from application to application. At one extreme, a cable may need to be only flexible enough to allow the connected equipment to be moved occasionally for maintenance. At the other extreme, the cable may be needed to supply a mobile crane from a cable reel, or a coal-face cutter from cable-handling gear.

Elastomeric-insulated and sheathed cable is used for all such applications. This may have flexible stranded conductors (known as 'class 5') or highly flexible stranded conductors (known as 'class 6'). Where metallic protection or screening is needed, this comprises a braid of fine steel or copper wires. For many flexible applications the cable is required to have a resistance to various chemicals and oils.

Although flexible cables will normally be operated on a 230/400 V supply, it is normal to use 450/750 V rated cables for maximum safety and integrity.

A number of cable types have been standardized in order to meet the range of performance requirements and the specification for these is incorporated into BS 6500. Guidance on the use of the cables is provided in this standard, and further information is available in BS 7540.

9.3.3 Wiring cables

The standard cable used in domestic and commercial wiring in the UK since the 1960s is a flat PVC twin-and-earth type, alternatively known as 6242Y cable. This comprises a flat formation of PVC-insulated live and neutral cores separated by a bare earth conductor, the whole assembly being PVC-sheathed to produce a flat cable rated at 300/500 V. Cable is also available with three insulated cores and a bare earth, for use on double-switched lighting circuits. These forms of cable are ideal for installation under cladding in standard-depth plaster. They are defined in BS 6004, which covers a large size range, only the smaller sizes of which are used in domestic and commercial circuits.

There are other cable types included in BS 6004 which have more relevance to non-domestic installations. These include cables in both flat and circular form, similar to the 6242Y type but with an insulated earth conductor. Circular cables designated 6183Y are widely used in commercial or light industrial areas, especially where many circuits are mounted together on cable trays. Also included in BS 6004 are insulated conductors designated 6491Y which are pulled into conduit or trunking, in those circuits where mechanical protection or the facility to rewire are key factors.

An alternative type of cable with outstanding impact and crush strength is *Mineral-Insulated Cable (MICC)* manufactured to BS 6207. This is often known by its trade name, Pyrotenax. In an MICC cable, the copper line and neutral conductors are positioned inside a copper sheath, the spaces between the copper components being filled with heavily compacted mineral powder of insulating grade. Pressure or impact applied to the cable merely compresses the powder in such a way that the insulation integrity is maintained. The copper sheath often acts as the circuit earth conductor. An oversheath is not necessary, but is often provided for reasons of appearance or for external marking. MICC cable has a relatively small cross-section and is easy to install.

In shopping and office complexes or in blocks of flats there may be a need for a distribution submain to feed individual supply points or meters. If this submain is to be installed and operated by the owner of the premises, then a 0.6/1 kV split-concentric service cable to BS 4533 may be used. This comprises a phase conductor insulated in PVC or XLPE, around which are a layer of copper wires and an oversheath. Some of the copper wires are bare and these are used as the earth conductor. The remainder are polymer covered and they make up the neutral conductor. For larger installations, 3-core versions of this cable are available to manufacturers' specification.

LSF versions of the above cable types are available from several manufacturers. Some of these have already been standardized according to BS 7211. In circuits supplying equipment for fire detection and alarm, emergency lighting and emergency supplies, regulations dictate that the cables will continue to operate during a fire. This continued operation can be ensured by measures such as embedding the cable in masonry, but there is a growing demand for cables that are fire resistant in themselves. Fire-resistance categories are set down in BS 6387.

The best fire resistance is offered by MICC cables, since the mineral insulation is unaffected by fire. MICC cable will only fail when the copper conductor or sheath melts and where such severe fires might occur the cable can be sheathed in LSF material to delay the onset of melting. MICC cable is categorized CWZ in BS 6387.

Alternatives to MICC cables for fire resistance have been developed by individual manufacturers. Some rely on a filled silicone rubber insulation which degrades during a fire but continues to provide separation between the conductors so that circuit integrity is maintained. Other types supplement standard insulation with layers of mica tape so that even if the primary insulation burns completely the mica tape provides essential insulation to maintain supplies during the fire. These cable types are assigned minimum fire-resistance categories BWX and SWX respectively and manufacturers may claim fire resistance up to CWZ for their own products. Both cable types are standardized in BS 7629.

Some circuits requiring an equivalent level of fire resistance need to be designed with larger cables than are found in BS 7629. Such circuits might be for the main emergency supply, fire-fighting lifts, sprinkler systems and water pumps, smoke extraction fans, fire shutters or smoke dampers. These larger cables are standardized in BS 7846, which includes the size range and LSF performance of BS 6724. These cables can be supplied to the CWZ performance level in BS 6387. However, there is an additional fire test category in BS 7846, called F3, which is considered to be more appropriate for applications where the cable might be subject to fire, impact and water spray in combination during the fire.

9.4   Parameters and test methods

There are a large number of cable and material properties which are controlled by the manufacturer in order to ensure fitness for purpose and reliable long-term service performance. However, it is the operating parameters of the finished installed cable which are of most importance to the user in cable selection. The major parameters of interest are as follows:

- current rating
- capacitance
- inductance
- voltage drop
- earth loop impedance
- symmetrical fault capacity
- earth fault capacity

These are dealt with in turn in the following sections.

9.4.1   Current rating

The current rating of each individual type of cable could be measured by subjecting a sample to a controlled environment and by increasing the load current passing through the cable until the steady-state temperature of the limiting cable component reached its maximum permissible continuous level. This would be a very costly way of establishing current ratings for all types of cable in all sizes, in all environments and in all ambient temperatures. Current ratings are therefore obtained using an internationally accepted calculation method, published in IEC 60287. The formulae and reference material properties presented in IEC 60287 have been validated by correlation with data produced from laboratory experiments.

Current ratings are quoted in manufacturers' literature and are listed in IEE Wiring Regulations (BS 7671) for some industrial, commercial and domestic cables. The ratings are quoted for each cable type and size in air, in masonry, direct-in-ground and in underground ducts. Derating factors are given so that these quoted ratings can be adjusted for different environmental conditions such as ambient temperature, soil resistivity or depth of burial.

Information is given in BS 7671 and in the IEE Guidance Notes on the selection of the appropriate fuse or mcb to protect the cable from overload and fault conditions, and general background is given in **sections 8.2** and **8.3**.

9.4.2 Capacitance

The capacitance data in manufacturers' literature is calculated from the cable dimensions and the permittivity of the insulation.

For example, the star capacitance of a 3-core belted armoured cable to BS 6346 is the effective capacitance between a phase conductor and the neutral star point. It is calculated using the following formula:

$$C = \frac{\varepsilon_0}{18 \ln[(d + t_1 + t_2)/d]} \quad [\mu F/km] \qquad (9.1)$$

where ε_0 = relative permittivity of the cable insulation (8.0 for PVC)
  d = diameter of the conductor   [mm]
  t_1 = thickness of insulation between the conductors   [mm]
  t_2 = thickness of insulation between conductor and armour   [mm]

Equation 9.1 assumes that the conductors are circular in section. For those cables having shaped conductors, the value of capacitance is obtained by multiplying the figure obtained using **eqn 9.1** by an empirical factor of 1.08.

The calculated capacitance tends to be conservative, that is the actual capacitance will always be lower than the calculated value. However, if an unusual situation arises in which the cable capacitance is critical, then the manufacturer is able to make a measurement using a capacitance bridge.

If the measured capacitance between cores and between core and armour is quoted, then the star capacitance can be calculated using **eqn 9.2**:

$$C = \frac{9C_x - C_y}{6} \quad [\mu F/km] \qquad (9.2)$$

where C_x = measured capacitance between one conductor and the other two connected together to the armour
  C_y = measured capacitance between three conductors connected together and the armour

9.4.3 Inductance

The calculation of cable inductance L for the same example of a 3-core armoured cable to BS 6346 is given by **eqn 9.3** as follows:

$$L = 1.02 \times \{0.2 \times \ln[2Y/d] + k\} \quad \text{[mH/km]} \tag{9.3}$$

where d = diameter of the conductor [mm]
        Y = axial spacing between conductors [mm]
        k = a factor which depends on the conductor make-up
          (k = 0.064 for 7-wire stranded
            0.055 for 19-wire stranded
            0.053 for 37-wire stranded
            0.050 for solid)

The same value of cable inductance L is used for cables with circular- or sector-shaped conductors.

9.4.4 Voltage drop

BS 7671 specifies that within customer premises the voltage drop in cables is to be a maximum value of 4 per cent. It is therefore necessary to calculate the voltage drop along a cable.

The cable manufacturer calculates voltage drop assuming that the cable will be loaded with the maximum allowable current which results in the maximum allowable operating temperature of the conductor. The cable impedance used for calculating the voltage drop is given by **eqn 9.4**:

$$Z = \{R^2 + (2\pi f L - 1/2\pi f C)^2\}^{1/2} \quad \text{[Ω/m]} \tag{9.4}$$

where R = ac resistance of the conductor at maximum conductor
         temperature [Ω/m]
        L = inductance [H/m]
        C = capacitance [F/m]
        f = supply frequency [Hz]

The voltage drop is then given by **eqns 9.5** and **9.6**:

For single-phase circuits:

$$\text{voltage drop} = 2Z \quad \text{[V/A/m]} \tag{9.5}$$

and for three-phase circuits:

$$\text{voltage drop} = \sqrt{3}Z \quad \text{[V/A/m]} \tag{9.6}$$

9.4.5 Symmetrical and earth fault capacity

It is necessary that cables used for power circuits are capable of carrying any fault currents that may flow, without damage to the cable; the requirements are specified in BS 7671. This assessment demands a knowledge of the maximum prospective fault currents on the circuit, the clearance characteristics of the protective device (as explained in **Chapter 8**) and the fault capacity of the relevant elements in the cable. For most installations it is necessary to establish the let-through energy of the protective device and to compare this with the adiabatic heating capacity of the conductor (in the case of symmetrical and earth faults) or of the steel armour (in the case of earth faults).

The maximum let-through energy (I^2t) of the protective device is explained in **Chapter 8**. It can be obtained from the protective device manufacturer's data. In practice the value will be less than that shown by the manufacturer's information because of the reduction in current during the fault which results from the significant rise in temperature and resistance of the cable conductors.

The fault capacity of the cable conductor and armour can be obtained from information given in BS 7671 and the appropriate BS cable standard, as follows:

$$k^2S^2 = \text{adiabatic fault capacity of the cable element} \qquad (9.7)$$

where $S =$ the nominal cross-section of the conductor *or* the nominal
                    cross-section of, say, the armour   [mm^2]
         $k =$ a factor reflecting the resistivity, temperature coefficient,
                    allowable temperature rise and specific heat of the metallic
                    cable element
                    ($k = 115$ for a PVC-insulated copper conductor within the cable
                       $= 176$ for an XLPE-insulated copper earth conductor external
                         to the cable
                       $= 46$ for the steel armour of an XLPE-insulated cable)

In practice it will be found that provided the cable rating is at least equal to the nominal rating of the protective device and the maximum fault duration is less than 5 seconds, the conductors and armour of cables to BS will easily accommodate the let-through energy of the protective device.

It is also important that the impedance of the supply cable is not so high that the protective device takes too long to operate during a zero-sequence earth fault on connected equipment. This is important because of the need to protect any person in contact with the equipment, by limiting the time that the earthed casing of the equipment, say, can become energized during an earth fault.

This requirement, which is stated in BS 7671, places restrictions on the length of cable that can be used on the load side of a protective device, and it therefore demands a knowledge of the earth fault loop impedance of the cable. Some cable manufacturers have calculated the earth fault impedance for certain cable types and the data are presented in specialized literature. These calculations take account of the average temperature of each conductor and the reactance of the cable during the fault. The values are supported by independent experimental results. BS 7671 allows the use of such manufacturer's data or direct measurement of earth fault impedance on a completed installation.

9.5 Optical communication cables

The concept of using light to convey information is not new. There is historical evidence that Aztecs used flashing mirrors to communicate, and in 1880 Alexander Graham Bell first demonstrated his photophone, in which a mirror mounted on the end of a megaphone was vibrated by the voice to modulate a beam of sunlight, thereby transmitting speech over distances up to 200 m.

Solid-state photodiode technology has its roots in the discovery of the light-sensitive properties of selenium in 1873, used as the detector in Bell's photophone. The Light Emitting Diode (LED) stems from the discovery in 1907 of the electroluminescent properties of silicon, and when the laser was developed in 1959, the components of an optical communication system were in place, with the exception of a suitable transmission medium.

The fundamental components of a fibre optic system are shown in **Fig. 9.9**. This system can be used for either analogue or digital transmissions, with a transmitter that converts electrical signals into optical signals. The optical signals are launched through a joint into an optical fibre, usually incorporated into a cable. Light emitting from the fibre is converted back into its original electrical signal by the receiver.

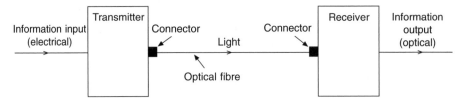

Fig. 9.9 Basic fibre optic system

9.5.1 Optical fibres

An optical fibre is a dielectric waveguide for the transmission of light, in the form of a thin filament of very transparent silica glass. As shown in **Fig. 9.10**, a typical fibre comprises a core, the cladding, a primary coating and sometimes a secondary coating or buffer. Within this basic construction, fibres are further categorized as multimode or single-mode fibres with a step or graded index.

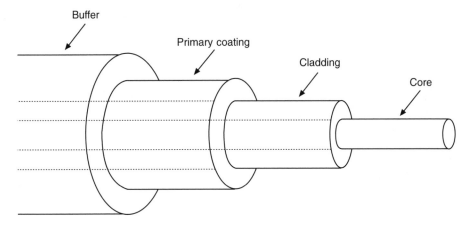

Fig. 9.10 Basic optical fibre

The core is the part of the fibre that transmits light, and it is surrounded by a glass cladding of lower refractive index. In early fibres, the homogeneous core had a constant refractive index across its diameter, and with the refractive index of the cladding also constant (at a lower value) the profile across the whole fibre diameter, as shown in **Fig 9.11(a)**, became known as a *step index*. In this type of fibre the light rays can be envisaged as travelling along a zig-zag path of straight lines, kept within the core by total reflection at the inner surface of the cladding. Depending on the angle of the rays to the fibre axis, the path length will differ so that a narrow pulse of light entering the fibre will become broader as it travels. This sets a limit to the

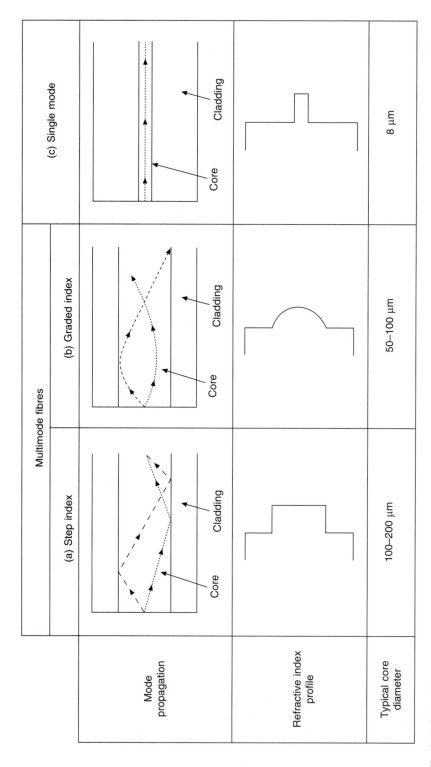

Fig. 9.11 Optical fibre categories

rate at which pulses can be transmitted without overlapping and hence a limit to the operating bandwidth.

To minimize this effect, which is known as *mode dispersion*, fibres have been developed in which the homogeneous core is replaced by one in which the refractive index varies progressively from a maximum at the centre to a lower value at the interface with the cladding. **Figure 9.11(b)** shows such a *graded index fibre*, in which the rays no longer follow straight lines. When they approach the outer parts of the core, travelling temporarily faster, they are bent back towards the centre where they travel more slowly. Thus the more oblique rays travel faster and keep pace with the slower rays travelling nearer the fibre centre. This significantly reduces the pulse broadening effect of step index fibres.

The mode dispersion of step index fibres has also been minimized by the more recent development of *single-mode fibres*. As shown in **Fig. 9.11(c)**, although it is a step index fibre, the core is so small (of the order of 8 μm in diameter) that only one mode can propagate.

Fibre manufacture involves drawing down a preform into a long thin filament. The preform comprises both core and cladding, and for graded index fibres the core contains many layers with dopants being used to achieve the varying refractive index. Although the virgin fibre has a tensile strength comparable to that of steel, its strength is determined by its surface quality. Micro-cracks develop on the surface of a virgin fibre in the atmosphere, and the lightest touch or scratch makes the fibre impractically fragile. Thus it must be protected, in line with the glass drawing before it touches any solid object such as pulleys or drums, by a protective coating of resin, acetate or plastic material, called the *primary coating*.

Typically the primary coating has a thickness of about 60 μm, and in some cases a further layer of material called the *buffer* is added to increase the mechanical protection.

9.5.2   Optical cable design

The basic aim of a transmission cable is to protect the transmission medium from its environment and the rigours of installation. Conventional cables with metallic conductors are designed to function effectively in a wide range of environments, as shown in **sections 9.2** and **9.3**. However, optical fibres differ significantly from copper wires to an extent that has a considerable bearing on cable designs and manufacturing techniques. The transmission characteristics and lifetime of fibres are adversely affected by quite low levels of elongation, and lateral compressions can produce small kinks or sharp bends which create an increase in attenuation loss known as *microbending loss*. This means that cables must protect the fibre from strain during installation and service, and they must cater for longitudinal compression that occurs, for example, with a change in cable temperature.

Fibre life in service is influenced by the presence of moisture as well as stress. The minute cracks which cover the surface of all fibres can grow if the fibre is stressed in the presence of water, so that the fibre could break after a number of years in service. Cables must be able to provide a long service life in such environments as tightly packed ducts which are filled with water.

The initial application of optical cables was the trunk routes of large telecommunications networks, where cables were directly buried or laid in ducts in very long lengths, and successful cable designs evolved to take into account the constraints referred to above. The advantages of fibre optics soon led to interest in other applications such as computer and data systems, military systems and industrial control. This meant that cable designs had to cater for tortuous routes of installation in buildings, the flexibility of patch cords and the arduous environments of military and industrial applications.

Nevertheless, many of the conventional approaches to cable design can be used for optical cables, with modification to take into account the optical and mechanical characteristics of fibres and their fracture mechanics.

Cables generally comprise several elements or individual transmission components such as copper pairs, or one or more optical fibres. The different types of element used in optical cables are shown in **Fig. 9.12**.

The primary-coated fibre can be protected by a *buffer* of one or more layers of plastic material as shown in **Fig. 9.12(a)**. Typically for a two-layer buffer, the inner layer is of a soft material acting as a cushion with a hard outer layer for mechanical protection, the overall diameter being around 850 μm. In other cases the buffer can be applied with a sliding fit to allow easy stripping over long lengths.

In *ruggedized fibres* further protection for a buffered fibre is provided by surrounding it with a layer of non-metallic synthetic yarns and an overall plastic sheath. This type of arrangement is shown in **Fig. 9.12(b)**.

When one or more fibres are run loosely *inside a plastic tube,* as shown in **Fig. 9.12(c)**, they can move freely and will automatically adjust to a position of minimum bending strain to prevent undue stress being applied when the cable is bent. If the fibre is slightly longer than the tube, a strain margin is achieved when the cable is stretched say during installation, and for underground and duct cables the tube can be filled with a gel to prevent ingress of moisture. Correct choice of material and manufacturing technique can ensure that the tube has a coefficient of thermal expansion similar to that of the fibre, so that microbending losses are minimized with temperature excursions.

Optical fibres can be assembled into a linear array as a *ribbon*, as shown in **Fig. 9.12(d)**. Up to 12 fibres may be bonded together in this way or further encapsulated if added protection is required.

In order to prevent undue cable elongation which could stress the fibres, optical cables generally incorporate a *strength member*. This may be a central steel wire or strand, or non-metallic fibreglass rods or synthetic yarns. The strength member should be strong, light and usually flexible, although in some cases a stiff strength member can be used to prevent cable buckling which would induce microbending losses in the fibres. Strength members are shown in the cable layouts in **Figs 9.13(b)** and **9.13(c)**.

The strength member can be incorporated in a *structural member* which is used as a foundation for accommodating the cable elements. An example is shown in **Fig. 9.13(c)**, where a plastic section with slots is extruded over the strength member with ribbons inserted into the slots to provide high fibre count cables.

A *moisture barrier* can be provided either by a continuous metal sheath or by a metallic tape with a longitudinal overlap, bonded to the sheath. Moisture barriers can be of aluminium, copper or steel and they may be flat or corrugated. In addition, other cable interstices may be filled with gel or water-swellable filaments to prevent the longitudinal ingress of moisture.

Where protection from external damage is required, or where additional tensile strength is necessary, *armouring* can be provided; this may be metallic or non-metallic. For outdoor cables, an *overall sheath* of polyethylene is applied. For indoor cables the sheath is often of low-smoke zero-halogen materials for added safety in the event of fire.

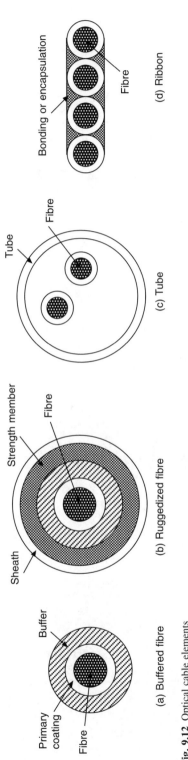

Fig. 9.12 Optical cable elements

(a) Buffered fibre

Primary coating

Buffer

Fibre

(b) Ruggedized fibre

Strength member

Fibre

Sheath

(c) Tube

Tube

Fibre

(d) Ribbon

Bonding or encapsulation

Fibre

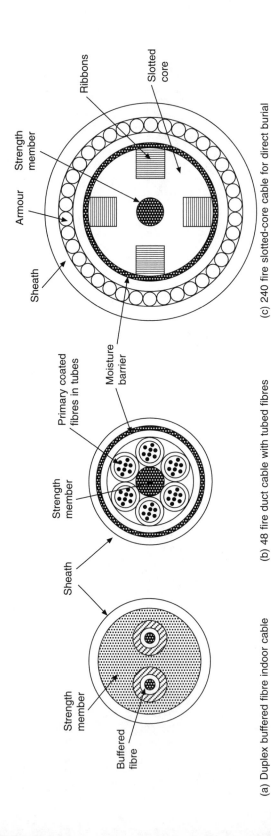

(a) Duplex buffered fibre indoor cable

(b) 48 fire duct cable with tubed fibres

(c) 240 fire slotted-core cable for direct burial

Fig. 9.13 Examples of optical cables

Although the same basic principles of cable construction are used, the wide range of applications results in a variety of cable designs, from simplex indoor patch cords to cables containing several thousand fibres for arduous environments, to suboceanic cables. **Figure 9.13** shows just a few examples.

9.5.3   Interconnections

The satisfactory operation of a fibre optic system requires effective jointing and termination of the transmission medium, in the form of fibre-to-fibre splices and fibre connections to repeaters and end equipment. This is particularly important because with very low loss fibres the attenuation due to interconnections can be greater than that due to a considerable length of cable.

For all types of interconnection there is a loss, known as the *insertion loss*, which is caused by Fresnel reflection and by misalignment of the fibres.

Fresnel reflection is caused by the changes in refractive index at the fibre–air–fibre interface, but it can be minimized by inserting into the air gap an index-matching fluid with the same refractive index as the core.

Misalignment losses arise from three main sources as shown in **Fig. 9.14**. Interconnection designs aim to minimize these losses. *End-face separation* (**Fig. 9.14(a)**) allows light from the launch fibre to spread so that only a fraction is captured by the receive fibre; this should therefore be minimized. Normally the fibre cladding is used as the reference surface for aligning fibres, and the fibre geometry is therefore important, even when claddings are perfectly aligned. Losses due to *lateral misalignment* (**Fig. 9.14(b)**) will therefore depend on the core diameter, non-circularity of the core, cladding diameter, non-circularity of the cladding and the concentricity of the core and cladding in the fibres to be jointed. *Angular misalignment* can result in light entering the receive fibre at such an angle that it cannot be accepted. It follows that very close tolerances are required for the geometry of the joint components and the fibres to be jointed, especially with single-mode fibres with core diameters of 8 μm and cladding diameters of 125 μm.

The main types of interconnections are fibre splices and demountable connectors.

Fibre splices are permanent joints made between fibres or between fibres and device pigtails. They are made by fusion splicing or mechanical alignment. In *fusion splicing*, prepared fibres are brought together, aligned and welded by local heating combined with axial pressure. Sophisticated portable equipment is used for fusion splicing in the field. This accurately aligns the fibres by local light injection and carries out the electric arc welding process automatically. Nevertheless, a level of skill is required in the preparation of the fibres, stripping the buffers and coatings and cleaving the fibres to achieve a proper end face. There are a number of *mechanical techniques* for splicing fibres which involve fibre alignment by close tolerance tubes, ferrules and v-grooves, and fixing by crimps, glues or resins. Both fusion and mechanical splicing techniques have been developed to allow simultaneous splicing of fibres which are particularly suitable for fibre ribbons. For a complete joint, the splices must be incorporated into an enclosure which is suitable for a variety of environments such as underground chambers or pole tops. The enclosure must also terminate the cables and organize the fibres and splices, and cassettes are often used where several hundred splices are to be accommodated.

Demountable connectors provide system flexibility, particularly at and within the transmission equipment and distribution panels, and they are widely used on patch cords in certain data systems. As with splices, the connector must minimize Fresnel

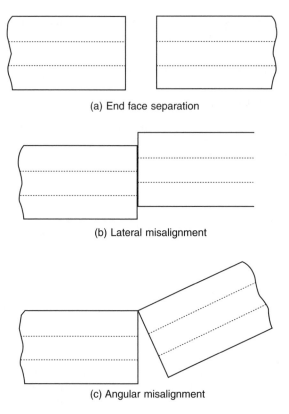

(a) End face separation

(b) Lateral misalignment

(c) Angular misalignment

Fig. 9.14 Sources of misalignment loss

and misalignment loss, but it must also allow for repeated connection and disconnection, it must protect the fibre end face and it must cater for mechanical stress such as tension, torsion and bending. Many designs have been developed, but in general the tolerances that are achievable on the dimensions of the various components result in a higher optical loss than in a splice. Demountable connectors have also been developed for multiple-fibre simultaneous connection, with array designs being particularly suitable for fibre ribbons. For connector-intensive systems such as office data systems, use is made of factory-predetermined cables and patch cords to reduce the need for on-site termination.

9.5.4   Installation

Optical fibre cables are designed so that normal installation practices and equipment can be used wherever possible, but as they generally have a lower strain limit than metallic cables special care may be needed in certain circumstances and manufacturer's recommendations regarding tensile loads and bending radii should be followed.

Special care may be needed in the following circumstances:

- because of their light weight, optical cables can be installed in greater lengths than metallic cables. For *long underground ducts* access may be needed at intermediate points for additional winching effort, and space should be allowed for larger 'figure 8' cable deployment.

- *mechanical fuses* and *controlled winching* may be necessary to ensure that the rated tensile load is not exceeded
- *guiding equipment* may be necessary to avoid subjecting optical cables to unacceptable bending stresses, particularly when the cable is also under tension
- when installing cables in trenches the footing should be free from stones. These could cause microbending losses.
- in buildings, and particularly in risers, cleats and fixings should not be overtightened, or appropriate designs should be used to prevent compression and the resulting microbending losses
- indoor cable routes should provide turning points if a large number of bends is involved. Routes should be as straight as possible.
- excess lengths for jointing and testing of optical cables are normally greater than those required for metallic cables
- where non-metallic optical cables are buried, consideration of the subsequent location may be necessary. Marker posts and the incorporation of a location wire may be advisable.

Blown fibre systems have been developed as a means of avoiding fibre overstrain for complex route installation and of allowing easy system upgrading and future proofing. It results in low initial capital costs and provides for the distribution of subsequent costs. Initially developed by British Telecom, the network infrastructure is created by the most appropriate cabling method, being one or a group of empty plastic tubes. As and when circuit provision is required, one or more fibres can be blown by compressed air into the tubes. Individual tubes can, by means of connectors, be extended within buildings up to the fibre terminating equipment. The efficient installation of fibres into the tube network often requires the use of specially designed fibres and equipment such as air supply modules, fibre insertion tools and fibre pay-offs. For installation it is necessary to follow the instructions provided by the supplier, taking into account the requirements for the use of portable electrical equipment and compressed air, and the handling, cutting and disposal of optical fibres. A novel variation of this system is a data cable used for structural wiring systems. In a 'figure-8' configuration one unit comprises a 4-pair data cable and the other an empty tube, so that when an upgrade is required to an optical system the appropriate fibre can be blown in without the need for recabling.

9.6 Standards

9.6.1 Metallic wires and cables

Most generally available cables are manufactured to recognized standards which may be national, European or international. Each defines the construction, the type and quality of constituent materials, the performance requirements and the test methods for the completed cable.

IEC standards cover those cables which need to be standardized to facilitate world trade, but this often requires a compromise by the parties involved in the preparation and acceptance of a standard. Where cables are to be used in a particular country, the practices and regulations in that country tend to encourage the more specific cable types defined in the national standards for that country. BS remains the most appropriate for use in the UK, and for the main cable types described in this chapter reference has therefore been made mainly to the relevant BS.

Some cables rated at 450/750 V or less have through trade become standard throughout the EU, and these have been incorporated into Harmonization Documents (HDs). Each EU country must then publish these requirements within a national standard. A harmonized cable type in the UK for instance would still be specified to the relevant BS and the cable would, if appropriate, bear the <HAR> mark.

The key standards for metallic wires and cables which have been referred to in the chapter are listed in **Table 9.1**.

Table 9.1 International and national standards for metallic wires and cables

IEC	HD	BS	Subject
	603	4553	600/1000 V PVC-insulated single-phase split concentric cables with copper conductors
60502-1		5467	Cables with thermosetting insulation up to 600/1000 V and up to 1900/3300 V
60227	21	6004	Non-armoured PVC-insulated cables rated up to 450/750 V
60245	22	6007	Non-armoured rubber-insulated cables rated up to 450/750 V
		6346	PVC-insulated cables
		6387	Performance requirements for cables required to maintain integrity under fire conditions
60055	621	6480	Impregnated paper-insulated lead sheathed cables up to 33 000 V
	621	(EA 09-12)	Paper-insulated corrugated aluminium sheathed 6350/11 000 V cable
60227 & 60245	21 & 22	6500	Insulated flexible cords and cables rated up to 450/750 V
60502-2	620	6622	Cables with XLPE or EPR insulation from 3800/6600 V up to 19 000/33 000 V
		6724	600/1000 V and 1900/3300 V armoured cables having thermosetting insulation with low emission of smoke and corrosive gases in fire
	22	7211	Non-armoured cables having thermosetting insulation rated up to 450/750 V with low emission of smoke and corrosive gases in fire
		7629	Fire-resistant thermosetting-insulated cables rated at 300/500 V with limited circuit integrity in fire
60364		7671	Requirements for electrical installations: IEE Wiring Regulations (16th edition)
60287		7769	Electric cables – calculation of current rating
		7835	Cables with XLPE or EPR insulation from 3800/6600 V up to 19 000/33 000 V with low emission of smoke and corrosive gases in fire
	603	7870-3	Polymeric-insulated cables for distribution rated at 600/1000 V
	620	7870-4	Polymeric-insulated cables for distribution rated from 3800/6600 V up to 19 000/33 000 V
	626	7870-5	Polymeric-insulated aerial-bundled cables rated 600/1000 V for overhead distribution
	604	7870-6	Polymeric-insulated cables for generation rated at 600/1000 V and 1900/3300 V
	622	7870-7	Polymeric-insulated cables for generation rated from 3800/6600 V up to 19 000/33 000 V

9.6.2   Optical communication cables

For communication systems and their evolution to be effective, standardization must be at an international level. Optical fibre and cable standardization in IEC started in 1979. In Europe ENs have been published; these generally use the IEC standards as a starting point but they incorporate any special requirements for sale within the EU.

Table 9.2 summarizes the main standards in the areas of optical fibres, optical cables, connectors, connector interfaces and test and measurement procedures for interconnecting devices.

Table 9.2 International and national standards for optical fibres, optical cables, connectors, connector interfaces and test and measurement procedures for interconnecting devices

IEC	EN	BS	Subject
		7718	Code of practice for installation of fibre optic cabling
60793			
60794			
	60794-3	EN60794-3	
60874			
	60874-17	EN60874-17	
	60874-19	EN60874-19	
61300	61300	EN61300	
61754	61754	EN61754	
	186000 to 186290	EN186000 to EN 186290	
	187000	EN187000	
	187100 to 187102	EN187100 to EN 187102	
	188000	EN188000	
	188100 to 188102	EN188100 to EN 188102	
	188200 to 188202	EN188200 to EN 188202	

References

9A.   Moore, G.F., *Electric Cables Handbook,* 3rd edn, Blackwell Scientific Publications Ltd, 1997.
9B.   Heinhold, L., *Power Cables and Their Application,* 3rd edn, Siemens AG, 1990.

Chapter 10

Motors, motor control and drives

Dr Norman N. Fulton
Switched Reluctance Drives Ltd
(A subsidiary of Emerson Electric Co.)

10.1 Introduction

Electric motors are used throughout industry, commerce and in the home in a wide variety of power ratings, speeds and duties. A typical home will easily have 40 motors in domestic appliances, hand tools and audio equipment; a modern car is unlikely to have fewer than 15 motors, ranging from the starter motor to adjusters for windows, seats and mirrors. Electric motors were invented well over 150 years ago and a number of distinct types of machine have developed, as described later in this chapter. Although usually regarded as a mature technology, the changing needs for motors and the emergence of new associated technologies, such as power semiconductors, continue to lead to changes in the types of machine that are used, in their applications and also in the way that motors are designed and built.

For instance, the field of adjustable speed drives was at one time dominated by dc machines because speed control could be readily managed without complicated high-speed switching of the power. With the emergence of power semiconductors, first in the form of thyristors and bipolar transistors, then MOSFETs and IGBTs, power switching technology has become commercially viable across a wide power range from watts to megawatts. One application of that technology has been in variable-frequency inverters, which enable the cheaper, more reliable ac induction motor to be used in many adjustable speed drives in place of the more complicated and expensive dc motor.

Advances in magnet technology have made permanent magnet motors more cost-competitive, especially in adjustable speed drives, and the so-called 'brushless dc drive' has already found widespread acceptance in applications in a wide variety of domestic and commercial products where a low-cost solution is required. In the industrial market, these drives are more commonly found in high-cost, high-precision drive systems, often where a positioning function is required in addition to speed control.

More recently, the switched reluctance motor has re-emerged. It was one of the earliest forms of electric motor but fell into disuse because of a lack of suitable control devices. Now incorporated into a switched reluctance drive using modern semiconductors, the switched reluctance motor is arguably cheaper and simpler than the induction motor, and it threatens to take over variable speed applications currently served by induction motors, as well as competing with dc motors and ac universal motors.

These types of motor and drive are described later in the chapter, but each of them relies on common basic principles and has to operate within certain constraints. These principles and constraints are explained before describing the various types of motor and drive.

10.2   General characteristics

This section covers the principles of torque production, the main components found in all motors and the related topics of losses, efficiency and heat dissipation.

10.2.1   The production of torque

The power delivered by a rotating shaft is the product of torque and speed. The mechanisms for the production of torque are therefore a critical part of motor operation and, as shown below, are sometimes used to distinguish the different types of motor. There are two basic principles for torque production in electric motors; these are known as *electromagnetic torque* and *reluctance torque*. Virtually all motors use one or the other, or a combination of both.

Figure 10.1 shows a conductor carrying a current. At right angles to the conductor there is a uniform magnetic field having a density of B webers per square metre or B tesla. If the conductor has a length l metres and the current is I amps, it can be shown that the force on the conductor in the direction shown (in newtons), is given by **eqn 10.1**:

$$F = B \times I \times l \tag{10.1}$$

This is termed *'electromagnetic' force* since it is the product of an electric current with a magnetic flux density. If this force acts on a conductor mounted on the rotor

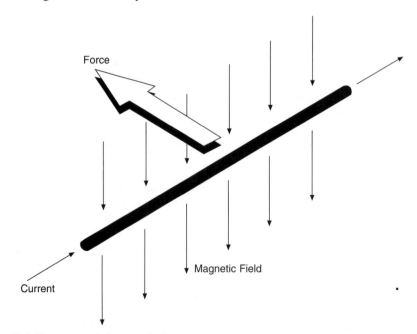

Fig. 10.1 Electromagnetic force production

of a motor at a radius r metres, then the torque acting on the rotor is $F \times r$ (in newton metres). This means of electromagnetic torque production is the most commonly used in electric motors, although different types of motors use different methods of passing the current I through the conductor and different ways of producing the flux density B. It is used in, for example, induction, synchronous and dc motors.

The second torque production method used in electric motors makes use of the fact that any magnetic circuit tends to move to a position at which its *reluctance* is a minimum (i.e. where the inductance of the exciting coil producing the mmf is a maximum). This is exploited by having a magnetic structure which has *saliency*. **Figure 10.2** shows a machine with a stator and a rotor both having salient poles: as the rotor turns, the air gap presented to the stator poles changes, resulting in a change in the reluctance of the magnetic circuit. By passing current into the winding around the stator poles, magnetic flux is set up in the magnetic circuit. The path of the flux is shown very approximately in **Fig. 10.2**. The flux produces a force which acts to reduce the reluctance of the magnetic circuit and the rotor will experience a reluctance torque as it attempts to align itself in a position of minimum reluctance (or maximum inductance) when the stator and rotor saliencies are aligned to give the minimum air gap between rotor and stator. This torque production method is used in stepper motors and switched reluctance motors; the torque can be produced by exciting a winding on only one member and it is not necessary to have a permanent magnet on either member.

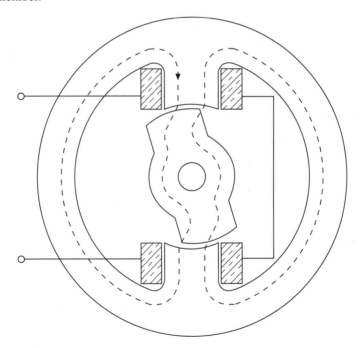

Fig. 10.2 Reluctance torque production

10.2.2 Main components and construction

Generally, motors comprise a stationary part, the *stator*, and a rotating part, the *rotor*. Typically, the stator is on the outside and the rotor rotates within it, although *inverted machines*, typically found in small fan drives, have the rotor on the outside. The main part of both stator and rotor is a *core* which is usually built from a stack of electrical steel laminations and has slots or salient poles to accommodate and support the windings. The laminations are normally 0.2 to 1.0 mm thick. In small and medium power motors they are punched from coils of electrical steel by a multi-stage press tool; in larger sizes they are produced on single-stage presses from sheet.

The general construction of small and medium power motors is illustrated by the sectional drawing of a cage induction motor in **Fig. 10.3**. The stator windings are embedded in slots around the inner bore of the stator core and the winding overhang outside the core is securely laced together. The outer surface of the rotor core is accurately machined to be concentric with the bearing journals and to have a diameter which, in conjunction with the stator bore, will produce the *air gap* of the machine.

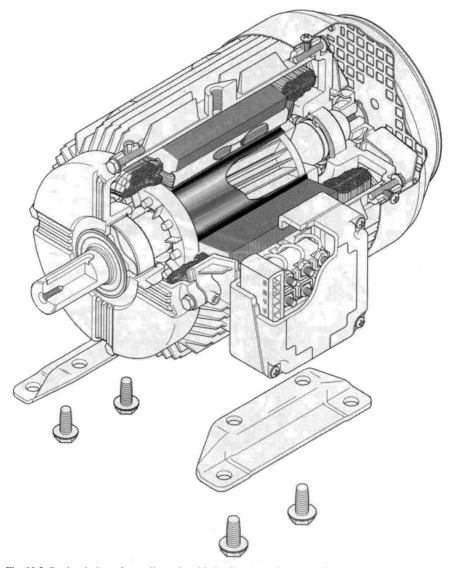

Fig. 10.3 Sectional view of a totally enclosed induction motor (courtesy of Invensys Brook Crompton)

The air gap will have a radial length of the order of 0.2 mm for very small machines and up to several cm for machines developing megawatts.

In some motors, the stator is contained in an external frame and the rotor is supported in bearings mounted in the *endshields* or *endbrackets*, which are located in the ends of the stator frame by spigots which ensure the rotor runs centrally within the stator. These are shown in **Fig. 10.3** and also more clearly in the exploded diagram **Fig. 10.4**. In many cases, the frame is finned to increase the surface area and improve dissipation of the heat generated by the electrical losses in the motor. The complete motor can be supported by feet on the stator (*foot mounted*) or by providing a flange on the drive-end endshield (*flange mounted*), or occasionally by both. By contrast, motors which are intended for incorporation into an enclosed appliance, such as a dishwasher pump, generally use a so-called *frameless* construction, where there is no separate frame for the core and skeletal bearing housings are mounted directly to the ends of the stator core. Larger high-voltage motors are generally mounted within fabricated steel frames and endshields.

One method of classification of machines, particularly for industrial applications, is to describe the degree of protection offered to the user and the mechanical protection afforded to the windings of the machine by the frame. Various levels of protection are used and these are defined in, for example, BS 4999 Part 105 (IEC 60034-5) in the form of various *IP classifications*. IP22, for instance, is a basic type of enclosure often referred to as drip-proof or *open ventilated* and this is commonly used for induction motors driving compressors. A much greater degree of protection is offered by an IP55 enclosure which is often known as *totally enclosed*. Special forms of enclosure are required for, for example, flameproof (EEx 'd') motors for hazardous areas and these requirements are described further in **Chapter 16**. An induction motor with an EEx 'd' enclosure is illustrated in **Fig. 16.3**.

As noted above, the rotor is normally mounted in bearings housed in the endshields. The rating of these bearings is dictated by, among other things, the speed of the motor and the side and end loads on the shaft. Simple sleeve bearings are suitable at low powers and loads, standard deep groove ball bearings are commonly used in the 1–200 kW range, roller bearings are often used for motors with high radial shaft loads (such as a belt drive with high tension in the belt) and sophisticated lubrication systems are used on large machines.

With the advent of adjustable speed drives which require a speed feedback signal, it is becoming more common to find some form of speed or position transducer mounted on the motor, typically at the non-drive end. The devices used range from simple, low-cost optical switches in switched reluctance motors through Hall-effect transducers in brushless dc machines to high-cost, high-resolution resolvers in dc servo drives. The devices may have their own terminal box and wiring loom or they may be connected through the main terminal box of the motor.

10.2.3 Losses and heat transfer

Losses in electrical machines are important first because they determine the efficiency of the machine and secondly because they generate heat in the machine. In many cases efficiency is not a primary concern to the user but it should be noted that a motor running continuously on full load will have an energy consumption which could equal its capital cost in a matter of months. The heat generated by the losses has to be dissipated effectively in order to ensure that the insulation in the machine does not operate at temperatures above its rated capability and that the surface

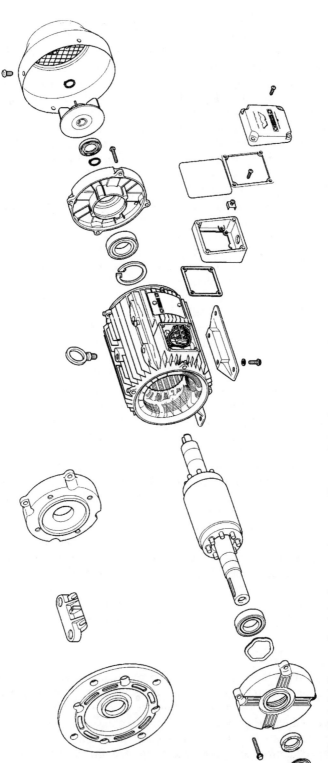

Fig. 10.4 Exploded diagram of an induction motor (courtesy of Invensys Brook Crompton)

temperature of the enclosure does not present a hazard to the user. It follows that the ability of the machine to dissipate its losses will directly affect its output rating.

The power loss in most electrical machines can be classified under four headings:

- resistive loss in the windings (*copper loss*)
- loss in the stator and rotor cores due to hysteresis and eddy current losses (*iron loss*)
- friction loss in the bearings
- windage loss associated with the rotor causing turbulence in the air surrounding it

Many machines operate at fixed speed and fixed voltage, and for these machines some components of the loss are independent of the power level at which they are operating; the friction and windage loss and many of the components of iron loss fall into this category. However, the copper loss in the windings generally varies as the square of the machine load. So although the losses *increase* with load, they *decrease* with load when taken as a proportion of power output; this means that the efficiency of an electrical machine generally increases as its power output increases and is likely to peak near full load.

The balance of the components of the loss varies with both the type and size of machine. In general, small machines have a much greater proportion of copper loss than iron loss, whereas the opposite holds for larger machines. Since the copper loss is generally load dependent, it follows that the efficiency of a small machine varies much more with load than a larger machine. In terms of absolute values of efficiency for different sizes of machine, induction machines of the same rated speed have full-load efficiencies of approximately:

0.1	kW	55%
1.0	kW	75%
10	kW	90%
100	kW	95%
1	MW	97%

The benefit of scale is clearly shown in these figures.

Because many of the power losses occur in the interior of the machine, arrangements have to be made to ensure that the heat resulting from these losses is conducted away effectively. The heat from the windings is normally dissipated into air flowing around exposed parts of the winding and/or into the core supporting the winding. To assist in the latter, the winding is thermally bonded to the core by impregnating the assembly with a varnish or epoxy compound which consolidates the winding, improves its electrical insulation and greatly improves the heat transfer to the core. Methods of impregnation include:

- dipping the wound core in varnish, draining the excess and baking it in an oven
- trickling a viscous gel (normally polyester based) onto the winding and allowing it to be drawn in by capillary action before setting
- evacuating the air from the winding in a vacuum chamber, flooding the chamber with varnish (normally an epoxy compound), increasing the pressure in the

chamber above ambient to force the impregnant into the winding, draining the excess and baking in an oven. This process is known as Vacuum Pressure Impregnation (VPI) and produces a very rugged assembly with good heat transfer from the winding to the core. A large VPI plant is shown in **Fig. 3.6**.

Heat from the magnetic losses in the cores can be removed either by direct cooling with an airstream or by conduction to the frame. In the latter case, the interference fit between the core and the frame needs to be carefully controlled to ensure good thermal conduction.

In small machines of less than a kilowatt rating, the heat transfer paths are short and dissipating the losses is relatively straightforward, particularly since the machine has a relatively high surface to volume ratio. In the medium range of power output up to several hundred kilowatts, many machines are of the totally enclosed type referred to in the previous section; these generally have a shaft-mounted fan outside the frame at the non-drive end, which is arranged to drive air over the outside surface of the frame. These are known as *Totally Enclosed Fan Ventilated (TEFV)* machines (see **Fig. 10.3**). Larger machines in the megawatt range present greater problems for heat transfer because of the absolute size of the losses involved (even though the efficiency is higher) and the fact that the surface to volume ratio falls as the machine size increases. Special arrangements have to be adopted, including dividing the cores into short packets with cooling air flowing between them, providing water cooling for the stator core, using closed air circulation within the machine in conjunction with a water-cooled heat exchanger etc. A large closed air circuit generator is shown in **Fig. 5.21** and large motors have very similar forms of construction. In the largest sizes, hollow conductors with de-ionized water flowing through them are used.

10.3 Main classes of machine

Electrical machines can be classified in many different ways, such as by:

- type of supply (ac, dc or switched)
- method of torque production (electromagnetic, reluctance or both)
- method of providing the 'working' flux (electromagnets or permanent magnets)
- fixed or variable speed
- number of phases

No one classification is entirely satisfactory, not least because the marriage of machines and electronic control systems has produced variable speed drives which do not fall easily into any of the main groups. In this chapter six main groups are identified which cover the most commonly found motors. These main groups, induction, synchronous, commutator, permanent magnet, switched reluctance machines and steppers, are shown in **Fig. 10.5**. Together, these cover the vast majority of all motors driving equipment in domestic, commercial and industrial applications. Each is described briefly in turn in the following sections.

10.3.1 Induction motors

Invented in the 1880s, the induction machine is by far the most common motor in domestic, commercial and industrial use, driving fans, pumps, compressors, conveyors,

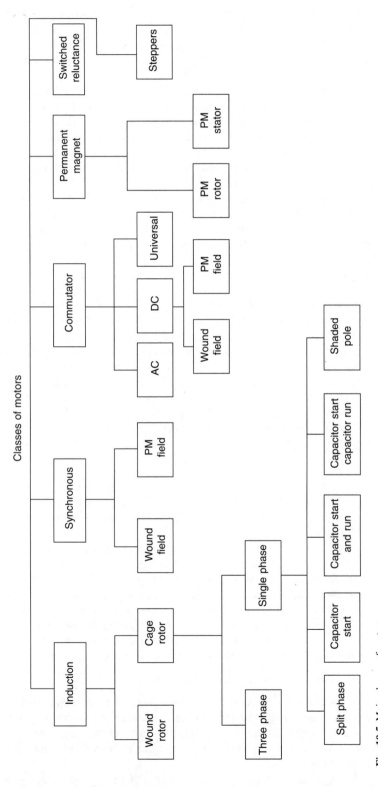

Fig. 10.5 Main classes of motors

machine tools and a wide range of other loads. Motors ranging from a few watts to several megawatts can be found, single-phase motors being commonly used up to around a kilowatt and three-phase machines for higher powers. Two main subdivisions are immediately apparent: the *cage rotor* induction motor and the *wound rotor* induction motor.

The simplest, commonest, most rugged and reliable type is the cage rotor induction motor, in which the rotor has a winding in the form of conducting bars embedded in slots and connected together at the ends of the rotor core by short-circuiting rings called *endrings*. In larger machines the cage is usually fabricated from copper bar and care has to be exercised to ensure the integrity of all the joints in the fabrication. In smaller motors, the cage is usually formed by casting aluminium into the slots of the rotor core and around the ends to form the cage. In these machines, the cage is often referred to as a 'squirrel cage' because of its shape. Sometimes the endrings have cooling fins cast onto them to act as rudimentary fans when the rotor is rotating, as shown in **Figs 10.3** and **10.4**. There is virtually no insulation between the cage and the rotor core, unless (exceptionally) particular steps are taken to insulate the cage during fabrication.

The wound rotor type, as its name implies, has a conventional insulated winding in the slots of the rotor core. Almost invariably it will be a three-phase winding, with its ends brought to shaft-mounted sliprings to provide a sliding connection. The winding can be connected either to a resistance (usually variable) or to a supply to modify the performance during starting or running, as will be described below.

Regardless of the type of rotor used, the stator carries the principal winding which, with the exception of some very small motors, is embedded in the slots of the stator core. The coils of the winding are connected so that, when fed from a balanced supply system (normally three phase, although two phase, six phase and 12 phase are also found in specialized applications) they produce a magnetic field of constant magnitude which rotates around the bore of the stator. It can be shown that this rotates at a speed of N_s given by

$$N_s = 120f/p \qquad (10.2)$$

where N_s is in rev/min, f is the supply frequency in Hz and p is an even integer describing the number of poles in the field pattern of the winding. For a 50 Hz supply and a 4-pole winding (the most commonly found type), the speed of the field is therefore 1500 rev/min.

This rotating field cuts the rotor conductors and *induces* (hence the name *induction motor*) voltages which drive currents around the cage. The magnetic field of the stator interacts with these currents to produce electromagnetic torque, as previously described in **section 10.2.1**. Note that the field only cuts the conductors if the speed of the rotating field is *not* equal to the rotor speed, so that when the rotor speed is synchronous with the magnetic field, the torque will be zero. The torque is therefore a function of the velocity of the rotor *relative* to the synchronous speed of the field. This relative speed is referred to as the *slip*, defined as the ratio:

$$s = (N_s - N_r)/N_s \qquad (10.3)$$

where N_s is the synchronous speed of the field and N_r is the rotor speed. The ratio is often expressed as a percentage, and full-load slip ranges from 10 per cent in small motors to less than 0.5 per cent on large motors. Because the induction motor cannot run at synchronous speed, it is sometimes (and particularly in continental Europe) called the *asynchronous motor*.

As the motor accelerates, the torque changes and a characteristic of the form shown in **Fig. 10.6** results. This is the torque–speed curve which is obtained by supplying the motor from a fixed voltage. At standstill the motor is connected to the supply and allowed to run up to full speed, and this torque–speed curve is followed regardless of the applied load. If the load has a characteristic shown by the broken line in **Fig. 10.6** then the difference between the motor torque and the load torque at any speed is the *accelerating torque* which is available to accelerate the load. If the torque–speed curve were to have dips in it which intercepted the load line, the motor would not accelerate through that point to reach full speed.

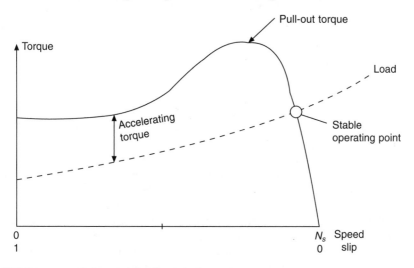

Fig. 10.6 Torque–speed characteristic of an induction motor

The shape of the torque–speed characteristic can be influenced considerably by, among other things, the depth and shape of the rotor conductors. To achieve high efficiency and power factor, the rotor resistance and reactance are minimized but unless the many design parameters are carefully chosen, the starting torque and current may be adversely affected. At starting, the slip is 100 per cent and the frequency of current in the rotor is the main supply frequency; under these conditions skin effect forces current to the top of the rotor bars, increasing their effective resistance and limiting the starting current. Special designs of rotor slot such as the *double cage* design have been developed to maximize this effect. Even when these designs are used, the starting current is typically 6 to 7 times the full-load rated current. This can create difficulties when a motor is to be started from a 'weak' supply with high internal impedance, causing a dip in the supply voltage which may affect other equipment and result in an unacceptably long run-up time for the motor.

The most common means of overcoming this problem for three-phase motors is to use a *star–delta start*, in which the stator windings are connected in *star* (sometimes called 'wye') for starting, and a timed contactor reconnects the windings to *delta* (sometimes called 'mesh') during run-up. The star and delta connections are shown in **Fig. 10.7**. The voltage appearing across each winding when in star is only 58 per cent of the full delta-connected voltage; the motor presents a higher impedance to the supply and the starting current (and, unfortunately, the torque) is limited to one-third of what it would have been in the delta connection. A second means of reducing starting current is a *soft starter*, which uses a simple device such as a triac or pairs

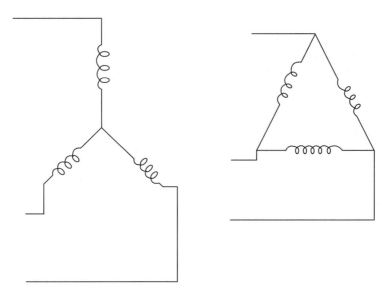

Fig. 10.7 Star and delta connections

of thyristors to delay the switching on of the voltage every cycle (see **section 11.4**), and this reduces the effective voltage applied to the motor during starting. High-voltage motors, which are usually connected in star for normal running in order to reduce the voltage across each winding and the level of insulation required, are often started through an autotransformer to reduce the motor voltage at starting.

When the motor has run up to speed on no load, the slip is very small, the torque produced being only just enough to overcome friction and windage losses. As the load on the motor is increased, the slip increases, the rotor current is increased and the torque is greater, reaching a peak at the *pull-out torque*. The pull-out torque is important because it determines the maximum temporary overload that the motor can withstand. The torque of an induction motor varies approximately as the square of voltage, so if the voltage drops to 90 per cent of its nominal value, the torque would be reduced to 81 per cent. For this reason it is important to ensure that the pull-out torque is adequate to cope with any short-term overloads even on lowest supply voltage.

The wound rotor variant was mentioned earlier, although it is now becoming relatively rare. It can be operated simply as a cage rotor, but the presence of a winding accessible to the user can be exploited in two main ways. First, a resistance can be added in series with each phase of the rotor winding. This resistance can comprise a group of fan-cooled resistors or a liquid resistor in the form of a tank of electrolyte into which electrodes are lowered. By using the maximum added resistance at standstill, the starting current is reduced to a minimum, and by selecting a particular value of resistance, pull-out torque can be achieved at standstill, with a good value of torque per amp. As the motor runs up, the resistance can be gradually reduced to zero, giving high efficiency at full speed. Secondly, the winding can be connected to a second supply which is able to inject currents to alter the torque–speed curve. Motors operated in this way are known as *doubly fed* motors and they were commonly used in speed-controlled drives. In recent times they have been supplanted by inverters feeding a standard cage rotor motor (see **section 10.4.4**).

It was noted above that the induction motor has to run at some level of slip even on no-load, since it has to supply losses to the rotor. If the rotor is driven faster, the slip will decrease to zero then become negative as the speed rises above synchronous speed and the machine will then naturally generate power back into the supply. This is a convenient way of braking an overhauling load, although the machine has a negative pull-out torque beyond which it cannot increase its braking torque and load control would then be lost.

10.3.1.1 Single-phase induction motors

Single-phase induction motors are common in domestic appliances such as refrigerators, freezers, fans and air conditioners. While they are necessary in situations where a three-phase supply is not available, they are typically limited to around 1 kW because of supply current limits and because of their inherent low efficiency and high torque ripple.

In **section 10.3.1**, reference was made to three-phase distributed windings in the stator and the way in which these windings produce a rotating field. In a single-phase motor the field pulsates, rather than rotates and it can be mathematically represented by two contra-rotating fields, each producing its own torque–speed curve, but in opposite directions, as shown in **Fig. 10.8**. The resultant field gives no starting torque so special arrangements have to be made for starting, and the motor has a much higher full-load slip (and hence lower efficiency) than the three-phase version.

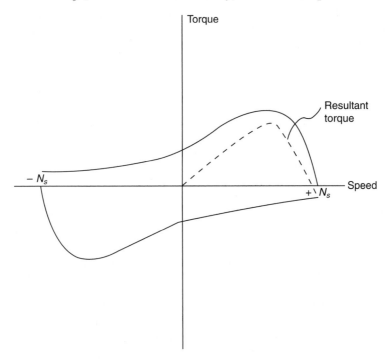

Fig. 10.8 Torque–speed characteristic of a single-phase induction motor

Single-phase motors are normally started by adapting them to be an approximation to a two-phase motor, in which two windings 90° apart in space around the bore of the machine have balanced emfs 90° apart in time applied to them. The 90° spacing between the windings is achieved by inserting the second winding (the *starting* or *auxiliary winding*) at the correct places in the stator, but providing balanced emfs 90° apart in time generally involves a greater approximation. There are four common arrangements, each having varying degrees of effectiveness.

In the *split-phase* or *resistance-split motor,* the starting winding uses fewer turns of a finer wire and so has a higher resistance and a lower reactance than the main winding. This results in the starting winding current leading the main winding current. The consequent phase difference is sufficient to provide reasonable starting torque. The starting winding is rated only for short periods because of its high current density and it is switched out when the motor reaches 60 to 70 per cent of full speed. Switching is usually done by a shaft-mounted centrifugal switch, although current-operated relays are sometimes used. The split-phase motor is best suited to infrequent starting with low-inertia loads, since its starting current is relatively high.

The *capacitor start motor* has a capacitor connected in series with the starting winding. The result is that the starting winding current leads the main winding current by a larger angle than in the split-phase case; this angle may approach 90° if a sufficiently large capacitance is used. A short-time rated electrolytic capacitor is normally used, and this is switched out of circuit when the motor reaches about 75 per cent of full speed. The capacitor start motor can deal with more frequent starting and higher inertia loads with higher starting torque such as pumps and compressors, and its starting current is lower.

In a *capacitor start and run motor*, a paper capacitor is connected permanently in series with the starting winding. The starting torque is low and this type is generally confined to fan drives, but running performance can approach that of a balanced two-phase motor if the capacitor is correctly chosen and it is generally quieter than a split-phase or capacitor start motor, with higher efficiency and power factor. These are also known as *Permanent Split Capacitor (PSC) motors.*

In the *capacitor start, capacitor run motor*, a large electrolytic capacitor is used for starting, but this is switched out before the motor reaches full speed and a smaller paper capacitor is in circuit for normal operation. There are alternative switching methods in which the paper capacitor is either permanently connected or switched into circuit when the electrolytic starting capacitor is switched out. In this way the good starting performance achieved with the large short-time rated electrolytic capacitor is combined with the good running performance achieved with the smaller paper capacitor.

Another variant of the single-phase induction motor is the *shaded pole motor.* These are to be found in sizes up to about 300 W output and comprise a standard cage rotor and a stator with a small number of salient poles, typically 2, 4 or 6. The poles each carry a coil, with the coils connected together to form the single main winding. At one side of each pole, near the air gap, a conducting ring is set into the lamination, the function of which is to distort the pulsating field and produce a crude approximation to a rotating field. Constructional variants abound, particularly in the so-called *uni-coil motors* using a single bobbin-wound coil to excite the magnetic circuit. Although very low in efficiency, these motors are often used for driving fans and pumps where an ac supply is readily available.

10.3.2 Synchronous motors

The construction of synchronous machines has been described in **Chapter 5** in relation to generators, so it is sufficient to note here that *wound field motors* are typically only found in the larger sizes of synchronous motor. However, where the wound field is replaced by a permanent magnet, the machine is (somewhat confusingly!) often called a *brushless dc machine* (see **section 10.3.4.2**). These abound in sizes below about 20 kW, particularly in small sizes where they are used, for example, in audio equipment and computer fan drives. Synchronous motors have higher efficiency than induction motors and for this reason are particularly found in the MW sizes in petrochemical and other pumping applications where operation is almost continuous.

All of these motors, whatever their size, share the characteristic that the rotor locks or *synchronizes* to the speed of the rotating field in the motor, so there is no variation of speed with load, unlike the induction motor. Various methods are used for starting and many of them use induction motor action to bring the motor to near synchronous speed, allowing it to lock onto the field. The synchronous running is exploited in applications where precise, constant speed is required, for instance in paper or textile making equipment.

10.3.3   Commutator motors

This category covers a variety of motors which share the feature of having a *commutator* mounted on the rotor shaft, to which the coils on the rotor are connected. **Figure 10.9** shows a typical commutator; it consists of copper bars (segments) set in insulation, with the ends of one or more coils connected to the riser portion. The surface of the commutator is accurately machined for concentricity and surface finish and carbon brushes provide sliding electrical connections. In all cases, the function of the commutator is to switch the polarity of one or more coils as the rotor rotates so that the direction of current in the coil is always correct with respect to the direction of the magnetic field, enabling electromagnetic torque to be produced in the desired direction.

Fig. 10.9 Commutator on a dc machine (courtesy of Invensys Brook Crompton)

10.3.3.1 AC commutator motors

These are now seldom produced, generally being variants of three-phase induction motors with a third or *tertiary* winding on the rotor. They were used in speed-controlled applications, particularly lifts and hoists, but are now superseded by inverter drives (see **section 10.4**).

10.3.3.2 DC motors

The dc machine, shown schematically in **Fig. 10.10**, is the classical commutator motor. The winding on the rotor is generally referred to as the *armature* and carries the main current. The magnetic field can be produced by a conventional winding, as shown in **Fig. 10.10**, which can either be supplied from a separate source (*separately excited*), connected in parallel across the armature supply (*shunt connected*), or connected in series with the armature to carry the same current (*series connected*). Alternatively, the field can be produced by permanent magnets housed in the stator, in which case the magnetic field strength cannot be easily varied.

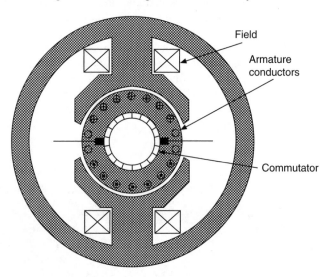

Fig. 10.10 DC motor in schematic form

The operation and control of the dc machine is, in principle, very simple. As explained in **section 10.2.1**, varying either the strength of the magnetic field or the magnitude of the armature current will directly vary the torque; varying the direction of either one will alter the direction of the torque. The ease of controlling the dc motor made it the obvious choice for controlled-speed drives before inverter-fed induction motors were available. It remains common in rail traction, steel rolling mills, winders, hoists and cranes, despite frequent forecasts of its demise.

The main disadvantage is the high cost of the machine and the life and reliability of the commutator and brushes, which also limit its operating speed. As the brushes wear, carbon debris is deposited on the winding insulation, potentially shortening its life. Overloads or fault currents can cause flashover on the commutator, often resulting in permanent damage to the surface of the commutator segments, necessitating a major overhaul.

Many of the developments in dc motors have concentrated on improving the commutation action, the best known being the introduction of *interpoles*. These are narrow poles situated between the main stator poles and carrying a winding with a few turns connected in series with the armature. The field of these interpoles is arranged to induce a motional emf in the coils undergoing commutation, thus enabling faster current reversal and preventing sparking.

Through-ventilated machines are the most common, with a fan driven either by the armature shaft or, more commonly, by a small auxiliary motor. Not only does this allow at least some of the carbon dust from brushes to be swept clear of the machine, it allows direct cooling of the armature winding and yields a higher output from a given motor size.

Before solid-state control became economic, series-connected dc motors were to be found in virtually every traction application. They exhibit high starting torque, with a falling torque which approaches zero at high speeds. By contrast, the shunt-connected machine operates substantially at constant speed, the speed being broadly set by armature voltage and its drop with load being relatively small. However, with the availability of modern controllers the separately excited machine is normally used; this can be programmed to give a range of torque–speed curves.

In small sizes, typically in automotive auxiliary drives for radiator fans and windshield wipers, the wound field is replaced by a cheaper permanent magnet, often using simple ferrite magnets. The dc supply is connected directly to the brushes, and reversal of direction is simply achieved by reversing this connection. Speed control is achieved by reducing the armature voltage by a variety of means.

10.3.3.3 Universal motors

These motors are so called because they will operate on either dc or single-phase ac supplies. This is because they are series connected, so the ac current reverses direction in the field and armature at the same time, leaving the direction of torque unchanged. Typically they are rated under 1.5 kW at several thousand revs/min. The highest speeds are normally found in vacuum cleaners (up to 25 000 rev/min is common) and the highest torques are in washing machines (up to 5 N m in horizontal axis models). They are normally of frameless construction to reduce cost. Being high-speed machines, the life of the carbon brushes is a major limitation (it may be only a few hundred hours in some cases) and commutator noise is often obtrusive.

Because of the high speeds, the specific power output of these machines is much higher than, say, induction motors running at 1500 or 3000 rev/min, so they are often used where space and/or weight is at a premium, with the final drive geared down from the motor shaft.

10.3.4 Permanent magnet motors

Motors with permanent magnets have been mentioned in two of the previous sections, illustrating the difficulty of using a simple classification method. Nevertheless it is worth summarizing here the types of motors with permanent magnets that are likely to be encountered.

Great care must always be exercised in working near or dismantling any permanent magnet machine. If the magnets are allowed to adhere to a surface suddenly there is a risk of the brittle magnet material chipping and firing out debris; the speed with which articles are attracted together often catches the user unaware and traps ends of fingers. In some cases removing the rotor from a machine can partially demagnetize the magnets. During manufacture or maintenance, metallic swarf tends to stick to the magnets, often causing a hazard.

The last two decades have seen huge strides in the quality and performance of magnetic materials (see **section 3.2.2**). The energy product of magnets now varies by an order of magnitude from the cheaper ferrites (typically used on cheap domestic appliance motors), through Alnico (traditionally used on loudspeakers) and samarium–cobalt to neodymium–iron–boron (see **Table 3.2** for a general comparison of properties). The range of materials now available to motor designers has enabled the performance boundaries of permanent magnet motors to be considerably extended.

10.3.4.1 Permanent magnets on the stator

In **section 10.3.3** it was noted that small dc motors sometimes use a permanent magnet instead of a winding to produce the field. These are usually referred to as *brushed permanent magnet* motors since the armature current is supplied through the carbon brushes as before. The motivation for adopting this construction is cost; in the smaller sizes it is more economical to use a magnet than a wound field. The stator normally comprises a steel shell, into which the blocks of magnet material (usually unmagnetized) are assembled. The blocks may be held in place mechanically or they may be bonded in place with a suitable adhesive. The armature and endframes are assembled and the motor is placed in a magnetizing fixture, where a very high pulse of magnetic field is applied to the motor to 'charge' the magnets. This system is amenable to volume production, but it can produce variable results in the motor performance due to differences in the field strength of the magnets.

Since the magnet field strength is approximately constant, the performance of these motors is similar to that of a separately excited motor supplied with constant field current and the speed regulation with load is relatively small. The speed is varied by controlling the armature voltage; this is done using external resistor(s) or by using a pair of brushes not diametrically opposite each other, or more commonly now by electronic control.

10.3.4.2 Permanent magnets on the rotor

In **section 10.3.2** it was noted that a synchronous motor may have the wound field on the rotor replaced by a permanent magnet system. This gives rise to two groups of motors, although in principle there is little difference between them.

The first group has a distributed stator winding embedded in a large number of slots, like an induction motor, and often has the rotor magnets embedded in the rotor core. When the stator is supplied with balanced polyphase voltages, the rotor field locks onto the resultant rotating field and synchronous running is achieved. Sometimes a rudimentary cage winding is also provided for starting the motor, or the frequency of the supply voltages can be linked to shaft position by providing rotor position feedback to the controller. These motors have a very low rotor loss and an overall efficiency which is significantly higher than induction motors. They have attracted much academic interest over the past few years but, in spite of their apparent promise, usage is limited and shows little sign of growth.

The second group is much more significant. Here the magnets are usually bonded to the surface of the rotor core and, particularly in smaller sizes, the stator has only a few slots. This is an inverted form of the brushed dc commutator motor described in **section 10.3.4.1**, since we now have stationary windings and a rotating magnet.

Instead of the mechanical commutator, the currents are electronically switched (commutated) at the correct moments by using position feedback from the rotor (often from the rotor magnets themselves). This class of motor is almost always described as the *brushless dc motor*. It is to be found in small sizes in medical and computer equipment and in larger sizes (typically up to 10 kW) in servo drives where absolute shaft position and shaft speed are of interest, for instance in weaving machines where many shafts have to move in precise relationship to each other.

10.3.5   Switched reluctance motors

These were known in primitive form from around the 1830s. However, since there was no convenient method of switching the currents in the inductive windings, they were overtaken by the advent of good quality commutators for dc motors and by the invention of the induction motor. In the 1970s the maturing capability of the power semiconductor rekindled interest in this separate class of machine which produces torque purely by reluctance action (see **section 10.2.1**). After intensive development at the academic level, products have been developed commercially and drives based on these motors are considered by many commentators to be at least the equal of drives based on motors producing electromagnetic torque.

The motors are characterized by having clearly visible poles on both stator and rotor (hence they are often described as *doubly salient*), but they only have windings on the stator. They are supplied not from a sinusoidal supply but from a dc voltage which is electronically switched to the appropriate winding, giving an essentially triangular phase flux. While it is possible to operate without rotor position feedback (so-called *open loop*), operation with simple position feedback (*rotor position switched*) is normal, and it is not possible to operate without an electronic drive system. Since the motor is supplied from a dc source, the number of independent phase circuits is a choice of the designer; systems with one to four phases are common and five phases or more have been seen in specialist applications. Similarly, the number of stator poles is flexible; 2 poles per phase gives a 2-pole field pattern (**Fig. 10.11(a)**) and 4 poles per phase gives a 4-pole pattern (**Fig. 10.11(b)**). The number of rotor poles is also variable. Using a number of rotor poles two different from the stator poles gives vernier action; two less makes the rotor move against the rotation of the field and two more makes it move with it. A common choice is the machine with 6 stator poles and 4 rotor poles shown in **Fig 10.11(a)**.

Torque is developed by the tendency for the magnetic circuit to adopt a configuration of minimum reluctance, that is for a pair of rotor poles to be pulled into alignment with an excited pair of stator poles, maximizing the inductance of the exciting coils. Continuous rotation in either direction is assured by switching the phases in the appropriate sequence, so that torque is developed continuously. In simple terms, the larger the current supplied to the coils, the greater the torque, although the design and analysis of these machines is complex because the magnetic circuit is generally operated above its linear region. The torque is independent of the direction of current flow, so unidirectional currents can be used. This permits a simplification of the electronic switching circuits compared with those required for most other forms of motor.

The motors are generally operated in *chopping mode* or in *single-pulse mode*. For the motor shown in **Fig. 10.11(a)** with each phase supplied by a switching circuit as shown in **Fig. 10.12,** current is established in a phase winding by connecting it to the dc supply by closing the switches S_1 and S_2 when the rotor poles are not aligned with

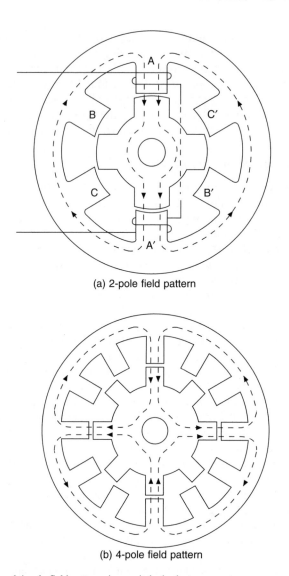

(a) 2-pole field pattern

(b) 4-pole field pattern

Fig. 10.11 2-pole and 4-pole field patterns in a switched reluctance motor

that phase. The current rises rapidly to the desired level as shown in **Fig. 10.13(a)**. S_1 and S_2 are now opened and the stored energy in the magnetic field ensures that the current continues to flow through D_1 and D_2. The voltage now impressed on the winding is negative, driving the current down. During this time some magnetic energy is being returned to the supply and, while the inductance continues to rise, some is being converted into mechanical output. When a lower specified current level is reached, S_1 and S_2 are closed and the current rises again. At the end of the required phase conduction period when the rotor poles are aligned with the stator poles, both switches are turned off and the current falls to zero. Torque is controlled by varying the level at which the current is chopped. This is only one of a number of methods of chopping control.

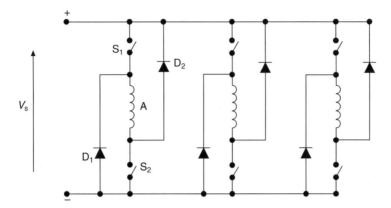

Fig. 10.12 Power converter for a switched reluctance motor

At higher speeds, the rise and fall times for the current will be such that the current is switched on and off only once in each conduction period and is never chopped, as shown in **Fig. 10.13(b)**. This is the single-pulse mode of operation in which torque is controlled through the switching angles.

Because of the flexibility of control of switched reluctance machines, their performance characteristics can be tailored to suit a wide range of applications. **Figure 10.14** shows motors developed for automatic door openers, rotary screw compressors and mining conveyors. Their operation is characterized by an ability to operate over very wide speed ranges, developing high efficiency over a wide range of both torque and speed. They can also be arranged to give extremely high overload torques in both motoring and braking.

The switched reluctance motor and controller are extremely robust. The motor has simple coils, with small endwindings and no overlapping of phase windings, and the rotor has no coils or magnets. The power converter has to supply only unidirectional currents and the maximum rate of rise of switch current is limited by the stator coils, thus avoiding the possibility of 'shoot-through' faults. More details on the design and control of these machines can be found in **Reference 10B**.

10.3.6 Stepper motors

Stepper motors are often considered as a separate class of machines because of the way in which they are operated, although constructionally they are similar to the other types discussed above and they produce torque by reluctance action. However, they are supplied from a source of discrete pulses, in response to which they move or '*step*' to a new angular position which is retained until the next pulse is sent. They are positioning devices rather than variable speed motors, although they are sometimes run in variable speed mode by simply increasing the pulse frequency so that they appear to move continuously. They are entirely dependent on the driving electronics, so they must be considered as part of a system, as shown in **Fig. 10.15**.

Two types are commonly encountered:

- a *variable reluctance stepper* produces torque purely by reluctance action and normally is constructed with a relatively small number of poles giving relatively large step angles. For instance a 6-stator/4-rotor pole machine will have a step angle of 30° giving 12 steps per rev. Switched reluctance machines are sometimes

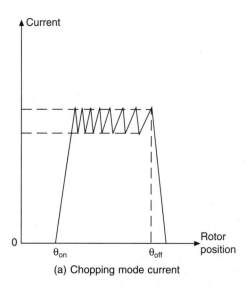

(a) Chopping mode current

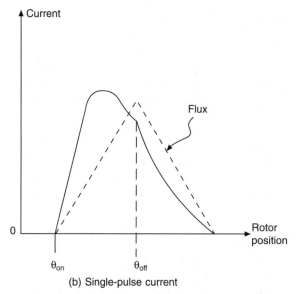

(b) Single-pulse current

Fig. 10.13 Typical chopping and single-pulse currents in a switched reluctance motor

thought of as large variable reluctance steppers because the construction is the same. These are normally found in the power range 0.1 kW to 2 kW.

- a *hybrid stepper* has on the rotor an axially magnetized permanent magnet, and it normally has a large number of rotor teeth. The stator poles are divided at the air gap to have several teeth per pole. These machines have relatively small step angles. A typical small hybrid stepper might have 8 stator poles each with 5 teeth, and a rotor with 50 teeth, giving a step angle of 1.8° or 200 steps/rev. The permanent magnet gives a *detent* or holding torque in the absence of any excitation on the stator poles. Hybrid steppers are found in disc drives, processing machinery and handling equipment.

Fig. 10.14 Selection of SR Drives (photo by courtesy of Switched Reluctance Drives Ltd)

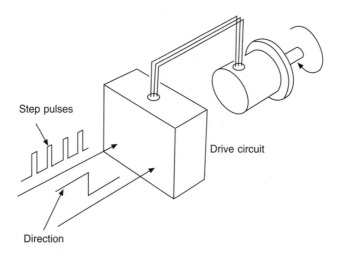

Fig. 10.15 Stepper motor system

There are many ways of operating steppers. The simplest is the *single-step* mode where a pulse is sent to a phase winding and the machine takes one step to the new detent position. By equally exciting two adjacent phases, *half-stepping* can be achieved, where the rotor takes up a position midway between the single step positions of the two phases. For very fine resolution such as for driving a print head in a printer, the currents in the phases are carefully controlled to be unequal, giving a mode known as *step division, mini stepping* or *micro stepping*. More complex schemes exist where a position transducer is used to give feedback, rather than relying on the motor to move to the correct position on demand. The transient behaviour of the rotor is of great importance during stepping and further details on this can be found in Chapter 8 of **reference 10A. Reference 10D** covers the entire subject in some detail.

10.4 Variable speed drives

Chapter 11 covers the details of different types of power electronic circuits which can be used in conjunction with the motors described in **section 10.3** to provide speed control and position control of the rotor shaft. A more detailed treatment can be found in Chapter 36 of **reference 10D**.

The concept of *quadrants* in the torque–speed plane is crucial to the specification for a variable-speed drive. A drive operating with positive torque and positive speed (or speed in the direction which the user defines as forwards) is said to be operating in the *first quadrant* of the torque–speed plane, as shown in **Fig. 10.16**. If the torque is reversed, braking or generating action takes place and power is extracted from the load. This is *second quadrant* operation. Many applications require the drive to operate in both motoring and braking modes; these are divided into drives that

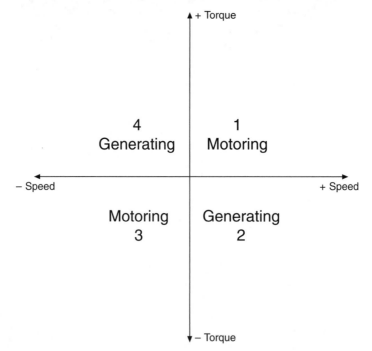

Fig. 10.16 Quadrants of operation

simply dissipate the regenerated power in resistors and drives that return power to the ac supply by a dedicated inverter. If the speed falls to zero under the braking action and the torque is maintained, the *third quadrant* will be entered in which both speed and torque are negative and the drive is motoring in the reverse direction. By reversing the torque once more, braking is achieved as the drive moves into the *fourth quadrant*.

Basic drive systems only operate in the first and third quadrants; the more sophisticated systems will operate in all four. The speed and smoothness of the transition between quadrant is a significant test of the quality of the drive; dc motor drives have excelled in this respect and much of the development of ac motor drives has been targeted at improving this aspect.

10.4.1   DC motor drives

These are probably the simplest motors to control, since separate control over field excitation and armature current is normally available. Typically the field is supplied through a single-phase controlled rectifier and the armature through a three-phase controlled rectifier. **Reference 10A** (especially Chapter 4) gives a full description of the operation and also discusses the control aspects of these drives.

10.4.2   AC motor drives

Section 10.3.1 explained the dependence of the speed of the induction motor on supply frequency (see **eqn 10.2**). It follows that by supplying the motor from a suitable source of variable frequency, the synchronous speed of the motor can be varied. For correct operation, the ratio of supply voltage to frequency also has to be kept sensibly constant (see Chapter 7 of **reference 10A**) so the voltage also has to vary with the flux. This gives rise to the use of a frequency inverter (see **section 11.6**) for controlling an ac induction motor. PWM inverters are now by far the most common in small and medium sizes, using the high switching speeds of IGBTs or MOSFETs; older designs of 6-step inverters are still found in very large sizes. A full discussion of different circuit topologies is given in Chapter 36 of **reference 10C**.

In recent years, much effort has gone into improving the transient performance of inverter drives in an attempt to take market share from dc drives where high control bandwidth is required. This has led to the development of *vector control, field-oriented control* and *direct torque control* by different manufacturers, all of which control the position of the rotor current in relation to the position of the rotating flux wave in the air gap. Such systems require rotor position feedback, which can be derived by hardware such as a shaft encoder, or software such as an observer.

In larger sizes, and commonly above 1 MW, synchronous machines are preferred to induction motors. These are normally driven by a different style of thyristor inverter, which uses the back emf of the machine itself to commutate the thyristors.

10.4.3   Switched reluctance drives

The switched reluctance motor cannot be operated without its power converter. While it is possible to operate it from a variable frequency inverter, with appropriate changes to the control, such complexity is not required. The most common configuration for the power converter is shown in **Fig. 10.12**, although many other variants exist, many of them offering a reduced number of switches at the expense of reduced control flexibility. In all cases, the power switches are in series with a phase winding, offering a number of advantages to the converter designer, not least the possibility of reduced switch size. Switching speeds are normally lower than in ac inverter drives, since there is no requirement for PWM at high speeds to synthesize a sinusoidal voltage.

10.4.4 Commissioning of drives

Most commercially available drives have control systems which allow the drive to be tailored to suit a variety of applications. This is especially true of general purpose drives sold from a catalogue; these generally have to be tuned to suit the parameters of the load, particularly if transient performance is an issue. Some systems have a degree of self-tuition and simply require the execution of a set-up program to allow the system to test the load so that correct responses can be stored for future operation. With the improvements accruing from cheap on-board signal processing, most systems now have comprehensive and intelligible fault signalling ability, and setting up new drives is now much simpler than with first-generation drives.

10.5 Ratings, standards and testing

National and international standards exist for specifying the rating and performance of several types of motor. Many of these also specify the test methods to be used, although these tests are often complex and can be done only by the manufacturer or a specialized independent test site. Contract testing can be undertaken by universities or consulting organizations. In the UK, Nottingham University Electrical Drives Centre is one of the best academic centres for testing machines and drives of all types on a contract basis, having testing capacity up to 750 kW and up to 15 000 rev/min at low power. Where the motor is a component in a drive system, the situation is much less clear, since there are few standards, other than EMC and Harmonic Limits (see **Chapters 11** and **15**) which apply to complete systems. For free circulation within the European market, CE marking under the Low Voltage and EMC Directives is required.

Table 10.1 indicates the most commonly encountered standards, with approximate national and international equivalents where appropriate. In spite of ongoing attempts at harmonization, there remains in some cases no simple correspondence between these standards and further advice (for instance from catalogues of national standards) should be taken for details of the equivalence of individual parts.

For motors other than standard induction motors, and particularly for drives, it is usually difficult to find a standard which is entirely relevant. In these cases, a detailed specification drawn up between the supplier and the customer is generally preferred, referring where appropriate to particular parts of other standards to cover specific aspects of construction and performance.

Table 10.1 National and international standards for electric motors and drives

Construction and dimensions:	BS:	4999, especially parts 105, 141 and 147
		5000 part 10
		EN 60035-6 and EN 60035-7
		EN 60529
	IEC:	60034-5
		60072
		60136
	NEMA	MG-13
Performance and rating:	BS:	4999 part 102
		5000 part 10
		EN 60034-1 and EN 60034-12
	IEC:	60034-2
		60072
	NEMA:	MG-10
Noise and vibration:	BS:	EN 60034-9 and EN 60034-13
		EN 60704
	IEC:	60034-2
		60072
	NEMA:	MG-3
Safety:	BS:	3456 (many parts now superseded)
		EN 60335
		5345
		5501 (potentially explosive atmospheres)
	IEC:	60335
		60079
	UL:	1004
	NEMA:	MG-2
EMC and harmonics:	BS:	EN 50081
		EN 50082
		EN 55014
		EN 55104
		EN 61800
	IEC:	1000
		50081
		50082
		55014
		55104
		61800

References

10A.   Hughes, A. *Electric Motors and Drives* (2nd edn), Butterworth-Heinemann, 1993.

10B.   Miller, T.J.E. Switched Reluctance Motors and Their Control, Oxford University Press, *Monographs in Electrical and Electronic Engineering,* No. 31, 0-19-859387-2.

10C.   Jones, G.R., Laughton, M.A. and Say, M.G. (eds), *Electrical Engineer's Reference Book* (15th edn), Butterworth-Heinemann, 1993.

10D.   Acarnley, P.P. *Stepping Motors: A Guide to Modern Theory and Practice* (2nd edn), Peter Peregrinus, 1984.

Chapter 11

Static power supplies

Dr C.D. Manning
Loughborough University

11.1 Scope

The primary function of a power supply is to draw electrical power from an existing electrical system and convert this power into a form suitable for a particular load. A wide variety of power supplies is used to meet the range of loads presented by common electrical equipment. This chapter covers the common power conversion circuits that use semiconductor switches. These are commonly known as static power supplies, and the overall requirements which influence their design and construction are described here. Systems that use rotating machines or which control the flow of power using metal-contact switches have been common in the past but are increasingly restricted to specialized applications; they are not covered.

The power that is handled by static power supplies can range from a few milliwatts, supplying low-voltage integrated circuits, to gigawatts in high-voltage dc links. This chapter concentrates on equipment of less than a few tens of kilowatts drawing ac power from a standard utility supply; this represents the majority of static power supplies manufactured and in use today.

11.2 General issues

11.2.1 Design optimization

The majority of static power supplies are manufactured to compete in markets where the purchase price is often the primary concern of the customer, given a minimum electrical specification and a basic level of reliability.

Other features such as size, weight, efficiency and acoustic noise are often secondary issues, although in some battery-charging applications efficiency may be critical and in aircraft weight is of paramount importance.

11.2.2 Thermal management

In static power supplies a significant proportion of the input power is dissipated as heat, and adequate cooling must be provided in order to limit the temperatures reached by the circuit components and to avoid damaging them. A maximum ambient temperature of 40°C to 60°C is often specified, and safe component temperatures must be maintained in these conditions. Component temperatures may be up to 70°C for resistors, 85°C or 105°C for capacitors, 90°C to 110°C for heat sinks, 100°C for wound components (inductors and transformers) and 120°C for power semiconductors.

At the lowest power levels, heat is removed from the components by natural convection and radiation. At higher power levels fan cooling is used, air speeds being typically in the range 0.5 to 5 m/s. Liquid cooling is normally used at much higher power levels than 10 to 20 kW.

11.2.3 Safety

Voltages in static power supplies are often at lethal levels (in excess of 55 V relative to earth). Prevention from electrocution is therefore necessary. Measures which are used to meet the requirements set down in standards are:

- the use of earthed metal enclosures in which any openings are below a specified size
- the referencing of output voltages to earth, which often requires the use of an isolation transformer
- design so that the insulation materials used to achieve safety isolation are maintained within their rated maximum temperatures

Protection against fire is also an important consideration and an input fuse is used to prevent sustained overheating and fire in the case of component failure.

11.2.4 Compatibility

Static power supplies must deal with various levels of disturbance in the utility supply (or other supplies such as standby generators and systems powered by renewable energy sources) and still maintain the required output to the load. The main disturbances to the utility supply which must be dealt with are:

- *blackout (outage)* – absence of utility supply for a cycle or more; this is often due to faults on the electrical supply system, and the operation of system protection equipment
- *brownout (sags)* – the utility supply voltage falls below the normal minimum rms level for more than one cycle. This is often due to the switching on of large loads on the utility system
- *overvoltage* – the utility supply voltage rises above the normal maximum rms level for more than one cycle; this is often due to the switching off of large loads on the utility system
- *voltage spikes* – high voltages of short duration, superimposed on the utility supply waveform; this is often caused by switching of inductive loads or by lightning strikes in the utility system
- *harmonic currents* – distortion of the sinusoidal supply waveform; this is due to harmonic currents (for instance from rectifiers, dc–dc converters and arc welding) flowing through the utility system impedances
- *Radio Frequency Interference (RFI)* – these arise typically from high-speed switching transients and from arc welding

Filters are commonly used at the input stage of static power supplies to protect against or limit the effect of these disturbances, and uninterruptible power supplies are used to protect against blackouts and brownouts.

11.2.5   Commissioning, testing, installation and maintenance

Commissioning of static power supplies is usually necessary only for equipment which is rated at tens of kilowatts or more. Below this power level, in the range which this chapter addresses, equipment is normally fully tested and set up by the manufacturer before despatch to the customer.

11.3   Supply-frequency diode rectifiers

Rectifiers convert ac to dc. Full-wave and half-wave rectifier circuits are available for both single-phase and three-phase utility supplies, and the choice usually depends upon the power to be delivered to the dc load.

The diodes used in rectifiers are often integrated into one component, which may be mounted on a heat sink. The need for a suitable safety margin often results in the use of 800 V-rated diodes for 230 V single-phase direct off-line power supplies, and 1200 V rating for 415 V three-phase direct off-line applications.

High-frequency filter capacitors are often connected in parallel with rectifier diodes. These capacitors protect the diodes from transient voltage spikes by acting as a low-impedance path to high-frequency voltages.

It is common to obtain the required output voltage using a transformer which is connected between the ac supply input and the rectifier. The use of an isolating transformer not only allows the output voltage to be set, but by isolating the rectifier circuit from the ac supply it also enables one dc output voltage rail to be referenced to earth. The earthing of a low-voltage output connection is necessary in many applications in order to prevent injury by electrocution.

11.3.1   Half-wave single-phase rectifiers

For low power levels, a single-phase half-wave circuit with a large dc output capacitor to reduce voltage ripple is normally used. The circuit is shown in **Fig. 11.1**. When the ac supply voltage $V_S(t)$ exceeds the voltage $V_L(t)$ at the output of the rectifier, the voltage difference causes a supply current $i_d(t)$ to flow. The rate of rise of this current is limited mainly by the supply inductance L_S. As the capacitance C is charged, $V_L(t)$ rises. When the supply voltage $V_S(t)$ decreases after passing its peak, it falls below $V_L(t)$ and the supply current $i_d(t)$ falls to zero. During the period when $i_d(t)$ is zero, $V_L(t)$ falls as the capacitor discharges. The decay in this voltage is approximately linear, but it will be exponential if the load is resistive and falls with an increasing rate when supplying constant power to a switching regulator.

When the circuit is first energized, the capacitor is initially fully discharged and there is a large surge of supply current as it is charged. To limit this surge to an acceptable level, thermistors with a negative temperature coefficient are often connected in the ac circuit. The resistance of the thermistor is initially high and the current is limited by this to the required level, but after a short period of conduction the thermistor temperature rises and its resistance drops, reducing the power loss and voltage drop in normal operation. This thermal characteristic requires a minimum time between the ac supply being switched off and reconnected in order that the thermistor temperature can fall sufficiently to limit the surge current on reconnection to an acceptable level. For applications requiring maximum efficiency, the thermistor may be replaced by a resistor connected in parallel with a relay contact or a semiconductor switch which closes after the initial charging surge.

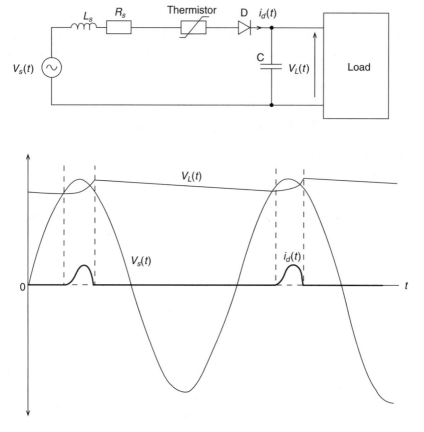

Fig. 11.1 Half-wave single-phase rectifier

11.3.2 Full-wave single-phase rectifiers

The three main circuits used are illustrated in **Fig. 11.2**.

The bridge rectifier circuit in **Fig. 11.2(a)** allows current to flow through diodes D1 and D3 when the supply voltage $V_S(t)$ exceeds the load voltage, and through D2 and D4 when $V_S(t)$ is more negative than the load voltage is positive. Supply current therefore flows in both polarities. Ideally this avoids a dc component of supply current and reduces the peak supply current levels. The rectifier delivers power during each half-cycle of the ac supply and this reduces the size of the capacitor required to limit the output voltage ripple.

The voltage-doubler rectifier shown in **Fig. 11.2(b)** can be regarded as two half-wave rectifiers, with D1 conducting during positive half-cycles of the ac supply and D2 conducting during negative half-cycles.

The dual-voltage rectifier of **Fig. 11.2(c)** has a link S which can be used to configure the rectifier as a bridge or as a voltage doubler. This arrangement is commonly used to deliver approximately the same dc voltage using either bridge rectification of 230 V ac or voltage-doubler rectification of 110 V ac. In this way, by changing the link S, the rectifier can be configured for use in most countries.

11.3.3 Three-phase bridge rectifiers

Three-phase bridge rectifiers are typically used above 1 kW to provide a single dc

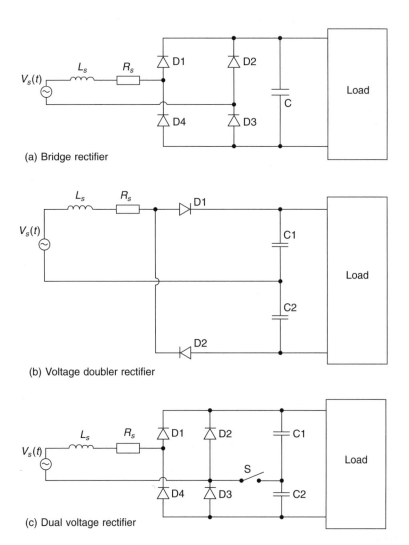

(a) Bridge rectifier

(b) Voltage doubler rectifier

(c) Dual voltage rectifier

Fig. 11.2 Full-wave single-phase rectifiers

supply by rectifying the line-to-line voltage. The scheme is illustrated in **Fig. 11.3(b)**. For completeness **Fig. 11.3(a)** shows the half-wave circuit, but this is not commonly used. The bridge rectifier produces six pulses of power near the peaks of the line-to-line voltages. It is common for the load of a three-phase rectifier to include a large inductance (shown as L_f) to create a dc current source. **Figure 11.3(c)** shows typical waveforms and it illustrates the time required for the supply current to rise and fall from the level of rectifier output current.

For high power outputs, two three-phase supplies are used. These have a phase difference of 30° and the supply is rectified to provide 12 pulses per cycle. A centre-tapped inductor is connected between the negative dc outputs of the two six-pulse rectifiers, and the negative supply to the load is connected from the centre tap of this inductor in order to limit sixth harmonic (300 Hz) currents. The main advantage of 12-pulse, compared with six-pulse, rectifiers is that the filtering requirements are reduced; the primary current waveforms have no fifth or seventh harmonic components,

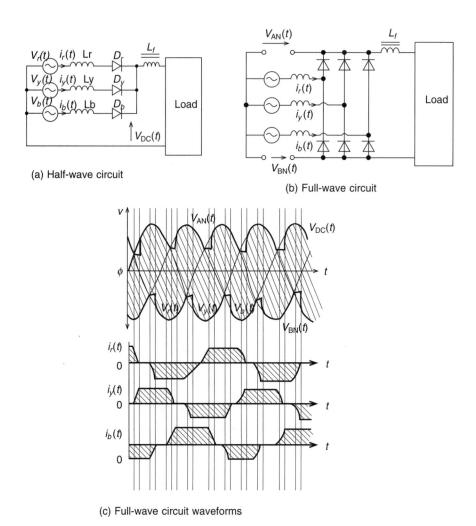

(a) Half-wave circuit

(b) Full-wave circuit

(c) Full-wave circuit waveforms

Fig. 11.3 Three-phase diode rectifiers

which reduce the filtering required in the ac supply, and the dc output has more pulses, which reduce the ripple contained in the dc output.

11.3.4 Voltage multipliers

These are rectifier circuits in which diodes and capacitors are used to deliver an output voltage which is an integral multiple of the peak ac input voltage. Different circuits are possible and a voltage doubler has already been shown in **Fig. 11.2(b)**.

A common circuit is shown in **Fig. 11.4** and the operation is:

- C1 charges to the negative peak supply voltage through D1
- when the ac source reverses, the voltage across C1 adds to the source voltage, and C2 is charged through D2 to twice the peak source voltage
- C3 and C4 charge in a similar way to C1 and C2, except that both these capacitors charge to twice the peak ac voltage

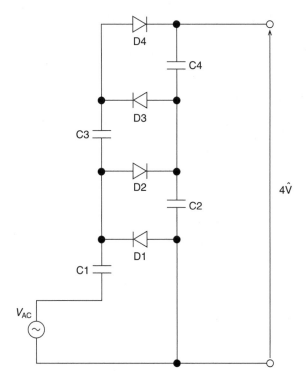

Fig. 11.4 Voltage multiplier

Further stages can be added to increase the output voltage, although the output voltage will not be the full multiple of the input voltage because of voltage drops in the components.

Voltage multipliers are suited to low-current high-voltage supplies, and they are rarely used for supplying more than a few milliamps.

11.4 Phase-controlled power converters

Phase-controlled power converters use thyristors, which begin to conduct when turned on by a gate signal. If the load current exceeds a 'latching current' level, a thyristor will latch in the conducting state and it will continue to conduct, irrespective of the gate signal. When the load current falls below a relatively small 'holding current' the latching action will stop and the thyristor will turn off provided the gate signal is in the off state. Power flow is controlled by timing the instant of turn-on of the thyristor to a set phase angle in the ac waveform.

Detection of ac supply zero-voltage crossings is used to ensure that the thyristor control signals are timed in correct phase with the supply voltage waveform. A ramp and a dc control voltage are taken as input to a comparator, which delivers an 'enable' signal to an oscillator. The oscillator output is amplified and passed through a pulse transformer (for electrical isolation) and through a current-limiting resistor to the thyristor terminals. A train of pulses is applied between gate and cathode terminals of the thyristor to minimize the size of the isolation transformer whilst ensuring that the thyristor cannot turn off for long during the conduction period set by the control circuit.

The switching of thyristors often requires snubbers. These prevent damage to the thyristors by limiting overvoltages, the rate of rise of voltage across the thyristor and the rate of change of thyristor current. A wide variety of snubber circuits are commonly used.

11.4.1 AC power controllers

Circuits for the control of single-phase ac power using phase-controlled thyristors are illustrated in **Fig. 11.5**. Anti-parallel thyristors (**Fig. 11.5(c)**), or a thyristor with an anti-parallel diode (**Fig. 11.5(b)**), or simply a triac (**Fig. 11.5(a)**) can be used.

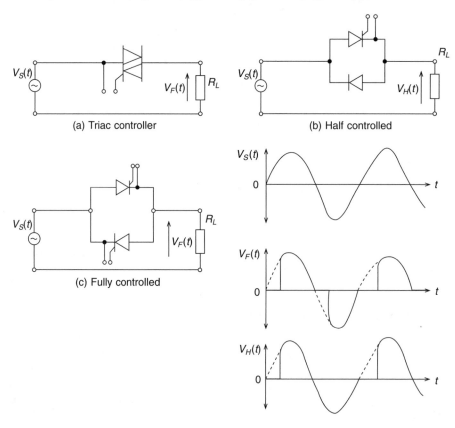

Fig. 11.5 Single-phase phase-controlled ac regulators

The resulting current waveform depends upon the load impedance, which is often resistive, for instance in the case of dimmers used for incandescent lights.

11.4.2 Phase-controlled rectifiers and inverters

A phase-controlled inverter uses the same circuitry as a phase-controlled rectifier, but with a different timing for the thyristor drive signal. The circuit is shown in **Fig. 11.6(a)**. The basic three-phase bridge arrangement is similar to that used in the diode rectifier circuit already seen in **Fig. 11.3(b)**; the main difference is that using the thyristors, the start of current conduction can be delayed for a controlled period after the zero-voltage crossing of the anode-to-cathode voltage.

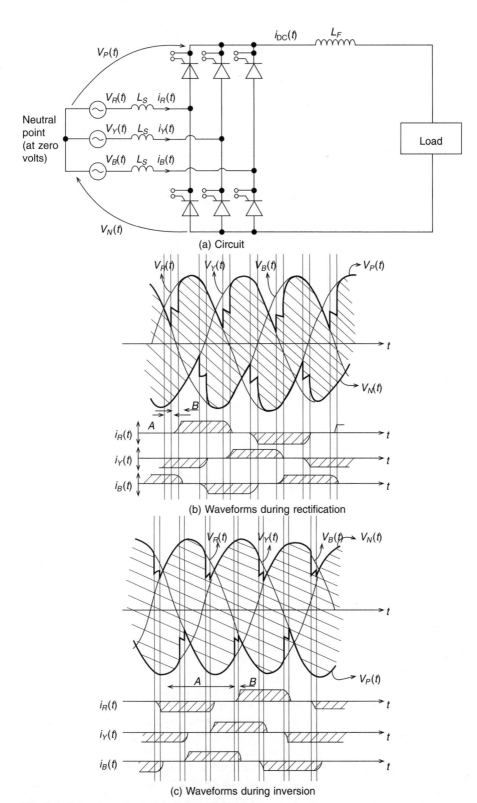

Fig. 11.6 Three-phase phase-controlled rectifier/inverter

Phase-controlled rectifiers and inverters are normally used at high power levels with three-phase supplies. A large inductance L_f is connected in series with the dc output in order to ensure a substantially ripple-free dc current.

The waveforms in a three-phase rectifier with constant-current output are shown in **Fig. 11.6(b)**. The conduction of each thyristor is delayed beyond the instant that a diode would conduct by a delay angle A. When conduction has begun, the rectifier output takes a time, represented by angle B to commutate from the previously conducting phase. This is illustrated by the gradual decay in the outgoing phase current $i_B(t)$ and the corresponding rise in the incoming phase current $i_t(t)$.

11.5   Linear dc voltage regulators

Linear regulators use a bipolar transistor connected in series with the load, and operating in the linear mode, in order to regulate the load voltage by controlling the transistor voltage drop. Basic schemes are shown in **Fig. 11.7**.

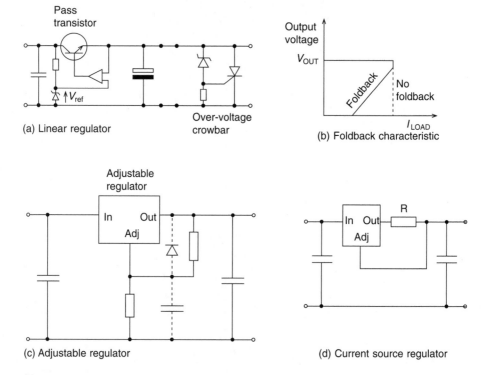

Fig. 11.7 Linear regulator schemes

Linear regulators are often implemented in a single three-terminal Integrated Circuit (IC), which requires only the connection of input and output decoupling capacitors and mounting on a heat sink. Fixed output voltages at standard values are delivered by regulators such as the 78XX series of positive voltage regulators and the 79XX series of negative voltage regulators. These regulator ICs usually contain additional features such as overtemperature and overcurrent protection circuits.

In cases where the heat dissipation in the internal pass transistor is high, a separate transistor is often used instead of an IC. This enables the control circuits to operate at an acceptably lower temperature while allowing a high temperature in the pass transistor itself. In such applications, the dissipation in the pass transistor, which is proportional to its voltage drop, is often minimized by the use of a 'low dropout' regulator. Standard linear regulators require the input voltage to be a minimum of 3 V higher than the output voltage for correct operation of the control circuitry. 'Low dropout' regulators operate satisfactorily with minimum voltage drop which may be less than 1 volt.

Output overvoltage protection is often used to safeguard the load. This protection works by short-circuiting the regulator output in the event of an overvoltage; the 'crowbar' for achieving this is shown in **Fig. 11.7(a)**. The resulting short-circuit current operates either the current-limit protection within the regulator or the input fuse. If the load current were limited by the regulator to its maximum value, then the pass transistor would dissipate maximum power with the output short-circuited, when the voltage across the regulator is also a maximum. Such high power dissipation would usually exceed the maximum acceptable for normal operation, and it would increase the required cooling. To avoid the need for this increased cooling, the output current is often reduced (or 'folded back') in cases where the load impedance is less than the full load impedance. A foldback characteristic is shown in **Fig. 11.7(b)**.

Adjustable regulators include an external resistive voltage divider to set the output voltage, as shown in **Fig. 11.7(c)**. A common IC used in adjustable regulators is the LM317.

Micropower linear regulators for use in battery-powered systems are optimized for very low quiescent current (the current that supplies the regulator control circuitry).

Dual-tracking regulators deliver equal positive and negative dc voltages. This is achieved by arranging that the pass transistor on the negative rail tracks the magnitude of the regulated output of the positive regulator.

11.6 Inverters

Although phase-controlled inverters require an existing ac system into which their output is delivered, a variety of other inverter types are used in order to create an ac supply-frequency power source.

In some applications, a dc source may be inverted to a square wave output by the use of four transistors which connect the dc input to the load in alternating polarity at the required frequency. This type of system is simple, efficient and has low noise; it also enables the use of slow switches for low-frequency applications.

Resonant schemes apply a square wave at the required output frequency to a resonant circuit which may be connected in series, in parallel or in series–parallel. The resonant circuit has a natural frequency which is equal to the required output frequency; it generates an approximate sine wave from the square wave, and acts as a low-impedance source.

Linear amplifiers are used to amplify a reference sine wave to the required ac output voltage, the output oscillating between positive and negative dc voltage rails. The linear amplifier has a low source impedance, and it offers low noise generation because of the avoidance of switching transients, but there are high power losses in the amplifier transistors.

Pulse-Width Modulated (PWM) inverters use high-frequency switching of the dc source voltage, in which the pulse width is proportional to the ac waveform amplitude.

Typically, a full bridge of transistors is controlled in such a way that a sinusoidal output voltage is achieved once switching frequency components have been filtered out. In comparison with linear inverters, PWM types are small, have higher efficiency and cope well with non-linear loads, but they generate substantial high-frequency noise.

11.7 Switching dc voltage regulators (dc to dc converters)

DC to dc converters are widely used for the regulation of dc voltage, and they are used in *Switched-Mode Power Supplies (SMPS)*. They are commonly treated as a single power converter, although isolated versions use a controlled inverter and a rectifier. The circuitry differs significantly from that used in AC supply-frequency power conversion.

Switching regulators use the duty cycle of a switch to regulate a voltage. The switch has low power loss because it alternates between an off-state, during which a negligible leakage current flows through the switch, and an on-state, during which the switch conduction voltage is low. The switched waveforms are averaged using low-pass filters which consist of capacitors and inductors. The size of these filter components reduces with increasing frequency, but the switching loss increases in proportion with frequency; typical commercial equipment operates at a switching frequency in the range 100–500 kHz.

The two types which are commonly used without isolation are the *buck converter* and the *boost converter*. These are shown in **Figs 11.8** and **11.9** respectively. The converters are said to operate in 'continuous mode' when the induction current $i_L(t)$ is continuous, and in the 'discontinuous mode' when the $i_L(t)$ falls to zero for a part of each cycle.

Buck converters deliver an output voltage which is less than the input voltage. The dc input is switched using S with a pulse width that is controlled in order that the average output voltage (filtered using the inductor L and capacitor C) has an accurately controlled dc value. Suitable ICs are available. These incorporate most of the components and simplify the design.

Boost converters deliver an output voltage which exceeds the input voltage. In this case the switch S connects inductor L across the supply voltage, ramping up the inductor current; when S is opened, the current is diverted through the diode D to the output, and the current is reduced linearly.

If electrical isolation between input and output is required, a wide variety of converters are possible. The introduction of a transformer into the circuit for electrical isolation creates a need to limit the dc magnetizing current in the transformer in order to minimize core saturation and avoid the consequent damage to components because of excessive currents. The most common types of circuit are the *flyback converter* and the *buck converter*. The flyback converter is used for low power levels and the buck converter is used generally for outputs higher than about 200 W, where the size and cost of the output capacitor in the fly back converter become prohibitive.

Circuits and waveforms for a typical flyback converter and a typical buck-derived forward converter are shown in **Figs 11.10** and **11.11** respectively.

The flyback converter uses a coupled inductor and operates in a similar way to the boost converter. The inductor current is built up in the primary winding $(i_s(t))$ when the switch S conducts, and is forced to flow in the secondary winding $(i_D(t))$ through the diode D when the switch is off.

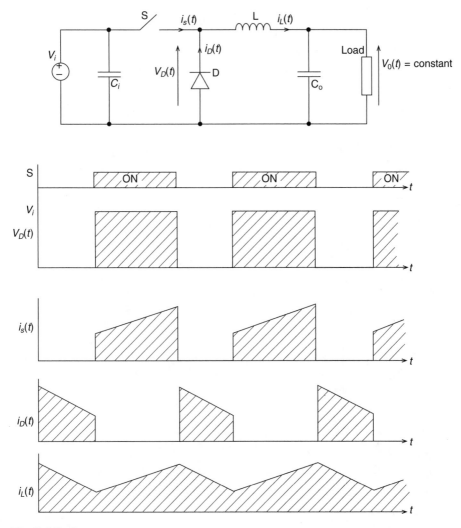

Fig. 11.8 Buck converter

The primary switches in the buck-derived converter invert the dc input voltage, and a PWM rectangular ac waveform $V_1(t)$ is applied to the primary winding of the isolation transformer. The secondary voltage is rectified by diodes D_1 and D_2 before being applied to the averaging filter which comprises inductor L and capacitor C. The output from the filter is a smoothed dc waveform with very little switching frequency ripple.

Stray inductances (mainly transformer leakage inductance) and stray capacitances (mainly in the switch and diodes) form resonant circuits which oscillate in response to switching transitions. These oscillations are damped in practical circuits by R–C snubbers which are connected across transformer windings and in parallel with diodes; the energy stored in stray and leakage inductances may be recovered using voltage clamps.

Auxiliary circuitry is necessary for the supply of power to the control circuit, for control circuit functions, to provide isolation in the feedback loop, and to drive power transistors.

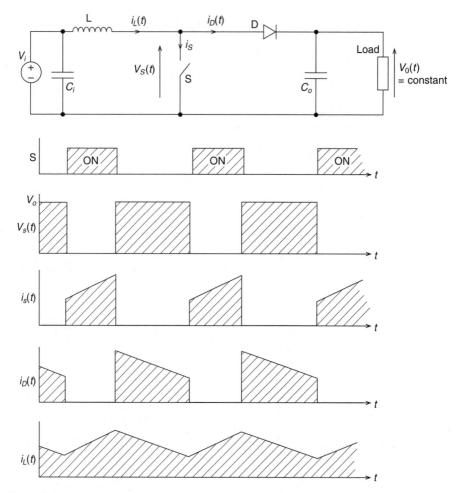

Fig. 11.9 Boost converter

11.8 AC supply harmonic filters

Filters are used to block and/or bypass selected frequency components of power circuit waveforms. Power circuit waveforms are usually composed of a dc component and an alternating component that is repeated each cycle. It can be shown that any repetitive waveform may be represented by the sum of a series of sinusoidal components, each of which has a frequency which is an integral multiple of the repetition frequency of the waveform. This series is known as the *Fourier series*. The component at the basic repetition frequency is known as the *fundamental* component, and other components are known as *harmonics*, the Nth harmonic having N times the frequency of the fundamental. The amplitudes of the harmonics generally decrease as their frequency increases. The representation of a square wave is shown in **Fig. 11.12**.

A filter incorporates inductors and capacitors, the impedance of which is proportional to frequency and inversely proportional to frequency respectively. These components are combined in a filter circuit to attenuate selectively the frequency components of the input waveform.

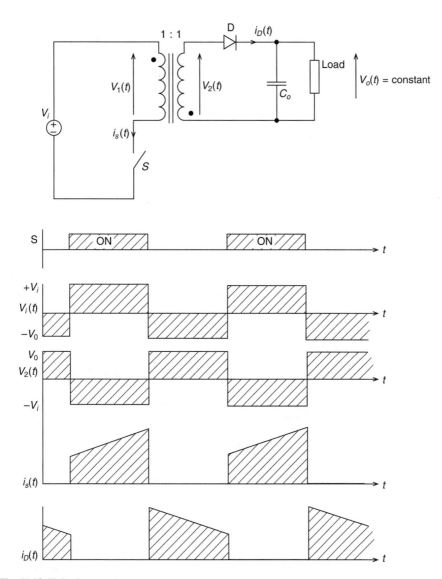

Fig. 11.10 Flyback converter

11.8.1 Supply current harmonics

Loads which are connected to the utility supply may disrupt the operation of other loads by drawing harmonic currents from the network. Power supply rectifiers, which charge large capacitors, are a main source of harmonic current flow in the utility distribution system.

Distorted current waveforms affect other loads connected to the utility system because of the harmonic voltages which are developed across common impedances in the distribution network. Return currents flowing in single-phase neutral wires are added at the three-phase supply neutral point. If the load in the three phases is balanced, with unity power factor, then the neutral currents cancel exactly and zero neutral current flows. Typical neutral currents in rectifiers do not cancel, and the

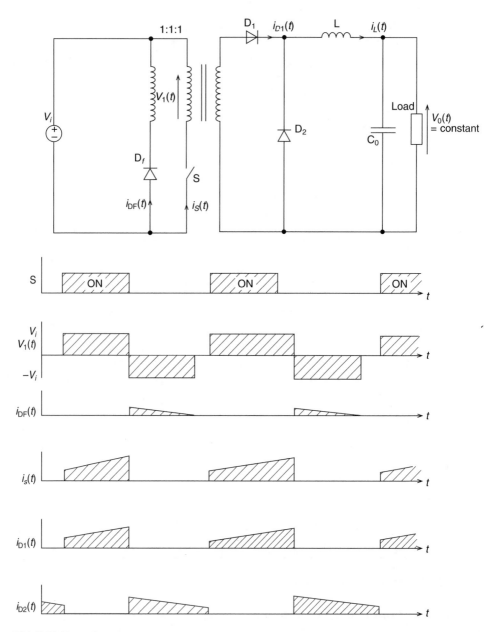

Fig. 11.11 Forward converter

resulting neutral current, being the sum of the line currents, requires a higher-rated cable.

Three-phase rectifiers usually deliver dc output. Load current does not flow through the neutral wire and the input current waveform is usually acceptable without special measures to reduce harmonics.

11.8.2 The generation of current harmonics

In diode rectifiers, which are used to convert ac mains to a dc voltage source, the dc

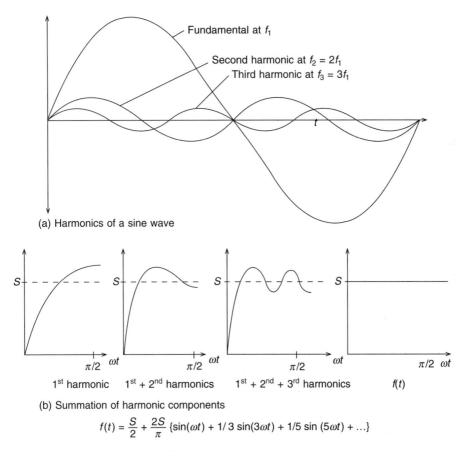

(a) Harmonics of a sine wave

(b) Summation of harmonic components

1st harmonic   1st + 2nd harmonics   1st + 2nd + 3rd harmonics   f(t)

$$f(t) = \frac{S}{2} + \frac{2S}{\pi} \{\sin(\omega t) + 1/3 \sin(3\omega t) + 1/5 \sin(5\omega t) + ...\}$$

Fig. 11.12 Frequency components of a square wave

voltage ripple at the rectifier output is limited by a large capacitor. In the half-wave rectifier shown in **Fig. 11.1**, the dc capacitor C draws a pulse of current during the mains voltage peak. A full-wave rectifier as shown in **Fig. 11.2** draws positive and negative pulses of current when the supply voltage is near its peak positive and negative values respectively.

The resulting supply current waveform is ac, and at the same fundamental frequency as the utility power supply, but it is not a pure sine wave. As previously explained, the waveform can be split into a Fourier series of pure sine wave frequencies which are harmonics of the fundamental frequency. The amplitudes of the harmonic components increase in magnitude if the pulse of current drawn by the capacitor is shorter; the pulse width depends upon the ac circuit impedance and the magnitude of the dc voltage ripple.

For a given dc voltage ripple requirement, which is set by the dc capacitance value and the dc load current, the width of the supply current pulse can be increased by adding series inductance in the rectifier circuit, either in the ac connections or on the supply side of the dc capacitance. This method is considered to be acceptable at low power levels, but above about 400 W output the inductor becomes prohibitively large and preference is given to active methods which allow the use of a smaller inductor.

11.8.3 Current harmonics and power factor

Power factor has been defined in **Chapter 2**. Simply restated here, it is the total power taken from the supply divided by the ac supply voltage and the ac supply current. Reactive power flow and current waveform distortion reduce the power factor from its ideal value of unity.

Rectifiers delivering a dc voltage output usually draw little reactive current, but as pointed out above, they draw significant harmonic currents. The amount of distortion in the supply current waveform is usually quantified using the ac supply power factor, since reactive power flow can be neglected. Typical single-phase full-wave rectifiers of the type described above have a power factor in the region of 0.65 due to distortion of the current waveform.

11.8.4 Reduction of current harmonics

In order to attenuate the harmonic currents drawn by a non-linear load, an inductor may be connected in series with the load. This has the effect of reducing the load impedance variation, this variation being responsible for generating the harmonic currents.

Alternatively, a circuit or *filter* which provides a low-impedance path at a harmonic frequency may be connected into the supply side of the load in order to bypass most of the harmonic current. At high power levels, a series resonant circuit is used; this comprises an inductor and a capacitor which are tuned to the harmonic frequency which is to be bypassed. In a 12-pulse inverter, for instance, a tuned filter is generally used to bypass the 11th and 13th harmonics and an *LC* filter is used for higher harmonics. Care is taken in the design of these filters to avoid high power loss at the fundamental frequency. Filters of this type are generally used in very high power applications such as HVDC transmission schemes.

Active power factor correction is often used in single-phase systems with a power output above about 400 W, where a passive system as described above would require an inductor of impractical size. Active power factor correction is usually based on a boost converter switching at a constant high frequency of 16 kHz or above, and using PWM control to achieve approximately unity power factor. The high-frequency switching enables a relatively small inductance value to be used. A typical scheme is shown in **Fig. 11.13**. In comparison with a non-corrected system, active power factor correction provides the benefits of reduced dc voltage ripple, reduced dc bulk capacitance and reduced rms supply current. It can also enable a universal input by working with low-voltage ac supplies at higher line currents. These benefits do not compensate for the increased cost, complexity and power losses associated with active power factor correction, but supply harmonic limits may be enforced by law, and this would lead to its widespread use.

A PWM-controlled inverter, when connected to the supply system as a load, can also be controlled to bypass current harmonics. In order to avoid high-frequency switching noise passing into the ac supply the PWM waveforms are filtered, and the cut-off frequency of the filter limits the maximum frequency of the harmonic currents which can be bypassed using this technique.

11.9 Filters for high-frequency noise

Switching converters generate high-frequency electrical noise which is conducted

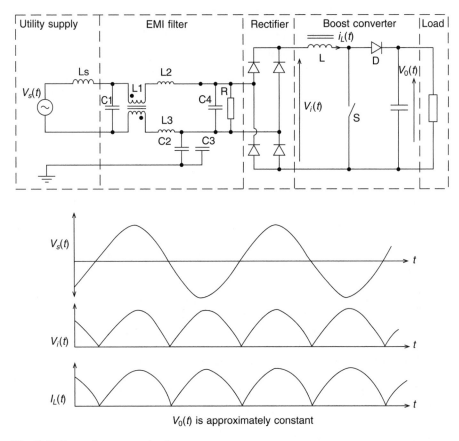

Fig. 11.13 Power factor correction boost converter

out of the equipment along electrical connections and emitted by the equipment in the form of electromagnetic radiation.

Conducted emissions are attenuated by an electromagnetic interference (EMI) filter as shown in **Fig. 11.13**. The EMI filter is a combination of *LC* filters for *common-mode* and *differential-mode* noise. In **Fig. 11.13**, C1 and C4 are X capacitors, which act with L2 and L3 to filter the differential-mode noise which flows between line and neutral connections. C2 and C3 are Y capacitors, which act with L1, L2 and L3 to filter the common-mode noise which flows through the earth connection and splits equally between line and neutral circuits. X capacitors have typical values in the range 10 nF to 2 μF; these are used in cases where failure of the capacitor would not lead to a risk of electric shock. Y capacitors have typical values in the range 0.5 nF to 35 nF. These are used where failure of the capacitor could lead to risk of electric shock. The mutual inductance of L1 typically ranges from 1.8 mH at 25 A to 47 mH at 0.3 A, and L2 and L3 may be provided by the leakage inductance of the windings of L1. The resistance R is required in order to discharge the X capacitors with a time constant of typically 1 second.

Power conditioners limit the disturbances which enter a load from the ac supply. They attenuate the high-frequency components of these disturbances using filters and by clipping voltage spikes. A typical single-phase power conditioner circuit is shown in **Fig. 11.14**. This circuit is similar to the filter circuit already shown in

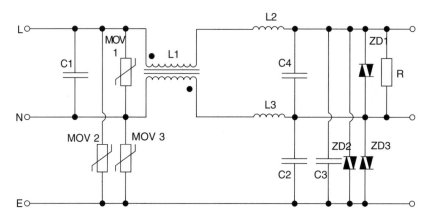

Fig. 11.14 Single-phase high-frequency power conditioner

Fig. 11.13, but it has non-linear voltage clamping components added; these are shown as the metal oxide varistors MOV1, MOV2 and MOV3 and suppressor diodes ZD1, ZD2 and ZD3.

Radiated emissions may be suppressed by various forms of screens, including earthed enclosures, screens between transformer windings, screens between components and their heat sinks, and screens around the joints between the halves of ferrite cores. The origin of high-frequency noise in switching power converters is the high rate of change of voltage and current during switching transitions. These high-frequency components in the waveform can be limited by minimizing the areas of the high-frequency current loops, and by using slow switching and snubbers. Components experiencing high rates of change of voltage propagate noise through capacitive coupling, and current loops with high rates of change of current propagate noise through radiated emissions.

The limits on *Radio Frequency Interference (RFI)* are usually required by law. This matter is dealt with in more depth in **Chapter 15**.

11.10   Systems with cascaded converters

The basic power conversion circuits which have been described are often combined in more complex applications. Some of the more common combinations are described in this section, and the merits of the different possibilities for some applications are set out.

11.10.1   The Uninterruptible Power Supply (UPS)

The UPS is used to provide a continuous stable supply in the event of a utility supply failure, this failure being typically defined as the ac voltage falling below 85 per cent of its nominal value. Typical applications are in medical and communication systems, in which the expense of a UPS is justified by the reduced chances of a system failure.

The UPS therefore incorporates a back-up power source, which is usually a battery. A typical scheme is outlined in **Fig. 11.15**. Such equipment is usually designed to provide 10–15 minutes' supply at full load, although to minimize the battery size some systems for computer supplies deliver the full load for just long enough for the hardware to shut down automatically and to save data. Diesel generators are often used where it is required to supply full load for longer periods.

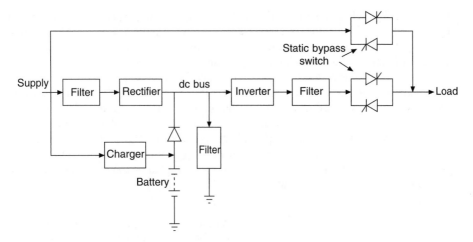

Fig. 11.15 Typical on-line UPS block diagram

Most UPS systems are either *on-line* or *standby* types.

An on-line UPS continuously rectifies the utility supply, charges the back-up battery as necessary and supplies power to the load by continuously inverting the dc. If the rectifier or inverter fails, a bypass switch is used to connect the load directly to the supply. When the utility supply fails, the back-up battery delivers dc power to the inverter and the ac supply to the load is maintained.

In a standby UPS, under normal operation the load is supplied directly from the utility network and the back-up battery is trickle charged. When a failure in the utility supply occurs, a changeover switch connects the output of an inverter to the load, and the inverter draws power from the battery. Standby systems are normally cheaper than on-line systems.

The *line-interactive* UPS is an off-line type which includes a system that boosts the supply voltage during line voltage sags. The voltage may be boosted by use of a tap-changing transformer. This system reduces the number of occasions when battery power is used; this helps to maintain a full charge in the battery for the event of a full loss of supply.

The inverter is the heart of a battery-powered UPS, and different types are used for different applications. The cheapest systems deliver a square wave, but these require severe derating of power when supplying such loads as switched-mode power supplies. Inverters delivering a sinusoidal output are generally used. The ac output of the inverter is stepped up to the required output voltage using a transformer.

11.10.2 Regulated dc power supplies

There is a large demand for power supplies that convert the utility ac supply to an isolated low-voltage dc output. This is mainly due to the near-universal use of transistor circuits for control, communication and computing equipment, these transistor circuits being powered by low-voltage dc.

For applications that require an accurately controlled output voltage, a switched-mode power supply or a linear regulator is necessary in order to compensate for fluctuations in the ac input voltage and to limit the variation of output voltage with load current. Switched-mode power supplies become economical above a few tens of watts output, but below this level linear regulators are often preferred, especially in low-noise, high-reliability applications. The dc output voltage is accurately controlled by comparing it with a voltage reference and by using negative feedback. The amplified voltage difference is used to control the regulator. This scheme requires an accurate voltage reference, which is usually temperature compensated, and an error amplifier; these are usually integrated with other components in an IC.

In a linear regulator a supply-frequency transformer is used at the input for isolation and to reduce the voltage level to that required by the load. The transformer output is rectified into a capacitor, and load current flows through the linear regulator to the output. A typical linear power supply scheme is shown in **Fig. 11.16**. These supplies require a large, heavy transformer and their efficiency is typically only 30–60 per cent. A large proportion of the losses occur in the regulator transistor, and they are dissipated using a large heat sink.

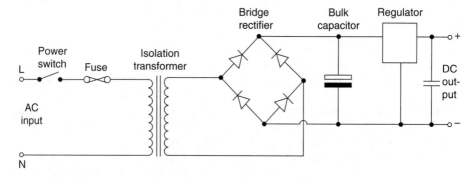

Fig. 11.16 Linear power supply scheme

Switched-mode power supplies use dc to dc converters. The ac supply is rectified into a capacitor, which results in some voltage ripple. The substantial dc power is passed through a dc to dc converter which regulates the output voltage using the switch duty cycle. The supply is isolated using a small high-frequency transformer. This avoids the high power dissipation and the large transformer of the linear power supply, and efficiencies of 70–90 per cent are usual. The circuit of a typical 1 kW to 10 kW off-line switched-mode power supply is shown in **Fig. 11.17**, and **Fig. 11.18** shows a typical 100 W off-line power supply. A switched-mode power supply requires more components than a linear power supply; it also produces significantly more electrical noise because of the high-frequency switching, and this requires careful layout of the power circuit with filters and screens to limit the noise to acceptable levels.

Commercial dc power supplies may incorporate the following control and monitoring functions:

- remote regulation of output voltage
- shutdowns for thermal overload, transformer primary overcurrent and output overvoltage
- external warning of power failure and external inhibit
- output current foldback
- parallel operation with forced current sharing

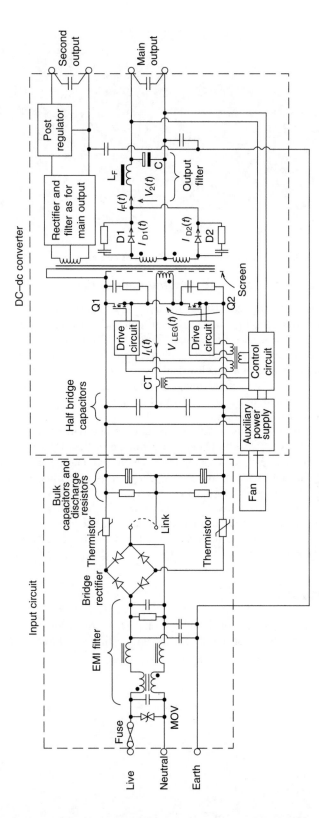

Fig. 11.17 A typical off-line switched-mode power supply power circuit

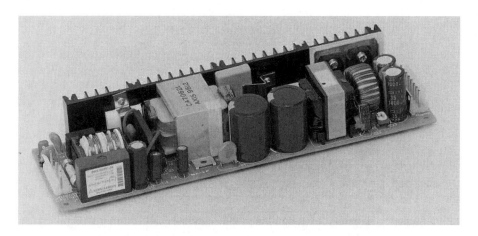

Fig. 11.18 A typical 100 W off-line power supply (courtesy of Coutant Lambda)

11.10.3 Power conditioners for photovoltaic arrays

A typical photovoltaic array consists of relatively large-area silicon semiconductor diodes which are exposed to light that passes through a window. The diode is designed such that photons absorbed within the semiconductor often produce charge carriers in the form of electrons and holes. The increased carrier densities give rise to a voltage, and the ability to deliver electrical power. Many diodes are connected in series in order to provide sufficient voltage. A photovoltaic array delivers maximum power at a given voltage and current, but this optimum voltage and current varies with light intensity and temperature.

Photovoltaic cells may be used without power control circuitry to supply families of integrated circuits that have wide operating voltage ranges. However, most large systems use a peak power tracking converter which delivers the maximum power available at any instant, at a controlled dc voltage; this controlled power may be supplied to a static load, a battery or to an inverter.

Power converters which deliver ac output power to the utility network using a PWM inverter have been proposed for widespread use, but such systems are rarely practical at present because of the special requirements for connection of power generating equipment.

11.11   Standards

The leading international, regional and national standards adopted by users and suppliers of power supplies for manufacturing and testing are shown in **Table 11.1**.

Table 11.1 International, regional and national standards relating to power supplies

IEC	EN	BS	Subject of standard	N. American
60065	EN 60065	EN 60065	Safety requirements for electronic apparatus for household use	
60335	EN 60335	EN 60335	Safety of household and similar apparatus	
60601-1	EN 60601-1	EN 60601-1	Medical electrical equipment – pt 1: general requirements for safety	
60950	EN 60950	EN 60950	Safety of IT equipment	CSA 22.2 no 950-95 UL 1950
61010-1	EN 61010-1	EN 61010-1	Safety requirements for electrical equipment for measurement, control and laboratory use Pt 1: general requirements	
61204			LV power supply devices, dc output – performance and safety	
			Industrial control equipment	UL 508
			Medical and dental equipment	UL 544
			Power units other than class 2	UL 1012
			LV video products without CRT displays	UL 1409
			Medical electrical equipment – pt 1: general requirements for safety	UL 2601-1
			Electrical equipment for laboratory use – pt 1: general requirements	UL 3101-1

References

The first two references given below are academic treatments, and the third gives a design engineer's approach to low-power dc power supplies. Useful information can also be obtained from the catalogues of companies that distribute power supplies; these catalogues often include general technical information as well as information about specific products.

11A.   Mohan, N., Undeland T.M. and Robbins, W.P. *Power Electronics: Converters, Applications and Design*, John Wiley & Sons Inc., 1995.
11B.   Williams, B.W. *Power Electronics,* Macmillan, 1992.
11C.   Billings, K.H. *Switchmode Power Supply Handbook*, McGraw-Hill, 1989.

Chapter 12

Batteries and fuel cells

Mr R.I. Deakin
Deakin Davenset Rectifiers

12.1   Introduction

The battery is at present the most practical and widely used means of storing electrical energy. The storage capacity of a battery is usually defined in ampere-hours (Ah); energy is strictly defined in kWh or joules, but since the voltage of the battery is nominally fixed and known, the Ah definition is more convenient. The terms *battery* and *cell* are often interchanged, although strictly a battery is a group of cells built together in a single unit.

Batteries can be classified into *primary* and *secondary* types.

A *primary* battery stores electrical energy in a chemical form which is introduced at the manufacturing stage. When it is discharged and this chemically stored energy is depleted, the battery is no longer serviceable. Applications for primary batteries are generally in the low-cost domestic environment, in portable equipment such as torches, calculators, radios and hearing aids.

A *secondary* (rechargeable) battery will absorb electrical energy, store this in a chemical form and then release electrical energy when required. Once the battery has been discharged and the chemical energy depleted, it can be recharged with a further intake of electrical energy. Many cycles of charging and discharging can be repeated in a secondary battery. Applications cover a wide range. In the domestic environment secondary batteries are used in portable hand tools, laptop computers and portable telephones. Higher-power applications in industry include use in road and rail vehicles and in standby power applications, the latter having already been described in **section 11.10.1**. The capacity of secondary battery systems ranges from 100 mAh to 2000 Ah. Their useful life ranges from two to 20 years; this will depend, among other things, upon the number of change–discharge cycles and the type and construction of battery used.

The *fuel cell* is a relatively new method of energy storage which is increasing in acceptance for certain applications. A fuel cell converts the energy of a chemical reaction into electrical energy in a way which is similar, in principle, to the battery, but a fuel cell is 'recharged' by continuous replenishment of the chemical materials.

12.2   Primary cells

The majority of primary cells now in use fall into one of eight types:

- zinc carbon
- alkaline manganese
- mercury oxide
- silver oxide
- zinc air
- lithium manganese dioxide
- lithium thionyl chloride
- lithium copper oxyphosphate

Each of these is briefly reviewed in turn in the following sections.

12.2.1 Zinc carbon

In its latest form, this has been in use since 1950. The basic cell operates with a voltage range of 1.2 V to 1.6 V, and cells are connected in series to form batteries such as AAA, AA, C, D, PP3 and PP9. Typical construction of a PP9 battery is shown in **Fig. 12.1**.

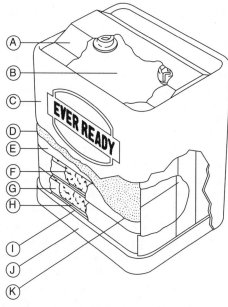

D Wax coating. This seals any capillary passages between cells and the atmosphere, so preventing the loss of moisture.

E Plastic cell container. This plastic band holds together all the components of a single cell.

F Positive electrode. This is a flat cake containing a mixture of manganese dioxide and carbon black or graphite for conductivity. Ammonium chloride and zinc chloride are other necessary ingredients.

G Paper tray. This acts as a separator between mix cake and the zinc electrode.

H Carbon coated zinc electrode. Known as a duplex electrode, this is a zinc plate to which is adhered a thin layer of highly conductive carbon which is impervious to electrolyte.

I Electrolyte impregnated paper. This contains electrolyte and is an additional separator between the mix cake and the zinc.

J Bottom plate. This plastic plate closes the bottom of the battery.

K Conducting strip. This makes contact with the negative zinc plate at the base of the stack and is connected to the negative socket at the other end.

A Protector card. This protects the terminals and is torn away before use.

B Top plate. This plastic plate carries the snap fastener connectors and closes the top of the battery.

C Metal jacket. This is crimped on to the outside of the battery and carries the printed design. The jacket helps to resist bulging, breakage and leakage and holds all components firmly together.

Fig. 12.1 Zinc carbon PP9 battery construction (courtesy of Energiser)

Zinc carbon batteries are low in cost, they have an operating temperature range from –10°C to 50°C, and the shelf life at a temperature of 20°C is up to three years. They are ideally suited to applications with intermittent loads, such as radios, torches, toys, clocks and flashing warning lamps.

12.2.2 Alkaline manganese

The alkaline manganese battery became commercially available in its present form in the late 1950s. The basic cell operates with a voltage range of 1.35 V to 1.55 V, and cells are connected in series to form the standard battery sizes N, AAA, AA, C, D and PP3. Typical construction of a single-cell battery is shown in **Fig. 12.2**.

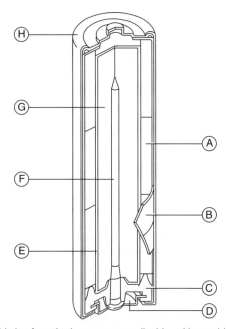

A Cathode pellet. This is of synthetic manganese dioxide, with graphite for conductivity, pressed into cylindrical pellets. The pellets make an interference fit with the can, ensuring good electrical contact. The graphite is the cathode current collector.
B Steel can. This acts as the cell container. The preformed stud at the closed end of the can functions as the cell positive terminal.
C Closure. The hard plastic cell closure forms the seal at the open end of the can and carries the bottom cover/current collector assembly.
D Bottom cover. This metal plate acts as the negative terminal of the cell. It is welded to the current collector to ensure good electrical contact.
E Separator. This is of non-woven synthetic material.
F Anode current collector. This is a metal 'nail'.
G Zinc paste anode. The anode is of amalgamated high purity zinc powder made into a paste with the electrolyte.
H Shrink label. This covers the outside of the cell and carries printed design, and provides insulation of the side of cell.

Fig. 12.2 Construction of an alkaline manganese battery (courtesy of Energiser)

Cost is in the medium–low range. The operating temperature range is –15°C to 50°C, and the shelf life at 20°C is up to four years. Alkaline manganese batteries are suited to a current drain of 5 mA to 2 A; this makes them ideally suited to continuous duty applications which can include radios, torches, cassette players, toys and cameras.

12.2.3 Mercury oxide

This has been in commercial use since the 1930s. The single cell operates with a narrow voltage range of 1.3 V to 1.4 V. This type is only available as a single button cell, the construction of which is illustrated in **Fig. 12.3**.

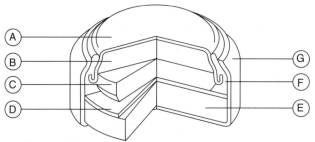

A Cell cap. The plated steel cap functions as the negative terminal of the cell. The inside of the cap is laminated with copper.
B Zinc anode. This is of high purity amalgamated zinc powder.
C Absorbent pad. This is of a non-woven material and holds the alkaline electrolyte.
D Separator. This is a synthetic ion-permeable material.
E Mercuric oxide cathode. The cathode is a pellet of mercuric oxide plus graphite for conductivity and sometimes manganese dioxide. It is compressed into the can.
F Sealing grommet. This plastic grommet both seals the cell and insulates the positive and negative terminals.
G Cell can. The can is of nickel plated steel and functions as the positive terminal of the cell.

Fig. 12.3 Mercury oxide button cell (courtesy of Energiser)

Cost is in the medium range. The operating temperature range is 0°C to 50°C, and the shelf life at 20°C is up to four years. Mercury oxide button cells are suitable for low current drains of 0.1 mA to 5 mA, with either continuous or intermittent loads. Their small size and narrow voltage range are ideally suited for use in hearing aids, cameras and electrical instruments.

12.2.4 Silver oxide

This has been in commercial use since the 1970s. The single cell operates with a very narrow range from 1.50 V to 1.55 V. Construction of a silver oxide button cell is shown in **Fig. 12.4**.

Cost is relatively high. They have an operating temperature range from 0°C to 50°C and a shelf life of up to three years. In button cell form, they are suitable for a current drain of 0.1 mA to 5 mA, either continuous or intermittent. Their small size and very narrow voltage range makes them ideal for use in watches, calculators and electrical instruments.

12.2.5 Zinc air

This battery has been in commercial use since the mid-1970s. A single cell operates

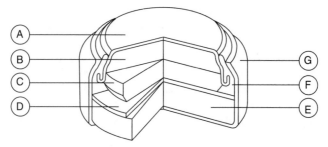

A Cell cap. The plated steel cap functions as the negative terminal
   of the cell. The inside of the cap is laminated with copper.
B Zinc anode. This is a high purity amalgamated zinc powder.
C Absorbent pad. This pad is of a non-woven material and holds
   the alkaline electrolyte.
D Separator. This is a synthetic ion-permeable membrane.
E Cathode. This is a pellet of silver oxide plus graphite for
   conductivity. It is compressed into the can.
F Sealing grommet. This plastic grommet both seals the cell and
   insulates the positive and negative terminals.
G Cell can. The nickel plated steel can acts as a cell container
   and as the positive terminal of the cell.

Fig. 12.4 Silver oxide button cell (courtesy of Energiser)

with a voltage range of 1.2 V to 1.4 V. The construction of a zinc air button cell is
shown in **Fig. 12.5**.

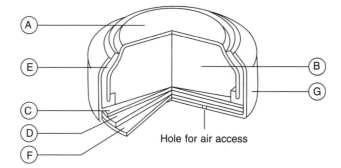

Hole for air access

A Cell cap. The plated steel cap functions as the negative terminal
   of the cell.
B Zinc anode. This is a high purity amalgamated zinc powder,
   which also retains the alkaline electrolyte.
C Separator. A synthetic ion-permeable membrane.
D Cathode. This is a carbon/catalyst mixture with a wetproofing
   agent on a mesh support, and with an outer layer of gas-
   permeable hydrophobic PTFE.
E Sealing grommet. This plastic grommet seals and insulates the
   positive and negative terminals, and seals the cathode to the base.
F Diffusion membrane. This permeable layer distributes air from
   the access holes uniformly across the cathode surface.
G Can. This plated steel can forms a support for the cathode, acts
   as a cell container and the positive terminal of the cell. The
   holes in the can permit air access to the catalyst cathode.

Fig. 12.5 Zinc air cell (courtesy of Energiser)

Because one of the reactants is air, the battery has a sealing tab which must be removed before it is put into service. Following removal of the tab, the cell voltage will rise to 1.4 V.

Cost is in the medium range. Operating temperature is from –10°C to 50°C and the shelf life is almost unlimited while the seal remains unbroken. They are produced as a button cell, and are suitable for a current drain from 1 mA to 10 mA, with either intermittent or continuous loads. The service life is relatively short due to high self-discharge characteristics, but end of life can be accurately estimated because of the flat discharge current. The general performance and the life predictability makes zinc air batteries ideal for use in hearing aids.

12.2.6   Lithium manganese dioxide

This type of battery has been available since 1975, but it has only become commercially available in the past few years. The single cell operates with a relatively high voltage, in the range 2.8 V to 3.0 V. A typical construction is shown in **Fig. 12.6**.

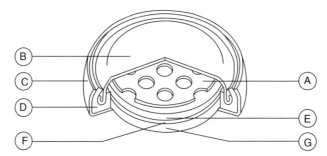

A   Current collector. This is a sheet of perforated stainless steel.
B   Stainless steel top cap. This functions as the negative
     terminal of the cell.
C   Stainless steel cell can. This functions as the positive
     terminal of the cell.
D   Polypropylene closure. This material is highly impermeable to
     water vapour and prevents moisture entering the cell after it
     has been sealed.
E   Lithium negative electrode. This is punched from sheet
     lithium.
F   Separator containing electrolyte. The separator is of non-
     woven polypropylene cloth and contains electrolyte, a
     solution of lithium perchlorate in a mixture of propylene
     carbonate and dimethoxyethane.
G   Manganese dioxide positive electrode. The cathode is made
     from a highly active electrolytic oxide.

Fig. 12.6 Lithium manganese dioxide cell (courtesy of Energiser)

Cost is relatively high. The operating temperature covers a wide range, from –30°C to 50°C, and the shelf life is four to five years. Lithium manganese dioxide batteries are suitable for continuous or intermittent loads; in the form of a button cell they are suitable for 0.1 mA to 10 mA current drain, or as a cylindrical cell they can deliver from 0.1 mA up to 1 A. They have a high discharge efficiency down to –20°C, and are suitable for applications in instruments, watches and cameras.

12.2.7 Lithium thionyl chloride

This type of battery has only recently become commercially available. The single cell has a high voltage in the range 3.2 V to 3.5 V, with an open-circuit voltage of 3.67 V. A cross-section of an example is shown in **Fig. 12.7**.

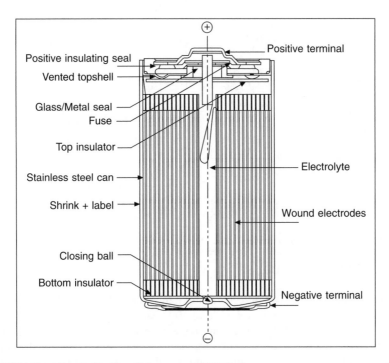

Fig. 12.7 Lithium thionyl chloride cell (courtesy of SAFT Ltd)

Cost is relatively high. The operating temperature range is one of the highest available, from −55°C to 85°C, and the shelf life is up to 10 years. These batteries are suitable for intermittent or continuous duties, with a current drain from 15 mA up to 1.5 A. Their high energy density, wide temperature range and high discharge efficiency makes these batteries particularly well suited to memory back-up applications in programmable logic controllers, personal computers and alarm systems.

12.2.8 Lithium copper oxyphosphate

This has only become commercially available recently. The single cell has a voltage range 2.1 V to 2.5 V, with an open-circuit voltage of 2.7 V. A sectional view of a typical unit is shown in **Fig. 12.8**.

These batteries are expensive because of their high energy density. They operate over a temperature range −40°C to 60°C and they have a shelf life of up to 10 years. They are suitable for continuous loads from 37 μA up to 25 mA. The high energy density and wide temperature range make these batteries specially suitable for applications in metering computers.

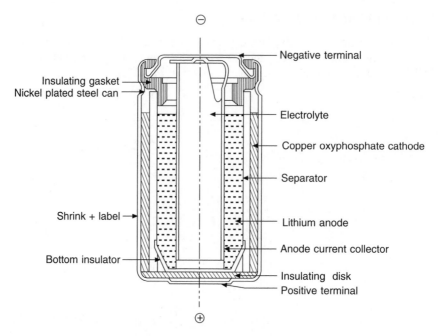

Fig. 12.8 Lithium copper oxyphosphate cell (courtesy of SAFT Ltd)

12.3 Secondary cells

The main forms of secondary or rechargeable battery are lead acid and nickel cadmium, both of which have been in use for about 100 years.

Nickel cadmium and lead acid batteries are formed by connecting a number of cells in series. Each cell consists of vertical plates which are connected in parallel and are divided into a positive group and a negative group. Adjacent positive and negative plates are insulated from each other by separators or rod insulators which have to provide the following functions:

- obstruct the transfer of ions between the plates as little as possible
- keep in place the active material
- enable the escape of charging gases from the electrolyte

All the plates and separators together form a *plategroup* or *element*.

The insulators in nickel cadmium batteries are vertical plastic rods, which may be separate or in the form of grids which are inserted between the plates. Corrugated or perforated PVC is also used as a separator. The active material is kept in place by steel pockets.

In lead acid batteries, the separators are normally microporous sheets of plastic, which are usually combined with corrugated and perforated spacers; porous rubber separators are also used. In some designs the separator completely surrounds the positive or negative plate. The selection of separator is very important, and it depends upon the plate design and the use for which the battery is intended. In batteries intended for short discharges, such as diesel engine starting, thin separators and spacers are used, whereas batteries designed for long discharges, as in standby power or emergency lighting, have thick separators and spacers. Batteries that require extra strength to withstand mechanical stresses often have a glass fibre separator against the positive plates in order to minimize the shedding of active material from these plates.

The outermost plates are of the same polarity on both sides. In lead acid batteries the outside plates are normally negative; in nickel cadmium batteries they are normally positive.

The plategroups are mounted in containers which are filled with electrolyte. The electrolyte is dilute sulphuric acid in lead acid batteries, and potassium hydroxide in nickel cadmium batteries. In a fully charged lead acid battery the electrolyte density is between 1.20 and 1.29. In nickel cadmium batteries it is between 1.18 and 1.30 irrespective of the state of charge.

The containers may house a single cell or they may be of a monoblock type, housing several cells. The top (or lid) may be glued or welded to the container, or it may be formed integrally with the container. In the lid is a hole for the vent, which is normally flame arresting, and the cells are topped up with electrolyte through the vent hole.

Below the plates there is an empty space which is known as the *sludge space*. This is provided so that shredded active material and corrosion products from the supporting structure can be deposited without creating any short-circuits between the plates. The sludge space in nickel cadmium batteries is smaller than in lead acid units because there is no corrosion of the supporting steel structure and no shredding of active material; it is usually created by hanging the plategroup in the lid. In lead acid batteries the sludge space may be created by hanging the plates from the lid or container walls (this is common in stationary batteries), by standing the plates on supports from the bottom (this is common for locomotive starting batteries) or by hanging the positive plates with the negative plates standing on supports (this is also used for stationary batteries).

From the plates, the current is carried through plate lugs, a connecting strap or bridge and a polebolt. The bolts pass through the lid and are sealed against the lid with gaskets. Above the lid, the cells are connected to form the battery. Monoblocks may have connectors arranged directly through the side of the partition cell wall, in which case the polebolts do not pass through the lid, or the connectors may be located between the cell lid and an outer lid. The monoblocks made in this way are completely insulated and easy to keep clean.

The main forms of secondary cell now commercially available are:

- nickel cadmium – sealed
- nickel cadmium – vented
- nickel metal hydride
- lead acid – pasted plate
- lead acid – tubular
- lead acid – Planté
- lead acid – valve-regulated sealed (VRSLA)

Each of these is described separately in **sections 12.3.1** to **12.3.7**, and a summary of the main features of the four lead acid types and nickel cadmium is given in **Table 12.1**.

12.3.1   Sealed nickel cadmium

This type of cell has been commercially available since the early 1950s. The single

Table 12.1 Comparison of features and performance of the main types of secondary battery

Battery type	Lead acid					Nickel cadmium
	Pasted plate		Tubular	Planté	VRSLA	
	Lead antimony	Lead calcium				
Cycle duty	Good	Poor	Very good	Suited to shallow discharge duty	Suited to shallow discharge duty	Suited to shallow discharge duty
Maximum temperature	45°C	35°C	45°C	45°C	40°C	45°C
Gas generation and maintenance	Low	Very low	Low	Very low	Negligible No topping up needed	Low to moderate
Volume/Ah indicator	50	50	45	100	35	55
Relative cost	70	70	90	100	80	200–300

cell operates across a voltage range 1.0 V to 1.25 V and is ideally suited to a heavy continuous current drain of up to eight times the nominal ampere-hour capacity of the battery. The low internal resistance of the cell makes it ideal for heavy current discharge applications in motor-driven appliances such as portable drills, vacuum cleaners, toys and emergency systems. Batteries of this type are capable of being recharged hundreds of times using a simple constant-current charging method.

The cells tolerate a wide temperature range from –30°C to 50°C, and they have a shelf life of up to eight years at 20°C.

Typical construction of a sealed nickel cadmium cell is shown in **Fig. 12.9**. The cell is available commercially in a variety of standard sizes including AAA, AA, C, D, PP3 and PP9.

12.3.2   Vented nickel cadmium

Using the electrochemistry of nickel and cadmium compounds with a potassium hydroxide electrolyte, vented cells use 'pocket' plates which are made from finely perforated nickel-plated steel strip filled with active material. The pocket plates are crimped together to produce a homogeneous plate. Translucent plastic or stainless steel containers are used. An example is shown in **Fig. 12.10**.

These batteries are robust and they have the benefits of all-round reliability; they are resistant to shock and extremes of temperature and electrical loading. They can be left discharged without damage and require very little maintenance. Their cycling ability is excellent and can offer a service life of up to 25 years.

Vented nickel cadmium batteries are used typically in long-life applications where reliability through temperature extremes in required and where physical or electrical abuse is likely. These applications include railway rolling stock, off-shore use, power system switch tripping and closing, telecommunications, uninterruptible power supplies (see **section 11.10.1**), security and emergency systems and engine starting.

12.3.3   Nickel metal hydride

The sealed nickel metal hydride battery has become commercially available in the

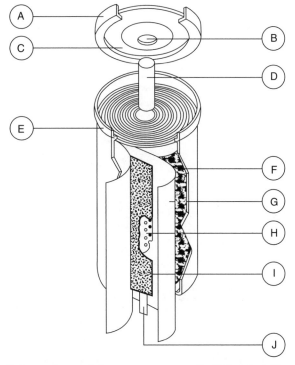

A   Nylon sealing gasket
B   Resealing safety vent
C   Nickel plated steel top plate (positive)
D   Positive connector
E   Nickel plated steel can (negative)
F   Sintered positive electrode
G   Separator
H   Support strip
I   Negative electrode
J   Negative connector

Fig. 12.9 Sealed nickel cadmium cell (courtesy of Energiser)

past few years. A single cell operates with a voltage range of 1.1 V to 1.45 V, and a nominal voltage of 1.2 V. The construction of a metal hydride cell is illustrated in **Fig. 12.11**.

These batteries are relatively expensive. Their energy density is in the region 44–46 Wh/kg, which is higher than the sealed nickel cadmium battery. The operating temperature range is –20°C to 60°C and the shelf life is up to five years. Nickel metal hydride batteries are suitable for applications with *mA current drains from 0.2 to 3 times the nominal ampere-hour capacity of the battery*. The high energy density makes this type of battery ideally suited to applications for memory back-up and portable communications.

Nickel metal hydride batteries will endure recharging up to 500 times, although the method of charging is slightly more complicated than for sealed nickel cadmium systems.

12.3.4   Lead acid – pasted plate

The positive plates of pasted plate cells are made by impressing an oxide paste into a current-collecting lead alloy grid, in which the paste then forms the active material of the plate. The paste is held in position by a long interlocking grid section. In addition to the long-life microporous plastic separator, a glass-fibre mat is used. This

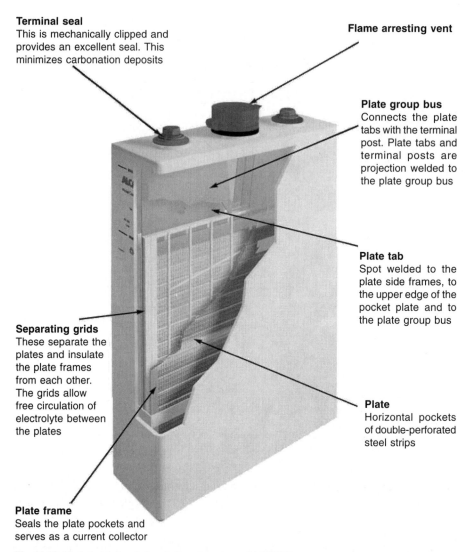

Terminal seal
This is mechanically clipped and provides an excellent seal. This minimizes carbonation deposits

Flame arresting vent

Plate group bus
Connects the plate tabs with the terminal post. Plate tabs and terminal posts are projection welded to the plate group bus

Plate tab
Spot welded to the plate side frames, to the upper edge of the pocket plate and to the plate group bus

Separating grids
These separate the plates and insulate the plate frames from each other. The grids allow free circulation of electrolyte between the plates

Plate
Horizontal pockets of double-perforated steel strips

Plate frame
Seals the plate pockets and serves as a current collector

Fig. 12.10 Vented nickel cadmium battery (courtesy of ALCAD)

becomes embedded in the face of the positive plate holding in place the active material and so prolonging the life of the battery.

Cells can be manufactured with lead antimony selenium alloy plates. These require very little maintenance, offer good cycling performance and are tolerant to elevated temperatures. Alternatively, lead calcium tin alloy plates may be used; these offer even lower maintenance levels.

Pasted-plate lead acid batteries are the lowest in purchase cost; their service life is in the range two to 20 years and the maintenance requirements are low, with extended intervals between watering. They are compact and provide high power density with excellent power output for short durations. Typical applications are in medium-to-long duration uses where initial capital cost is a primary consideration. These include telecommunications, UPS, *power generation transmission*, switch tripping and closing, emergency lighting and engine starting.

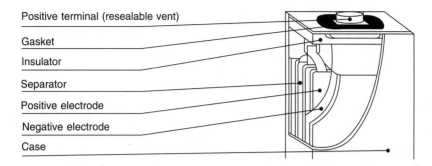

Positive terminal (resealable vent)

Gasket

Insulator

Separator

Positive electrode

Negative electrode

Case

Fig. 12.11 Nickel metal hydride battery (courtesy of SAFT Ltd)

12.3.5 Lead acid – tubular

In a tubular cell, an example of which is shown in **Fig. 12.12**, the positive plate is constructed from a series of vertical lead alloy spines or fingers which resemble a comb. The active material is lead oxide; this is packed around each spine and is retained by tubes of woven glass fibre which are protected by an outer sleeve of woven polyester or perforated PVC. This plate design enables the cell to withstand the frequent charge–discharge cycles which cause rapid deterioration in other types of lead acid cell.

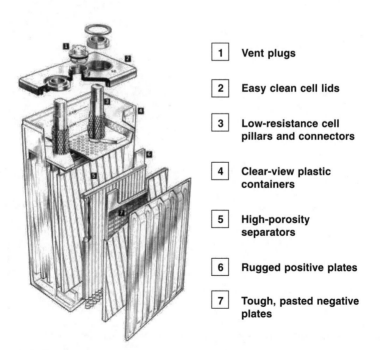

1 Vent plugs

2 Easy clean cell lids

3 Low-resistance cell pillars and connectors

4 Clear-view plastic containers

5 High-porosity separators

6 Rugged positive plates

7 Tough, pasted negative plates

Fig. 12.12 Tubular cell lead acid battery (courtesy of Invensys)

The advantages of the tubular cell are excellent deep cycling characteristics, service life up to 15 years and a compact layout which gives a high power per unit volume. The low antimony types also offer extended watering intervals.

Typical applications are where the power supply is unreliable, and where discharges are likely to be both frequent and deep. These include telecommunications, UPS, emergency lighting and solar energy.

12.3.6 Lead acid – Planté

The distinguishing feature of the Planté cell is the single pure lead casting of the Planté positive plate. The active material is formed from the plate surface, eliminating the need for mechanical bonding of a separately applied active material to a current collector. This makes the Planté cell the most reliable of all lead acid types. An example of the construction of Planté batteries is shown in **Fig. 12.13**.

1. Vent plugs
2. Cell lids
3. Cell pillars and connectors
4. Bar guard
5. Negative plates
6. Separators
7. Planté positive plates
8. Plastic containers

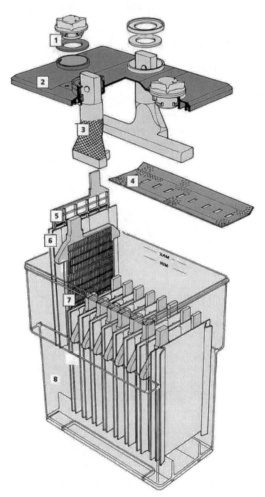

Fig. 12.13 Lead acid Planté cells (courtesy of Invensys)

Ultra High Performance Planté cells have a thinner plate. This gives the highest standards of reliability for UPS and other high-rate applications.

The advantages of the Planté cell are extremely long life (which can be in excess of 20 years for High Performance cells and 15 years for Ultra High Performance cells) and the highest levels of reliability and integrity. Constant capacity is available throughout the service life, and an assessment of condition and residual life can readily be made by visual inspection. Usage at high temperatures can be tolerated without significantly compromising the life expectancy, and maintenance requirements are very low.

Typical applications are long-life float duties in which the ultimate in reliability is required. These include critical power station systems, telecommunications, UPS, switch tripping and closing, emergency lighting and engine starting.

12.3.7 Lead acid – valve regulated sealed (VRSLA)

In sealed designs, lead calcium or pure lead is used in the grids. The separator is a vital part of the design because of its influence on gas recombination, and the amount of electrolyte that it retains; it often consists of a highly porous sheet of microfibre. Examples of sealed cells are shown in **Fig. 12.14**.

Beside these sealed units, there is a large variety of so-called maintenance-free batteries which rely more upon carefully controlled charging than upon the gas recombination mechanism within the cells. Non-antimonial grids and pasted plates are used, and in most units the electrolyte is immobilized or gelled. The batteries are provided with vents which open at a relatively low overpressure.

The benefits are long life, no topping up, no acid fumes and no requirement for forced ventilation. There is no need for a separate battery room and the units can be located within the enclosure of an electronic system. In addition these batteries are lighter and smaller than traditional vented cells.

Typical applications include main exchanges for telecommunications, PABX systems, cellular radio, microwave links, UPS, switch tripping and closing, emergency lighting and engine starting.

12.4 Fuel cells

A fuel cell converts the energy of a chemical reaction directly into electrical energy. The process involves the oxidation of a fuel, which is normally a hydrogen-rich gas, and the reduction of an oxidant, which is usually atmospheric oxygen. Electrons are passed from the fuel electrode to the oxidant electrode through the externally connected load, and the electrical circuit is completed by ionized particles that cross an electrolyte to recombine as water.

The fuel cell has a number of advantages over the conventional heat engine and shaft-driven generator for the production of electrical power. The generation efficiency is higher in a fuel cell, it has a higher power density, lower emission and vibration characteristics, and reduced emission of pollutants.

An individual fuel cell operates at a dc voltage of about 1 V. Cells must therefore be connected in stacks of series and parallel connection in order to deliver the voltage and power required for many applications.

Another complication is that many fuel cells cannot operate effectively using a raw hydrocarbon fuel; most types require a reformer, which converts the hydrocarbon into a hydrogen-rich gas suitable for passing directly into the fuel cell.

The two main types of fuel cell at present being considered are the *Solid Polymer Fuel Cell (SPFC)* and the *Solid Oxide Fuel Cell (SOFC)*. These two main classes are distinguished by the type of electrolyte they use. This in turn determines their operating temperature.

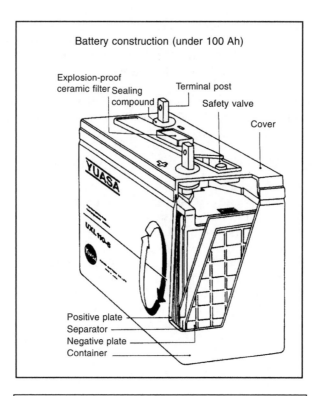

Battery construction (under 100 Ah)

Explosion-proof ceramic filter Sealing compound
Terminal post
Safety valve
Cover

Positive plate
Separator
Negative plate
Container

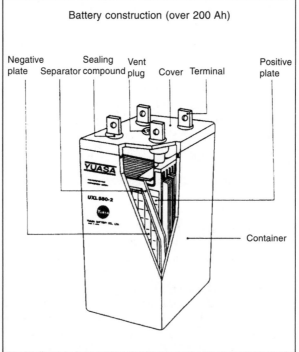

Battery construction (over 200 Ah)

Negative plate   Sealing   Vent plug   Cover   Terminal   Positive plate
Separator compound

Container

Fig. 12.14 Examples of sealed lead acid batteries (courtesy of Yuasa)

12.4.1 The Solid Polymer Fuel Cell (SPFC)

An example of an SPFC is shown in **Figs 12.15** and **12.16**.

The SPFC operates at temperatures below 100°C. It is a potentially clean method of generating electricity which is silent, robust and efficient. Applications for the

Fuel cell structure

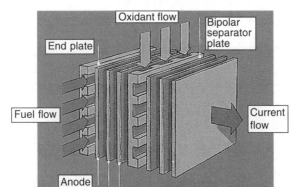

Fig. 12.15 The solid polymer fuel cell (courtesy of ETSU)

Fig. 12.16 The solid polymer fuel cell (courtesy of Loughborough University)

SPFC cover the domestic, commercial and industrial range, but it will probably be best suited to small-scale Combined Heat and Power (CHP) applications, and to transport duties.

The SPFC is often described by reference to power density, which ranges from 0.25–1.0 kW/litre. Expected life is in the range 8–20 years.

12.4.2   The Solid Oxide Fuel Cell (SOFC)

The SOFC is illustrated in **Fig. 12.17**.

Fig. 12.17 The solid oxide fuel cell (courtesy of ETSU)

The SOFC operates at much higher temperatures than the SPFC, in the region 850°C to 1000°C. It offers the possibility of high electrical efficiency together with high-grade exhaust heat; applications for the SOFC are therefore in the combined heat and power field and in other generation applications where a significant heat load is present.

Advantages of the SOFC are a high electrical efficiency, and the facility to produce steam or hot water from the high temperatures that are available. Typical power ratings are in the range 150–250 kW and life expectancy is 5–20 years, with adequate maintenance.

12.5   Battery charging

12.5.1   Small commercial batteries (up to 10 Ah)

The rechargeable batteries used in portable equipment are mainly nickel cadmium or nickel metal hydride. Recharging of these batteries can be carried out by the following methods:

- transformer–rectifier
- switched-mode power supply unit
- capacitive chargers

The transformer–rectifier circuit is used to reduce ac mains voltage to a lower dc voltage. The charge delivered to the battery can be regulated using a constant current, a constant voltage, constant temperature or a combination of these. The transformer in this system provides the inherent advantage of isolation between the mains and low-voltage circuits.

In the switched-mode power supply unit, the mains voltage is rectified, switched at a high frequency (between 20 kHz and 1 mHz), converted to low voltage through a high-frequency transformer, rectified back to dc which is then regulated to charge the battery in a controlled manner. Although more complex than the transformer–rectifier circuit, this technique results in a smaller and more efficient charging unit.

In a capacitive charger the mains is rectified and then series coupled with a mains-voltage capacitor with current regulation directly to the battery pack. This makes for a small and cheap charging unit, but it has to be treated with care, since the capacitor is not isolated from the mains.

In addition to these three methods, several manufacturers have responded to the need for shorter recharge periods and better battery condition monitoring by designing specific integrated circuits that provide both the charging control and the monitoring.

12.5.2   Automotive batteries

The main types of requirement for recharging automotive batteries are:

- single battery, out of the vehicle
- multiple batteries, out of the vehicle
- starter charging, battery in the vehicle

Transformer/reactance is predominantly used in each of these applications, with either a ballast in the form of resistance, or resistance to control the charging current. More sophisticated chargers have constant voltage and current-limiting facilities to suit the charging of 'maintenance-free' batteries. A typical circuit is shown in **Fig. 12.18**.

Vented automotive batteries are usually delivered to an agent in the dry condition. They have to be filled and charged by the agent. A multiple set of batteries connected in series is usually charged for a preset time in this case, and this requires the use of a bench-type charger. Bench chargers are rated from 2 V to 72 V, with a current capability ranging from 10–20 A dc (mean). Examples are illustrated in **Fig. 12.19**. The bench-type charger may also be used for charging a single battery out of a vehicle, or this battery may alternatively be left on charge indefinitely using a constant-voltage, current-limited charger.

Starter charging is used to start a vehicle which has a discharged battery. A large current is delivered for a short period, and starter chargers usually have a short-term rating. They can be rated at 6, 12 or 24 V, delivering a short-circuit rated starting current from 150 A to 500 A dc (rms) and a steady-state output of 10 A to 100 A dc.

Output voltage is from 1.8 V to 3.0 V per cell. Voltage ripple may be typically up to 47 per cent with a simple single-phase transformer–rectifier, but the ripple will depend upon output voltage, battery capacity and the state of discharge. Voltage regulation is important in starter charging applications since high–voltage excursions can damage sensitive electrical equipment in the vehicle.

Since these chargers are usually short-time rated, some derating may be required if they have to operate in a high ambient temperature.

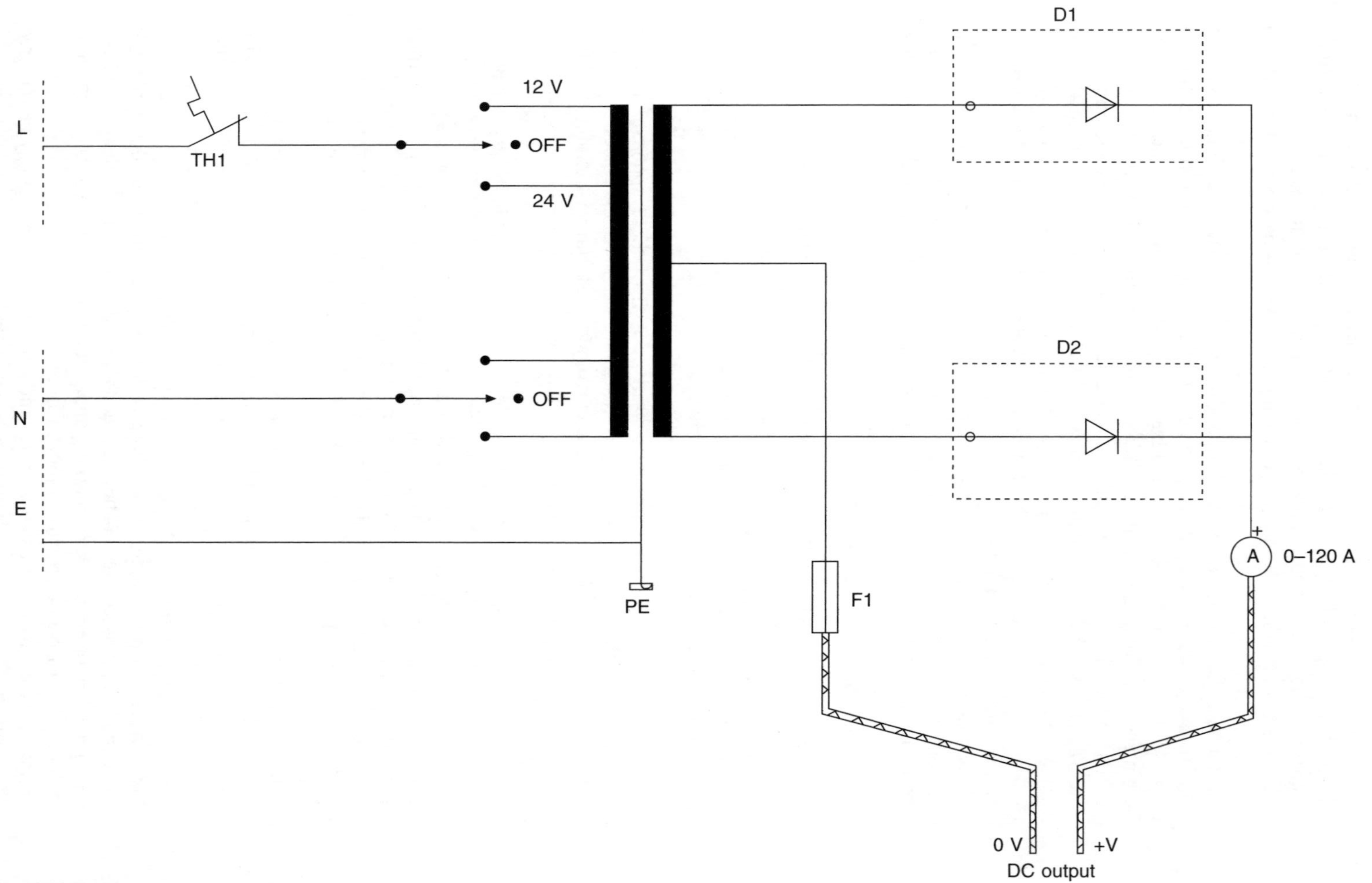

Fig. 12.18 Typical circuit for automotive battery charging

Fig. 12.19 Examples of bench-type battery chargers (courtesy of Deakin Davenset Rectifiers)

Basic charger circuit

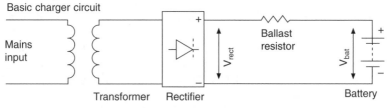

Transformer to match battery voltage and provide isolation
Rectifier converts ac to dc
Ballast resistor to limit current

$$I = \frac{V_{rect} - V_{bat}}{Ballast}$$

Fig. 12.20 Modified constant potential, or taper charger

Safety features are built into automotive battery chargers in order to avoid the risk of damage to the battery, to vehicle wiring or to the operator. The normal safety features include:

- reverse polarity protection
- no battery – short-circuit protection
- thermal trip – abuse protection

12.5.3 Motive power

Motive power or traction battery chargers are used in applications where the batteries provide the main propulsion for the vehicle. These applications include fork lift trucks, milk floats, electrical guided vehicles, wheelchairs and golf trolleys. The requirement is to recharge batteries which have been discharged to varying degrees, within a short period (7–14 hours). Both the battery and the charger may be subject to wide temperature variations. A well-designed charger will be simple to operate, will automatically compensate for fluctuations in mains voltage and for differences between batteries arising from such factors as manufacture, age and temperature, and will even tolerate connection to abused batteries which may have some cells short-circuited.

The most common type of charger is the *modified constant potential* or *taper charger* shown in **Fig. 12.18**. In all but the smallest chargers, the ballast resistor shown in this circuit is replaced by a reactance; this reactance may be in the form of a choke connected in series with the primary or secondary windings of the transformer, or more usually it is built into the transformer as leakage reactance.

While the battery is on charge, the voltage rises steadily from 2.1 V per cell to 2.35 V per cell, at which point the battery is approximately 80 per cent charged and gassing begins. Gassing is the result of breakdown and dissipation of the water content in the electrolyte. Charging beyond this point is accompanied by a sharp rise in voltage, and when the battery is fully charged the voltage settles to a constant voltage, the value of which depends upon a number of factors, including battery construction, age and temperature.

During the gassing phase (above 2.35 V per cell) the charging current must be limited in order to prevent excessive overheating and loss of electrolyte. The purpose of the ballast resistor in the charger circuit is to reduce the charging current as the battery voltage rises, hence the name *taper charger*. By convention, the current output from the charger is quoted at a voltage of 2.0 V per cell, and the proportion of this current which is delivered at 2.6 V per cell is defined as the *taper*. A typical current limit recommended by battery manufacturers is one-twelfth of the battery capacity (defined as the 5-hour rate in ampere-hours) at the mean gassing voltage of 2.5 V per cell. A disadvantage of the high-reactance taper charging system is that the output may be very sensitive to changes in the input voltage; the charge termination method and the required recharge time must therefore be taken into account in sizing the charger.

Because of the inefficiencies of energy conversion, particularly due to the heating and electrolysis during the gassing phase, the energy delivered by the motive power charger during recharging is 12–15 per cent higher than the energy delivered by the battery during discharge.

To recharge a battery fully in less than 14 hours a high rate of charge is necessary and the termination of charging when the battery is fully charged must be controlled. The two types of device for termination of charge are:

- voltage–time termination
- rate-of-charge termination

A *voltage–time controller* detects the point at which the battery voltage reaches 2.35 V per cell, and then allows a fixed 'gassing' time for further charging, which is usually 4 hours. This method is not suitable for simple taper chargers with nominal recharge times of less than 10 hours because of the variation in charge returned during the time period as a result of mains supply fluctuations. If a short recharge time is required with a voltage–time controller then a *two-step taper charger* is used, in which a higher charging current is used for the first part of the recharge cycle. When the voltage reaches 2.35 V per cell, the timer is started as above, and the current during the further charging period is reduced by introducing more ballast in the circuit.

The *rate-of-charge* method of termination has predominated in large chargers during the past decade, offering benefits to both the manufacturer and the user. In the rate-of-charge system the battery voltage is continuously monitored by an electronic circuit. When the battery voltage exceeds 2.35 V per cell, the rate of rise of the battery voltage is calculated and charging is terminated when this rate of rise is zero, that is when the battery voltage is constant. This method can be used with single-rate taper chargers with recharge periods as short as 7 hours because of the higher precision of termination. In order for a rate-of-charge termination system to operate satisfactorily, there must be compensation for the effects of fluctuations in mains supply. A change in mains voltage results in a proportional change in the secondary output voltage from the charger transformer; if uncompensated this will cause a change in charging current and therefore in battery voltage. For a 6 per cent change in mains voltage, the charging current may change as much as 20 per cent and the battery voltage may change by 3 per cent.

Many chargers have the facility for *freshening* or *equalizing*.

Freshening charge is supplied to the battery after the termination of normal charge in order to compensate for the normal tendency of a battery to discharge itself. A freshening charge may be a continuous low current, or trickle charging, or it may be a burst of higher current applied at regular intervals.

Equalizing charge is supplied to the battery in addition to the normal charge to ensure that those cells which have been more deeply discharged than others (due, for instance, to tapping off a low-voltage supply) are restored to a fully charged state.

A *controlled charger* is a programmable power supply based on either thyristor phase angle control or high-frequency switch mode techniques. The main part of the recharge cycle is usually at constant current and the power taken by the charger is therefore constant until the battery voltage reaches 2.35 V per cell. Many options are available for the current–voltage profile during the gassing part of the recharging cycle, but all of these profiles deliver a current which is lower than the first part of the cycle.

Voltage drop in the cable between the charger and the battery is important because the charge control and termination circuitry relies upon accurate measurement of the battery voltage. It is not normally practical to measure the voltage at the battery terminals because the measuring leads would be either too costly or too susceptible to damage, and it is common practice to sense the voltage at the output of the charger and to make an allowance for the voltage drop in the cables. Alteration to the length or cross-section of these cables will therefore cause errors, especially with low-voltage batteries.

Motive power chargers are typically available from 6 V (three cells) to 160 V (80 cells) with mean dc output currents from 10 A to 200 A. A typical circuit is shown in **Fig. 12.21**.

The dc output current is rated at 2.0 V per cell; the taper characteristic set by the transformer reactance then results in 25 per cent output current at typically 2.65 V per cell. Rating is not continuous, and derating to 80 per cent is typical to take into account the taper characteristic. Consideration must be given to this if a multiple shift working pattern is to be adopted. Typical charger units, together with their ratings, are shown in **Fig. 12.22**.

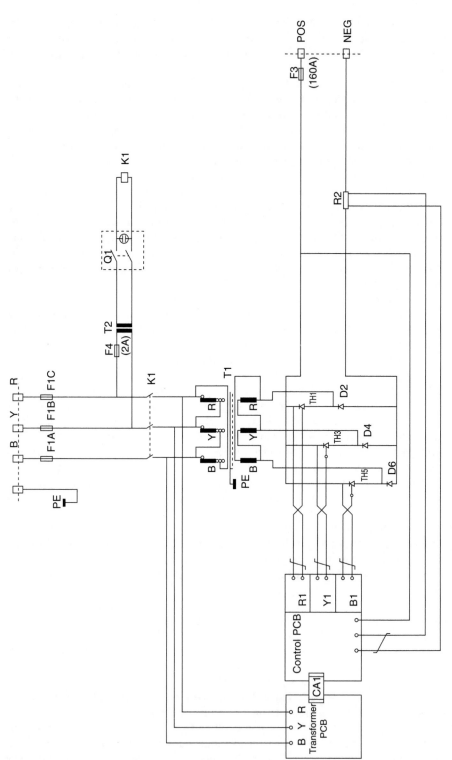

Fig. 12.21 Typical circuit diagram for a motive power charger

EURO 92
Motive power chargers

Charger selection table (based on 80% discharge at 2.0 V per cell)

Charger rating (amps)	Recharge times (hours)						
	7.5	8	9	10	11	12	13
15	84	88	96	103	112	120	130
20	112	118	128	138	149	160	173
25	139	147	159	172	186	200	216
30	167	176	191	206	223	240	259
35	195	205	223	240	260	280	302
40	222	235	255	275	297	320	345
45	250	264	286	309	334	360	389
50	278	293	318	343	371	400	431
60	334	352	382	412	445	480	517
70	390	410	445	480	519	560	603
80	445	469	509	549	593	640	690
90	501	527	572	617	668	720	776
100	556	586	636	686	742	800	862
110	617	644	699	754	816	880	948
120	668	703	763	821	890	960	1034
130	723	761	826	891	964	1040	1120
140	778	820	890	960	1038	1120	1206
160	890	937	1017	1097	1186	1280	1378
180	1001	1054	1144	1234	1334	1440	1550
200	1112	1171	1271	1371	1483	1600	1723

Fig. 12.22 Examples of motive power chargers, with their ratings (courtesy of Deakin Davenset Rectifiers)

Voltage ripple is typically 15–25 per cent for single-phase chargers and 5–15 per cent for three-phase chargers. The precise level of ripple will vary with time. It will depend upon mains voltage, battery capacity and the depth of discharge of the battery.

12.5.4   Standby power applications

Typical standby power applications include emergency lighting, switch tripping, switch closing and telecommunications. The main functional requirements for the battery charger in these cases are:

- to ensure that the state of charge of the battery is maintained at an adequate level, without reducing battery life or necessitating undue maintenance
- to ensure that the output voltage and current of the complete system are compatible with the connected electrical load
- to ensure after a discharge that the battery is sufficiently recharged within a specified time to perform the required discharge duty
- to provide adequate condition monitoring, to the appropriate interface standards

Assuming initially that the battery is fully charged, the simple option is to do nothing. A charged battery will discharge if left disconnected from a load and from charging equipment, but if the battery is kept clean and dry this discharge will be quite slow. For some applications, open-circuit storage is therefore acceptable.

For most applications, however, there is a need for battery charge to be maintained. The current–voltage characteristic is not linear, and a small increase in charging voltage will result in a large increase in current. Nevertheless, it is always possible to define a voltage which, when applied to a standby power battery, will maintain charge without excessive current, and the charging current flowing into the battery has only to replace the open-circuit losses in the battery, which are usually small.

Once these open-circuit losses have been made up, any additional current flowing is unnecessary for charging purposes and is normally undesirable. In vented cells it causes overheating and gassing and eventually, if not checked, damage and loss of capacity of the cell. In sealed cells there can be overheating, in extreme cases expulsion of gases through the pressure vent, and ultimately a loss of capacity.

On the other hand, if the battery voltage is allowed to fall too much, the open-circuit losses will not be replaced and the battery will slowly discharge.

The charging voltage has therefore to be controlled carefully for best battery maintenance. The usual limits are within ±1 per cent of the ideal voltage, which is normally termed the *float voltage*. The float voltage has a negative temperature coefficient which must be accounted for when batteries are to operate in exceptionally hot or cold environments. Float voltages for the major types of standby power cell are shown in **Table 12.2**.

For vented cells there are, however, circumstances under which the float voltage should be exceeded. Batteries which are new or have suffered abuse will benefit from a vigorous gassing up to the *boost voltage* shown in **Table 12.2**. Batteries which have stood on float charge with no discharge–charge cycle for many months will benefit from a *refresh charge* with gassing, at the refresh voltages which are also shown in **Table 12.2**.

Few dc standby power systems can be designed without taking account of *limits on the load voltage*. For good battery operation it is necessary to charge at the float voltage (or sometimes a higher level), but it is also necessary to discharge the battery

Table 12.2 Typical cell voltages for systems with limited load voltage excursions

Cell type	Cell voltage (V)			
	End of discharge	Float	Refresh	Boost
Vented nickel cadmium	1.1	1.45	1.55	1.7
Vented lead acid	1.8	2.25	2.45	2.7
Sealed lead acid	1.8	2.27	N/A	N/A

to a sufficiently low voltage if the full capacity is to be released from the cells. These considerations impose fundamental limits which define the maximum excursion of the system output voltage. **Table 12.3** shows the minimum and maximum voltages which are reasonable in a 50 V standby power system for each for the three major battery types. Systems with other voltages will require excursions which are in direct proportion.

Table 12.3 Load voltage ranges in a 50 V standby power system

Function	Minimum volt/cell (V)	Maximum volt/cell (V)	50 V system			
			No. of cells	Minimum voltage (V)	Maximum voltage (V)	± voltage excursion (%)
Vented lead acid cells						
Float	1.80	2.25	25	45.0	56.3	11.3
Refresh	1.80	2.45	24	43.2	58.8	15.6
Boost	1.80	2.70	22	39.6	59.4	19.8
Vented nickel cadmium cells						
Float	1.10	1.45	39	42.9	56.6	13.7
Refresh	1.10	1.55	38	41.8	58.9	17.1
Sealed lead acid						
Float	1.80	2.27	25	45.0	56.8	11.8

It can be seen from **Table 12.3** that, allowing for the discrete steps in voltage when changing the number of cells, it is not possible to achieve a voltage excursion of less than about ± 12 per cent under conditions of float alone; the excursion limits are larger in refresh operation, and still larger with boost.

Table 12.3 shows that the sealed lead acid cell offers minimum overall voltage variation because of the absence of the larger excursions due to refresh and boost charges. If it is not possible to use a sealed lead acid system, another alternative is to disconnect the load for the full boosting operation; this should normally be necessary only at the time of system installation in any case.

A voltage regulator should be included in the system if closer limits of voltage variation are required. Diode regulators are now reliable and widely used; they operate by switching banks of series-connected diodes in and out as the battery voltage slowly varies. It is important to ensure that switching in the regulator occurs only as a result of changes in battery voltage, and not as a result of load changes; if the regulator responds to changes which occur as a result of a load change, excursions outside the specified voltage limits may occur because of the delays in the process of switching the diode bank. Ensuring that this distinction is made normally requires a computer simulation for all but the simplest regulators. Control of the switching of the diode bank is best achieved by a programmable controller, especially if additional complicated relay-type logic is required in the system. Diode regulators are large and their heat dissipation is substantial. Higher-efficiency regulators using actively switched devices are becoming available, but they are at present limited to relatively low power applications.

Condition monitoring is now included in most dc systems in order to warn of excessive battery voltage excursions. The applications are diverse, but the main features are:

- *high voltage detection:* this is necessary in order to prevent a fault on the supply system from damaging the battery or load circuit
- *low voltage detection:* this warns of load failure due to insufficient voltage, and it is also needed to trigger the disconnection of sealed batteries which may be damaged by excessive discharging
- *charge failure:* this is needed in order to stimulate action to restore the ac supply or to prepare for disconnection of the load
- *earth leakage:* this is needed where an unearthed load system is used, for safety and for the avoidance of 'double faults'

Communication of a fault signal is through a 'volt-free' contact on a relay, which usually signals to the monitoring centre using a 110 V or other voltage supply.

For *charging of the battery*, it has been seen in **Table 12.3** that there are restrictions on the choice of charging voltage. The preference for many applications is to limit the voltage to the float voltage, and while it may be necessary to increase the charging voltage to speed up the recharging, the voltage should be returned to the float voltage as soon as possible. At this float voltage level, all the cell types discussed will be recharged to about 80 per cent of their nominal capacity. Assuming the charger current is at the adequate level shown in **Table 12.4**, recharge to 80 per cent capacity will be achieved in the times indicated in the table.

Table 12.4 Charging time and current necessary to recharge to 80 per cent of capacity using float voltage

Cell type	Charging current	Charging time
Vented lead acid	7% of capacity	14 hours
Vented nickel cadmium	20% of capacity	6–8 hours (depending on type)
Sealed lead acid	10% of capacity	9 hours

To recover the remaining 20 per cent of the charge is more difficult, and different techniques are necessary for the three cell types. A vented lead acid battery will be fully recharged at float voltage in about 72 hours, but if the charging voltage is boosted to 2.7 V per cell, a full recharge will take about 14 hours. For a vented nickel cadmium cell full charge will never be achieved without increasing the voltage above the float level. Exact times vary between cell types, but typically a refresh charge at 1.55 V per cell will give full charge in about 200 hours on the highest-performance cells, and boosting to 1.7 V per cell will reduce this time to 9–10 hours. Sealed lead acid cells will reach full charge after about 72 hours at float voltage; an increase of charge voltage to 2.4 V per cell will reduce the charging time to about 48 hours, but there is no way in which this can be significantly reduced further.

If a fast recharge is essential, an alternative is to oversize the battery; for instance if 100 Ah capacity is required with a full recharge within 8 hours, then a battery with 125 Ah capacity could be installed.

Most cells will protect themselves from excess charging current, provided that the voltage is limited to the float voltage, but above this level excessive charging current can damage the cell. For vented lead acid cells the 7 per cent of capacity shown in **Table 12.4** is recommended as an upper limit. For sealed lead acid cells, the recommended upper limit is 50 per cent of capacity. Nickel cadmium cells normally require a minimum charge current of 20 per cent of capacity, as indicated in **Table 12.4**, but a lower limit of 10 per cent of capacity is recommended when boosting in order to avoid excessive gassing and electrolyte spray.

Standby power chargers are available in a wide range of capacities to suit many applications. Typical dc outputs are 6, 12, 24, 30, 48, 60, 110, 220 and 440 V, with dc mean output current ranging from 1 A to 1000 A. DC output current is rated at 100 per cent of the output current at the full specified voltage. Regulation of the output is generally within ±1 per cent for an input voltage change of ±10 per cent and a load current change of 0–100 per cent. Typical circuits are shown in **Figs 12.23** and **12.24**. Although standby power systems are continuously rated, some derating may be necessary for operation in tropical climates if this was not originally specified.

Differing levels of output smoothing can be incorporated into the charging system, depending upon the application. General applications require a maximum of 5 per cent ripple, but for telecommunications supplies, specifications are based on CCITT telecommunication smoothing, which requires 2 mV phospometrically weighted at 800 Hz. The key components of an installation are shown schematically in **Fig. 12.25**.

12.6 Battery monitoring

The monitoring of battery condition is becoming more important as remote operation and reduced maintenance requirements are increasingly specified in standby power systems. The options which are available are summarized in the following sections.

12.6.1 Load-discharge testing

The basis of this test is to discharge the battery into a selected load for a preset time, after which the battery voltage is checked. In the majority of cases this will require that the battery is taken off line.

The load-discharge test is very reliable, but it requires a special resistance load bank, and it is labour intensive and time consuming. **Figure 12.26** shows an example of the type of equipment that is required.

12.6.2 *dV/dt*-load testing

The technique here is to connect a fixed known load to the battery for a short period and to record the battery voltage over this period.

By comparing the recorded load voltage with data from the battery manufacturer, estimations can be made regarding the condition of the battery and its ability to perform.

12.6.3 Computer monitoring of cell voltages

In this case, measurement leads are taken from each cell to a central monitoring computer.

The voltage of each cell is recorded through charge and discharge of the battery. Although cell voltage is not a direct indication of residual capacity, trends can be observed and weak cells can be identified.

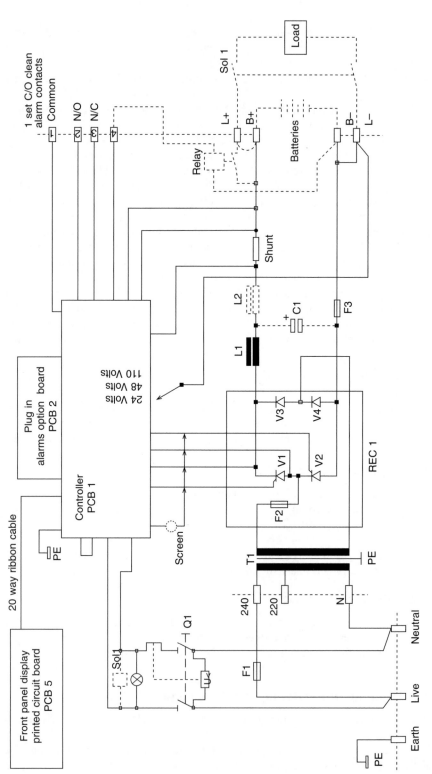

Fig. 12.23 Typical circuit for a standby charger

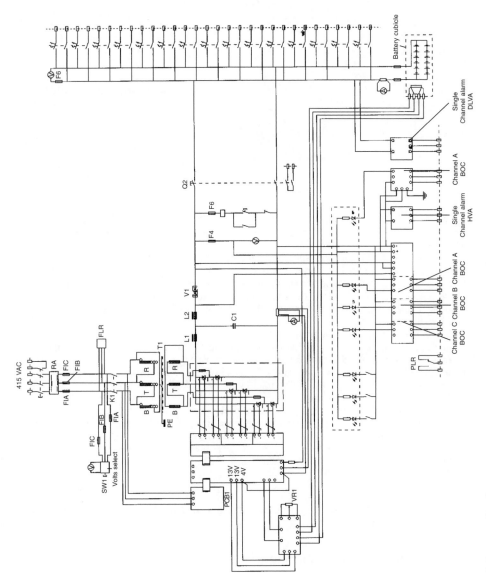

Fig. 12.24 Typical circuit for a standby charger

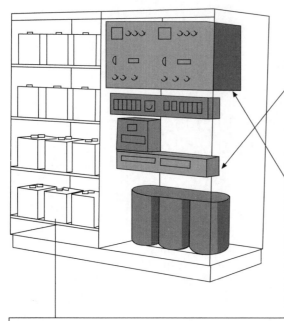

The rectifier/battery charger

The rectifier/charger converts incoming ac mains into dc to provide a stable, constant voltage output with automatic current limiting. The rectifier–charger automatically floats and boost charges the battery simultaneously providing the required dc output to the load.

Control and distribution

Distribution panels typically provide several fused outlets, monitoring of busbar voltage and current, earth fault status together with busbar and interbusbar controls. Depending upon application, the control and distribution equipment may be contained within the charger cubicle (as shown) or may be housed in a separate matching enclosure.

The battery

The battery is a key element in a secure power supply and must be selected with care. The type of battery selected will depend on many factors including: reliability, operating tempera-ture, cost, life, standby time required, maintenance parameters, ventilation and available space. Depending on the type of battery, it can be supplied on a stand or in a ventilated cabinet. The battery should always be located as close as possible to the dc power supply to minimize line voltage drop.

The main types of battery used are:

*   valve regulated lead acid; minimal gassing so can be used in electronics enclosure, maintenance free, relative installed cost medium, typical life 10 yrs
*   flat plate lead acid: gas, with adequate ventilation and periodic maintenance required. Relative installed cost low, typical life 10–15 yrs.
*   Planté lead acid: gas, with adequate ventilation and periodic maintenance required. Highest levels of reliablility. Relative installed cost high, typical life 20–25 yrs.
*   nickel cadmium: gas, with adequate ventilation and periodic maintenance required. Ideal for arduous conditions. Relative installed cost very high, typical life 20–25 yrs.

Fig. 12.25 Key elements of a standby power battery unit

An example of this type of system is illustrated in **Fig. 12.27**, which includes a typical screen display.

12.6.4 Conductance monitoring

This method has increased in popularity in recent years and several manufacturers now offer a standard monitoring product based upon the technique. **Figure 12.28** shows typical equipment.

Monitoring the conductance of individual cells over their working life can give an indication of impending failure, allowing preventive maintenance to be carried out. A particular advantage is that the readings can be reliably used whatever the state of charge of the battery.

Fig. 12.26 Equipment for load-discharge testing (courtesy of Deakin Davenset Rectifiers)

Fig. 12.27 Computer monitoring of cell voltages (courtesy of Invensys)

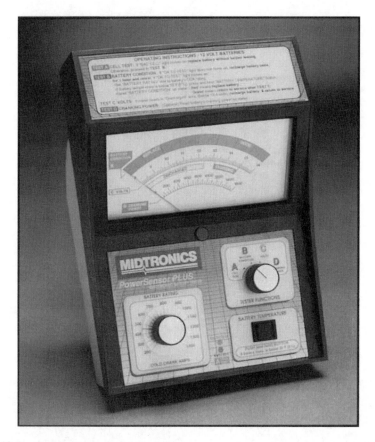

Fig. 12.28 Conductance monitoring equipment

12.7   Installation, testing and commissioning

All battery equipment should be installed according to manufacturers' recommendations.

In particular, larger lead acid or nickel cadmium installations should be sited in a correctly constructed room, vehicle, cubicle or rack which allows ready access for maintenance and ensures adequate ventilation of batteries and battery chargers. The battery should not be located close to a source of heat such as a transformer or heater. Battery racks should be assembled to manufacturers' instructions, adhering to recommendations regarding spacing between cells, and drip trays and insulators should be fitted where applicable. All cell connections should be cleaned, coated with no-oxide grease and tightened to the suppliers' specification. A typical installation is shown in **Fig. 12.29**.

The rating of the ac mains input cable to the battery charger should be based on the worst-case condition, which is the peak current that will be drawn by the charger at the lowest mains voltage. The rating of the dc cables from the battery charger and in the battery should be no lower than the maximum protection fuse rating, and attention should be paid to the overall voltage drop in the system, especially on lower voltage dc systems. The cables should be adequately protected, and segregation may be necessary on systems with a large dc current. All equipment should be earthed with a conductor of the correct cross-sectional area.

Fig. 12.29 Typical battery installation (courtesy of Invensys)

Battery systems should be commissioned in accordance with procedures recommended by the manufacturer. It may be necessary to charge batteries for several days before their full capacity is reached, and in some cases a special initial commissioning charge level may be required.

Tests should be set up so that the results enable the operation of the system to be compared with the original specification. For larger systems, the manufacturer should propose a set of witness tests to be performed at the works, followed by final tests to prove capacity and duty on site. A typical test schedule should be designed to establish:

- the ability of the battery charger to support the standing load, where applicable
- the ability of the battery charger to recharge a fully discharged battery in the required time, whilst simultaneously time supporting the standing load, again where applicable
- the ability of the battery charger to maintain regulation within specified limits throughout the required range of load and input conditions
- correct operation of the monitoring features and instrumentation
- the ability of the battery and charger system to supply the load for the required period during a power failure

12.8 Operation and maintenance

12.8.1 Primary cells

Primary cells cannot be recharged and once their energy is depleted they should be disposed of in the manner recommended by the manufacturer. The cells should be examined occasionally during their life for signs of leakage or physical deformation. Chemical hazards can arise if batteries are misused or abused and in extreme cases if there is a risk of fire or explosion.

12.8.2 Secondary cells

Small secondary cells should be examined periodically under both charging and discharging conditions. Their cycle life is not indefinite, and as the battery ages its ability to accept recharge and hold its capacity will decrease. This ageing is usually shown by the dissipation of heat during charging, as a result of the inability to accept charge, whilst on discharge a reduction of capacity will be observed. Excessive overcharging will cause leakage of electrolyte or deformation of the cell.

Cells should be handled carefully and according to manufacturers' instructions, with the operator wearing the correct protective clothing. When handling vented cells, supplies of saline solution (for eye washing) and clean water should be readily available. If a cell is to be filled, only purified water should be used when mixing the electrolyte.

The maximum storage time for cells which are filled and nominally charged is 6 months for nickel cadmium, 12 months for valve-regulated lead acid batteries (at 20°C) and 8 weeks for vented lead acid batteries. Cells which are stored for longer than this should be periodically charged and the electrolyte level should be checked, where applicable.

Whether filled or empty, the battery may contain explosive gases, and no smoking, sparks or flames should be permitted in the vicinity of the installation.

Although the voltage at any point in a battery system can be reduced by the removal of inter-cell connectors, the cells are electrically live at all times and cannot be de-energized or isolated in the conventional sense. When connecting cells together, insulated tools should be used wherever possible to avoid accidental short-circuits and sparks. Tools should be cleaned before use if they are to be used on both lead acid and nickel cadmium batteries, since acid will destroy a nickel cadmium battery.

On large installations it is good practice to operate a system of 'cell log sheets'. Such sheets or books are normally supplied by the cell manufacturer. Completion of these sheets will require the measurement of voltage, ambient temperature and, if applicable, the specific gravity of each cell at regular intervals. Cell electrolyte level (where applicable) and the tightness of cell connections should also be checked at these intervals.

Other regular operational and maintenance checks on the charger system should include:

- ac mains input voltage is correct and within tolerance
- dc charger output voltage and current are correct and within tolerance, after compensation for ambient temperature
- fuses are intact
- protection devices operate correctly
- all connections are tight
- no components are operating at excessive temperature

Faults to be watched for are:

- loose connections in the cells or in the charger system

- low or high electrolyte level (where applicable)
- debris or electrolyte spillage on the top of cells which may lead to short-circuits
- excessive loss of electrolyte (where applicable) due to overcharging or battery ageing
- cells overheating because of their inability to accept charge
- loss of capacity due to undercharging (specific gravity should be checked where applicable)
- loss of charger regulation due to a control circuit fault
- ac or dc fuses operated
- excessive loading on the battery or on the charger

12.9 Standards

There are many standards covering various types of battery and battery charging system. The key IEC recommendations together with equivalent BS and EN standards and related North American standards are summarized in **Table 12.5**.

Table 12.5 Comparison of international, regional and national standards for batteries and battery chargers

IEC	EN	BS	Subject	N. American
428		5142	Specification for standard cell	
896	60896-1	6290-1	General specification for lead acid cells	
896	60896-2	6290-2	Planté cells	IEEE 450/484
896	60896-3	6290-3	Pasted cells	IEEE 450/484
896	60896-4	6290-4	Valve-regulated sealed cells	IEEE 1188
1056	61056-1	6745	Portable valve regulated lead acid cells	
		3031	Sulphuric acid for use in lead acid batteries	
		4974	Specification of water for lead acid batteries	
		7483	Specification for lead acid batteries in light vehicles	
254	60254	2550	Specification for lead acid traction batteries	
		6287	Code of practice for safe operation of traction cells	
95	60095	3911	Lead acid starter batteries	
		6133	Code of practice for safe operation of lead acid stationary batteries	
623	60623	6260	Nickel cadmium single cells	IEEE 1106
285	60285	5932	Nickel cadmium cylindrical cells	
622	60622	6115	Nickel cadmium single cells (prismatic)	ANSI C 18.2 M
		6132	Code of practice for safe operation of alkaline secondary cells and batteries	
1044	61044	EN61044	Operation charging of lead acid traction batteries	
	DIN 41774		Traction battery chargers – taper characteristics	
	DIN 41773		Traction battery chargers – characteristics	
952	60952	EN60952	Aircraft batteries	
335-2-29	60335-2-29	3456-2-29	Domestic battery chargers	UL 1564 ANSI 1564
146	60146	4417	Specification for converters	UL 458 UL 1012 UL 1236 UL 1310
950	60950	7002	Information technology equipment	UL 1950

References

12A.   May, G.J., *Journal of Power Sources,* **42**, 1993, pp. 147–153.
12B.   May, G.J., *Journal of Power Sources,* **53**, 1995, pp. 111–117.
12C.   *Rechargeable Batteries Handbook*, Butterworth-Heinemann, Oxford, UK, 1992.

Chapter 13

Electroheat

Dr Peter L. Jones
Pealjay Consultants

Mr I. Harvey

13.1 Introduction

Much of industry depends on heat processes to refine, shape, modify, dry, bake or carry out any one of a number of operations. Traditionally, much of this heat has been provided by fossil fuel; in earlier days this was coal and its derivatives and more recently it has been oil and gas. Electricity was used to provide drives for machinery and auxiliaries such as blowers, or for very specialized applications. The large-scale use of electricity was limited to such processes as electrochemical extraction of the chemicals originating from salt, or the extraction of aluminium from bauxite. In part, the explanation for the limited use of electroheat has been an economic one, and in some areas this continues to be the case. At the point of delivery, electricity in most countries is much more expensive than fossil fuels. This is inevitable because most electricity has been generated using fossil fuel, with a conversion efficiency that cannot be better than that of the Carnot cycle. Furthermore, the equipment necessary for the generation and distribution of electricity requires very significant capital investment. As a result, the case for the use of electricity in process heating needs to take into account other aspects of manufacturing costs and process benefits, such as improvements in product quality, reduced manning levels and increased throughput for a given capital investment.

Nevertheless, electroheat technologies are now widely established in processing and manufacturing industries ranging from metals, glass and ceramics to textiles, paper, food and drink. The heating techniques use frequencies which cover a significant part of the electromagnetic spectrum, from dc to terahertz, and at powers which range from a few watts to many megawatts. The different processes which have been developed, and the frequency and power ranges which they use, are summarized in **Table 13.1**.

The relative importance of the different techniques has changed as industry moves away from heavy engineering operations. In particular, some electroheat processes related to the iron and steel industry have decreased in relative importance whilst others, aimed at materials with higher added value, are becoming more widely accepted. It is also apparent that even in the heavy industries such as metals and ceramics, certain traditional and novel electroheat techniques are of great importance since they bring with them better process control, leading to higher product quality and

Table 13.1 Electric heating processes

Technique	Frequency range	Power range
Direct resistance	0–50 Hz	<1 kW–30 MW
Indirect resistance	50 Hz	0.5 kW–1 MW
Arc melting	50 Hz	1–100 MW
Induction heating	50 Hz–1 MHz	2 kW–30 MW
Dielectric heating	5–100 MHz	1 kW–5 MW
Microwave heating	0.9–25 GHz	1–100 kW
Plasma torch	4 MHz	1 kW–1 MW
Laser CO_2	30 THz	0.1–60 kW
Ultraviolet	750–1500 THz	1 kW

reduced manufacturing costs. Further details of most of these processes can be found in **reference 13A**.

It is possible to categorize electroheat techniques in a variety of ways. The following are examples:

- by the part of the frequency spectrum in which they operate, for example infrared, microwave or ultraviolet
- by the material being processed, such as metal melting or metal heat treatment
- by the generic name of the process, such as induction heating, dielectric heating or laser

No single categorization is generally accepted and a mixture of these is adopted in the structure of this chapter. Metal melting by various methods is dealt with in **section 13.2**. The route to a final component may also involve heat treatment, either for size reduction, to modify properties or to work into final form. A number of electroheat techniques are used and these are grouped under direct resistance heating, in **section 13.3**, and indirect resistance heating, in **section 13.4**. The later sections deal with more recently accepted technologies.

13.2   Metal melting

Electroheat is used in a number of ways in the metals industry, from the initial refining through to the formation into finished shape. In the case of refining, perhaps the greatest use is for the extraction of aluminium from bauxite. Because this is a very energy-intensive process, it is normally carried out in places where cheap hydroelectric power is available, such as Canada. There are relatively few installations in Europe.

Metal melting may take place for secondary refining, for recycling of scrap or for the production of castings. It is normally considered separately from metal heating for hot working, because the implementation of the methods for liquid metals is usually quite different.

The electroheat furnaces which are established for the melting of metals include arc, electron beam and coreless induction furnaces. The choice is usually based on the demand schedule for liquid metal which is required by the casting process. The requirements of a small foundry producing a few specialist castings each day are quite different from those of a large automated foundry. The seven main types of furnaces commonly used in the melting of metals are described in the following sections.

13.2.1 The arc furnace

This is mainly used for the production of liquid steel by the melting of scrap. It consists of a squat, refractory-lined cylindrical vessel with a movable domed refractory roof, as shown in **Fig. 13.1**. The roof is swung open to allow the furnace to be charged with scrap metal, and is closed prior to the start of the melting process. The three graphite electrodes pass through holes in the roof, and each is supplied through flexible cables and bus tubes from one of the three phases of a transformer. The metal charge forms the star point of the three-phase load circuit.

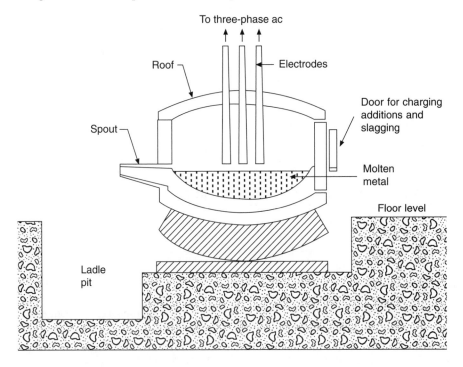

Fig. 13.1 Arc furnace

A variant of this conventional scrap melter is the vacuum arc furnace, which is used for the refining of metals. A partial vacuum is maintained in the vessel to minimize oxidation during melting.

The substation for a large arc furnace is normally adjacent to the furnace itself, and it contains the furnace transformer, which normally has a star-connected primary and (in the UK) an input voltage of 33 kV. The transformer must withstand the very large electromechanical forces which are produced by the high short-circuit currents. It is oil cooled and the power is varied using on-load tapchangers. Further comments on this class of transformer can be found in **section 6.3.10**.

As an electrical load, the arc furnace is less than ideal. The current can vary from zero, when the arc is extinguished at current zero, to a short-circuit level when scrap contacts the electrodes. The physical movement of the pieces of solid charge and of the melt causes variation in the arc length which can fluctuate many times within a second. In addition, the arc length varies due to mechanical vibration of the electrode and its supporting structure. These effects combine to prevent a constant power level being delivered by each of the phases, and this results in a number of problems which affect the performance of the furnace, the local electricity supply and the acoustic noise level.

13.2.2 The submerged arc process

This is not essentially an arc process, since heating occurs mainly by direct resistance effects, with perhaps some limited heating from arcs and sparks during interruption of the current path. The general arrangement is shown in **Fig. 13.2**.

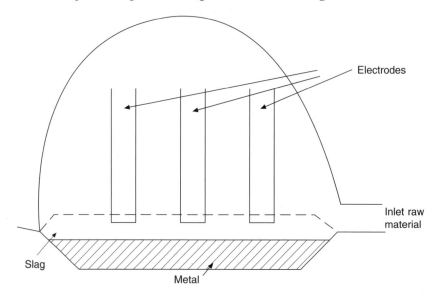

Fig. 13.2 Submerged arc furnace

The principal application is in the reduction of ferroalloys such as ferromanganese, nickel, chrome, silicon, tungsten and molybdenum, which have a high melting point. These metals are subsequently remelted in an arc furnace to produce special alloys. Fused oxides may also be produced by the submerged arc process.

13.2.3 Electroslag refining

This is an alternative to vacuum arc processes for materials which are not unduly reactive in air. A high degree of refining can be obtained, since the droplets of molten metal penetrate the molten slags, enabling desuphurization to be carried out and oxide inclusions to be removed. As in the vacuum arc furnace, a molten pool is formed on the solidifying ingot, and there are similar advantages.

The general arrangement of an electroslag refining furnace is shown in **Fig. 13.3**. The current path varies according to the construction of the furnace and varies with the state of the melt. In general terms, current passes from the electrodes, through the slag and from the slag to the baseplate, through the vessel walls or the ingot, or both.

13.2.4 The electron-beam furnace

Electron beams can be used for melting, as well as for welding and the production of evaporated coatings. The beam is obtained from a heated filament or plate, and is accelerated in an electron gun by a high electric field which is produced by one or

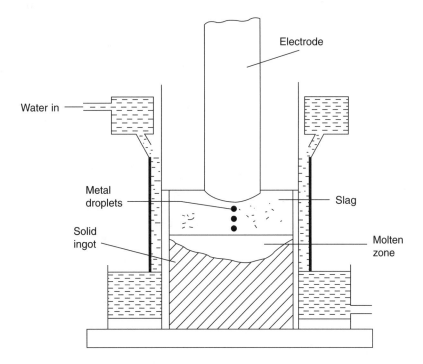

Fig. 13.3 Electroslag refining furnace

more annular electrodes. Electrons on the axis of the gun pass through the final anode at very high speeds, typically 85×10^6 m/s at 20 kV. The electron gun and chamber are kept at a low pressure of around 0.001 Pa. Since little energy is lost from scatter or from the production of secondary electrons, practically all of the kinetic energy of the beam is converted into heat at the workpiece. The conversion from electrical energy input to thermal energy in the workpiece is therefore very high.

The electron-beam furnace is shown diagrammatically in **Fig. 13.4**. It utilizes a cooled ingot mould in the same way as the vacuum and electroslag furnaces.

13.2.5 The coreless induction furnace

This consists of a refractory crucible encircled by a solenoid coil which is excited from a single-phase ac supply. A diagrammatic view is shown in **Fig. 13.5**.

The fluctuating axial magnetic field links the charge within the crucible and causes I^2R heating in the charge. The power induced in the charge and the efficiency of the furnace can be calculated according to the equations given later in **section 13.6**, using appropriate material property values. Depending upon the resistivity of the material in the charge, the coreless furnace converts the electrical input power to heat in the charge with an effciency of 65–80 per cent. This efficiency is largely independent of frequency, but is improved by the high charge permeability found in ferromagnetic materials such as steel scrap.

13.2.6 The channel induction furnace

During the 1970s the channel furnace became accepted as a melting unit for cast iron, and largely replaced the coke-fired cupola. Its acceptance was due mainly to

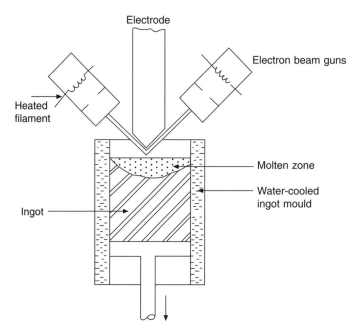

Fig. 13.4 Electron-beam melting furnace

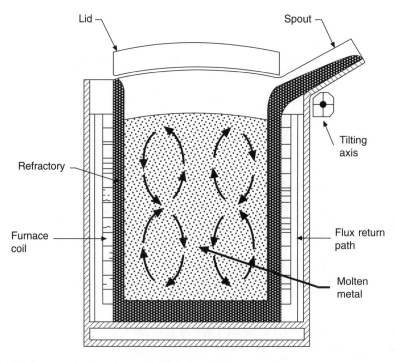

Fig. 13.5 Diagrammatic view of a coreless induction furnace

improved refractory technology, which allowed useful levels of power to be applied.
Until that time the furnace had mainly found application in iron foundries as a
holding unit, in which previously melted liquid metal is maintained at a temperature
suitable for pouring into the casting moulds.

In the channel induction furnace, a mains-frequency solenoid coil encircles one leg of a lamination pack, and the resultant alternating magnetic field induces joule heating in a loop of molten metal surrounding the coil, this metal loop acting as a single-turn secondary of the transformer. The arrangement is illustrated in **Fig. 13.6**.

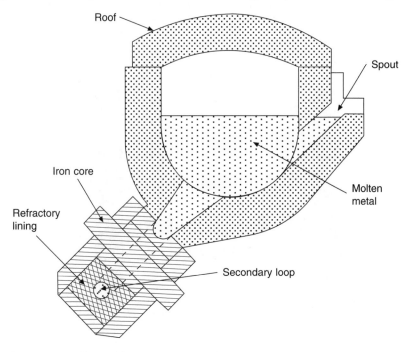

Fig. 13.6 Diagrammatic view of a channel induction furnace

The close coupling between the induction coil and the metal loop results in a higher power factor than is achieved with the coreless induction coil, and the efficiency of converting electrical power into heat is therefore also higher, being typically above 95 per cent. The molten metal loop must be continuous, and it is not allowed to solidify while energized in case the continuity is broken during contraction.

The use of channel furnaces for aluminium melting has been constrained by problems of oxide build-up in the metal loop. This necessitates special designs to allow easy cleaning. For bulk melting applications, European furnace companies and researchers have in recent years produced high-power designs of up to 1.3 MW per inductor, with freedom from loop blockage. This enables the high efficiency of this type of furnace to minimize melting costs.

13.2.7   The resistance furnace

Electric radiant heating techniques for metal melting are most widely used in the zinc and aluminium foundry sectors. There is much less use in the copper base alloy market which requires higher melting temperatures. In general, resistance heating is used where induction furnaces are not suited to the metal demand patterns. In the case of aluminium the application of resistance heating ranges from crucible bale-out furnaces of 25 kg capacity, where metal is scooped out of the container as required for pouring into moulds, to well-insulated box-like receivers of over 10 tonnes capacity.

Crucible furnace designs comprise a steel shell which contains insulating material, heating elements and a carbon-based crucible which is located in the central chamber. The general arrangement is shown in **Fig. 13.7**. The highest outputs are around 120 kW, which gives a maximum melting rate for aluminium of 230 kg/hour in a 600 kg capacity furnace. This can be compared with a coreless induction furnace of similar capacity which, with its higher power density, can achieve melting rates about three times greater.

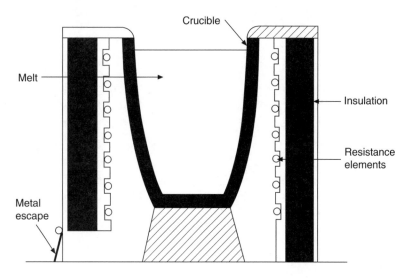

Fig. 13.7 Diagrammatic view of a resistance crucible furnace

Crucible furnaces exist in both tilting and bale-out configurations, and great strides have been made to improve performance figures in the light of competition between manufacturers and from other fuels. As an example, a typical 180 kg bale-out furnace has a holding requirement of about 4 kW, with advantages of automatic start-up, good working environment and close temperature control. Typically, the metal content is about eight times the optimum melt rate achieved when these large structures have reached thermal equilibrium.

An alternative high-efficiency heating mode for metals with a low melting point is provided by immersion heating, using silicon carbide or wire elements in silicon carbide-based tubes.

13.3　Heat treatment by direct resistance heating

This is used in the iron and steel industry for heating rods, wire and billets prior to rolling, forging or other hot working. It is also used for annealing of ferrous and non-ferrous materials. The process works through direct contact with the material, and a schematic of the general arrangement is shown in **Fig. 13.8**.

Direct resistance heating is also used for melting glass, in electrode boilers for water heating or steam raising, and in salt baths for the surface heat treatment of metallic components. In these applications it may be used alone or in combination with other fuels.

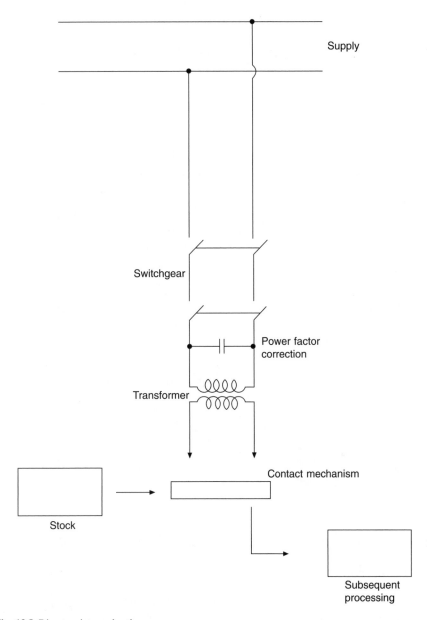

Fig. 13.8 Direct resistance heating

13.3.1 Metals

The resistivity of several common metals is shown in **Table 13.2**, and the variation of some of these resistivities with temperature is illustrated in **Fig. 13.9**.

Because of its relatively high resistivity, steel can be heated efficiently in billets of up to 200 mm² section, provided that the length of the billet is several times greater than its diameter. Heating time is from seconds to a few minutes, so heat losses (for instance radiation from the surface and thermal conduction through the contacts) are small. The efficiency of the process can be 90 per cent or better. The

Table 13.2 Electrical resistivity of typical metals

Metal	Electrical resistivity at 20°C ($\mu\Omega$ cm)
Copper	1.6
Aluminium	2.5
Nickel	6.1
Iron	8.9
Mild steel	16.0
Stainless steel	69.0

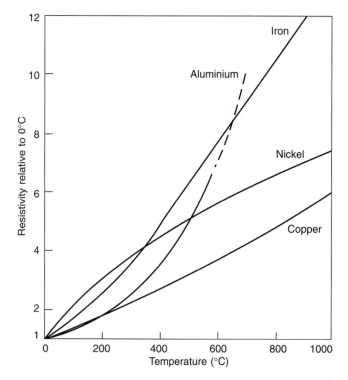

Fig. 13.9 Variation of resistivity with temperature for aluminium, copper, nickel and iron

workpiece resistance is low, and for these high efficiencies to be achieved the supply resistance must be much lower.

For copper and other low-resistivity metals, the length-to-diameter ratio of the billet should be considerably greater than 6 if the process is to be successful, and consequently it is usually applied to the annealing of wire and strip.

In all cases the cross-sectional area of the current path must be uniform, otherwise excessive heating will occur at the narrower sections, with the possibility of melting. In billets of larger cross-section, a non-uniform current density can arise as a result of skin effect. This leads to higher heating rates at the surface, but these are counteracted by increased heat loss at the surface.

The direct resistance heating of bar and billets is a single-phase load that is switched at frequent intervals. This results in transient voltage disturbances and in voltage unbalance at the point of common coupling. The load can be phase balanced by inductive and capacitive components, and in large units the switch-on disturbances may be compensated by a soft-start arrangement.

13.3.2 Glass

At temperatures above 1100°C, glass has a low viscosity and a resistivity which is low enough for direct resistance heating to be considered at acceptable voltages. The variation of the resistivity of glass with temperature is shown in **Fig. 13.10**. In the UK, electricity is used in mixed melting units where electrodes are typically added to a fuel-fired furnace to increase the output for a relatively low additional capital investment.

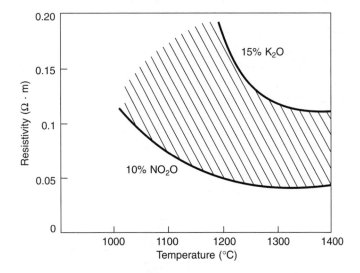

Fig. 13.10 Variation of resistivity with temperature of glasses

Current is passed between electrodes which are immersed in the molten glass. The electrodes must withstand the high temperatures and forces which arise from the movement of molten glass across their surface. It is also necessary to protect the electrodes from exposure to the atmosphere. Molybdenum or tin oxide electrodes are used in order to avoid contamination of the glass by pick-up of electrode material, and the current densities are of the order of 1500 A/m^2. Single-, two- or three-phase electrode systems may be employed. Scott-connected transformers (as described in **section 6.2.3**) are used for the two-phase connection. The three-phase arrangement produces electromagnetic forces in the glass which, together with the thermal forces, lead to a significant movement in the melt; this improves quality and melt rate provided it is not excessive. Three phase is also preferred for phase balance and low cost.

Existing electric furnaces have usually followed a traditional design pattern with a rectangular tank, but circular designs have now been adopted and it is claimed that these give improved stirring in the melting zone.

The power input to the furnace, and hence the melting rate, is controlled by varying the input voltage with tapped transformers or saturable reactors; an alternative method is to change the effective surface area of the electrodes by raising them from the melt.

13.3.3 Salt baths

These can be used for the heat treatment of metal components. The heated salt reacts chemically with the surface layer of the workpiece to give the required surface properties. At temperatures above about 800°C, direct resistance heating is the only practical method.

Although the salt is a good conductor when molten, the bath must be started up from cold by using an auxiliary starting electrode to draw a localized arc. The electrodes have to withstand the corrosive effects of the salt, and are manufactured from graphite or a corrosion-resistant steel alloy. Currents of up to 3 kA at 30 V are required, and both single-phase and three-phase units are available.

13.3.4 Foods and other fluids

In recent years the principles of direct resistance heating have been applied to a range of other pumpable fluids, typically food and aggressive liquids such as zinc phosphate.

The equipment in these applications consists of a pipeline interspaced with four electrodes which are each built into an insulated housing and connected to a three-phase supply in delta. The general arrangement is shown in **Fig. 13.11**. The pipeline itself is made from stainless steel and is insulated internally by a polymeric coating to ensure that the current flows through the fluid itself.

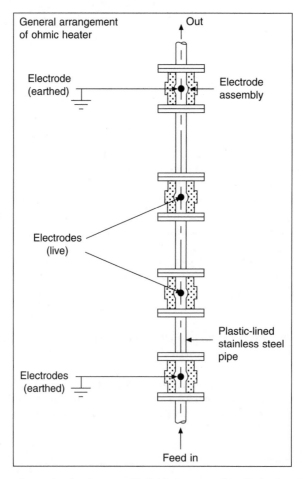

Fig. 13.11 Direct resistance heating for pumpable fluids (courtesy of EA Technology Ltd)

In the food industry the technique is known as 'ohmic heating'. Equipment rated up to 100 kW and 3.3 kV has been used in this industry to heat up to 120°C liquids and pastes which often contain solid particulates with characteristic dimensions up to 25 mm. This results in the cooking and sterilizing of the food in a single, very rapid operation of a few tens of seconds. When linked into an aseptic packaging line it gives an almost indefinite shelf life in ambient temperature storage.

13.4 Indirect resistance heating

In this method, electricity is passed through a suitable conducting material which forms a heating element, and heat is transferred to the workpiece or the medium to be heated by conduction, convection or radiation, or a combination of these. The element is the hottest component of the system, and the factors which determine the choice of element materials will depend on the nature of the heat transfer process as well as the physical and chemical characteristics of the process environment.

Infrared heating falls under the generic heading of indirect resistance heating, but for clarity it is covered separately in **section 13.5**.

13.4.1 Metallic elements

Metallic elements traditionally take the form of wire, strip or tape. For a given operating temperature, the shortest operating life tends to be found with the lowest cross-section of wire or tape, because failure occurs through progressive oxidation of the surface and the consequent reduction in mechanical strength.

The choice of metal composition for the element will depend upon the required operating temperature, the material resistivity, the temperature coefficient of resistance, corrosion resistance at high temperature, mechanical strength, formability and cost. Many metals and alloys are used, but the most common for industrial applications are based on nickel–chromium, iron–nickel–chromium or iron–chromium–aluminium alloys.

13.4.2 Sheathed elements

The elements may be protected from the working environment by an insulation layer and an outer sheath.

In the heating elements of many domestic appliances (such as cooker rings, immersion heaters and kettle elements) a purified magnesium oxide powder separates the helical element coil from a sheath which may be of copper, stainless steel or a nickel-base alloy. These elements are often rated in watts per square centimetre of sheath.

Such *mineral-insulated* elements are also used in cartridge heaters, radiant panels and immersion heaters for industrial applications. Also available are *thin-strip* or *band heaters*, which have mica insulation between the element and the sheath.

13.4.3 Ceramic elements

A range of ceramic materials exist which have a sufficiently high electrical conductivity to act as element material. These materials include silicon carbide, molybdenum disilicide, lanthanum chromite and hot zirconia. Graphite is another recognized non-metallic element material, but this can only be used in inert atmospheres such as steam and carbon dioxide.

Ceramic elements usually have a hot zone which is created by a thin section or a spiral cut; this hot zone is supported by two or more cold ends which are of a thicker section or have been impregnated with a metallic phase to lower the resistance locally.

13.4.4 Electric ovens and furnaces

These are used for a great variety of purposes, ranging from sintering ceramic materials at temperatures up to 1800°C to drying processes which are close to ambient temperature. Power ratings may be from a few kilowatts up to more than a megawatt. An *oven* usually has an upper temperature limit of about 450°C, and a *furnace* is usually used for higher temperature processes. Most modern ovens and furnaces incorporate lightweight insulation with low thermal conductivity to reduce thermal inertia and surface losses.

Metal resistance-heating elements for furnaces are usually in the form of wire, strip or tube. Elements for low voltage and high current have a heavy-section construction in an alloy casting or with corrugated or welded tube. Helically wound wire heating elements are made with a mandrel-to-wire diameter ratio of between 3 : 1 and 8 : 1. This ratio is limited by the tendency of the helical winding to collapse under its own weight at high operating temperatures. Coiled-wire or strip elements may be inserted in ledges or grooves, supported at intervals by pegs of nickel alloy or ceramic. The end connections of the elements are normally of a different material in order to reduce attack from oxidation and chemical reaction with the refractories, and they have a lower resistance in order to reduce the heat dissipation where the leads pass through the furnace wall.

Ovens may use natural or forced convection, and are widely employed in a range of applications including curing, baking, the annealing of glass and aluminium, and the drying and preheating of plastics prior to forming. Coiled nickel–chrome wire or mineral-insulated metal-sheathed elements are distributed around the oven in order to obtain a uniform temperature distribution. The heat transfer rate may be increased by using a fan to circulate air over the heating elements and onto the workpiece, the air being recirculated through ducts. An important advantage of convective ovens is that the operating temperature is normally the element temperature, and the maximum temperature is never exceeded since the process is self-limiting. This prevents overheating if the material is left in the oven too long, and it is particularly important for temperature-sensitive materials such as plastics.

High heating rates can be achieved by direct radiation from heating elements in infrared ovens. Here the oven walls are made of sheet metal which reflects or re-emits the radiation. Infrared processes are usually associated with surface heating which is applied, for example, to paint and other coatings. The efficiency in these applications may be very high in comparison with other methods which also heat the substrate.

Forced-convection furnaces allow high heating rates and with careful design good temperature uniformity can be achieved, but they are normally limited to a maximum temperature of 700–900°C.

Higher operating temperatures can be achieved with the *pit furnace*, in which radiation is the dominant mode of heat transfer and, by using a retort, the process can be carried out in a controlled atmosphere.

The *bell furnace* may be used as a hot-retort vacuum furnace. The layout of a bell furnace is shown in **Fig. 13.12**. By reducing the pressure inside the bell retort,

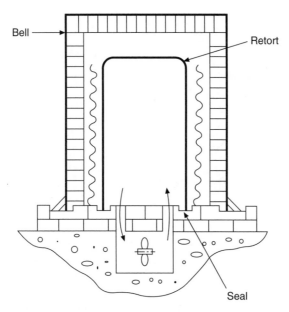

Fig. 13.12 Bell furnace

oxidation of the product is minimized. Since heat losses by convection are greatly reduced at low pressure, the bell can be raised when the required temperature is reached, and one bell may be used to heat several retorts.

Further information on furnace types and rating methods is available in **reference 13A**.

13.5 Infrared heating

Infrared is one of the most widely used electrical process heating techniques in industry, finding application in all sectors. Using electromagnetic energy as a source, it provides the potential for high input power densities and non-contact heating.

The peak wavelength of an infrared source depends upon its temperature. As the temperature is increased, the peak wavelength is shorter, proportionately more of the power is radiated at shorter wavelengths and the power densities are higher. This effect is illustrated in **Fig. 13.13**.

Typical input power densities for different types of infrared sources are given in **Table 13.3**.

Lower-temperature infrared sources (long and medium wave) are normally used for surface heating applications such as the drying of coatings, curing of paints and powder finishes, reflow soldering and browning of foods. Short-wave heaters are used where higher product temperatures are required (for instance in metal treatment) or where high intensities are needed; they may also be used for the through-heating of certain materials, for example in the moisture profiling of paper and board.

The efficient use of infrared radiation for heating is assisted by matching the spectral output of the infrared source to the absorptive properties of the product in the infrared range. The optimum choice may not always be apparent at first. For example, in the drying of paper, the first choice would be medium wave due to the very strong absorption, but this would provide surface drying only; by using a short-

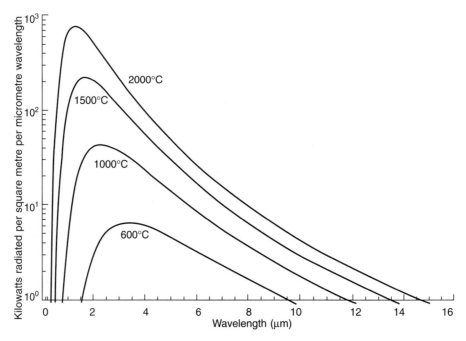

Fig. 13.13 Black-body radiation characteristics

Table 13.3 Characteristics of infrared heaters

Heater type	Maximum surface temperature (°C)	Maximum power density (kW/m²)	Wavelength
Embedded ceramic	600	68*	long
Mineral-insulated element	800	40*	long
Tubular quartz-sheathed element	900	50*	medium
Circular heat lamp (375 W)	2100	30*	short
Linear heat lamp (1 kW)			
• parabolic reflector	2100	50*	short
• elliptical reflector	2100	90**	short
Quartz-halogen linear heat lamp (12 kW)			
• parabolic reflector	2700	200*	short
• elliptical reflector	2700	3000**	short

* Average power density.
** Power density at focus.

wave source, some of which is transmitted and reflected by a polished surface behind the paper, penetration is achieved and through-drying is obtained.

Control of infrared heating may be effected in a number of ways. Infrared elements are generally of less than 10 kW rating, and an installation will often use many heaters. These can be switched on and off individually or in groups, or the voltage and current to each heater or group can be separately adjusted. One aspect of altering the power input to an infrared heater is that its temperature and hence its spectral output will be changed. This can have implications for the efficient heating of a few materials which show a variation in infrared absorption with wavelength. Recent improvements in the speed of response of lower-temperature infrared sources have extended the previous limitations in control of infrared heating.

The sensing of product parameters such as temperature or moisture content can provide information for control systems.

13.6   Induction heating

When an electrically conducting body is placed in an alternating magnetic field, eddy currents are induced in it and it is heated as a result of I^2R losses. The induced current density is greatest at the surface of the workpiece and decreases through its thickness. This phenomenon is known as the *skin effect*. The *skin depth*, δ, is defined as

$$\delta = (2\rho/\omega\mu)^{1/2}$$

δ is given in metres, where ρ is the electrical resistivity of the workpiece ($\Omega \cdot$ m), ω is the angular frequency $2\pi f$ of the coil current (rad/s) and μ is the absolute magnetic permeability of the workpiece (H/m). μ is given by the product $\mu_r \times \mu_0$, where μ_0 is $4\pi \times 10^{-7}$ and μ_r is known as the relative permeability. The relative permeability is a function of the applied magnetic field strength for magnetic materials, and it has the value of 1 for non-magnetic materials such as copper and aluminium. If 'thickness' is defined as the diameter of a round billet or the depth of a plate or slab, the ratio of thickness to skin depth is an important yardstick for the performance of an induction heating system. As this ratio increases, a greater proportion of the total power is dissipated near the surface of the workpiece, and the efficiency of the conversion of electrical energy into heat also increases.

The distribution of heating in the body of the workpiece can be controlled by the choice of frequency and the effect this has on the skin depth. If the frequency is high, most of the heat is dissipated in a thin surface layer, while lower frequencies give a more uniform distribution.

In ferrous metals, additional heating arises from hysteresis loss. This is normally small in comparison with the eddy current loss, but it is exploited in the heating of metal powders at high frequencies.

Induction techniques are used for both through heating and surface heating of metallic materials at frequencies across the range from 50 Hz to 1 MHz. They are used for metal melting and at very high frequencies they are applied in the manufacture of semiconductor materials and in the hot working of glass.

For regular-shaped workpieces such as cylindrical rods and wide rectangular slabs, the power generated in the workpiece and the induction heating efficiency can be calculated from analytical solutions to the diffusion equation for the induced current, supplemented by empirical factors. Details are given in **reference 13B**. The analytical solutions assume constant material properties throughout the workpiece, where in practice the resistivity and specific heat vary with temperature and the relative permeability of magnetic materials is a function of both field strength and temperature, reducing to unity above the Curie temperature, which is about 750°C for steel. Computer-based numerical solutions are now commonly used to take account of these variations.

The power induced in a round solid billet, illustrated in **Fig. 13.14**, is given approximately by:

$$P_W = (N \times I \times K_C)^2 \times (\pi \times d \times \rho)/(\delta \times L_W) \times (L_W/L_C)^2 \times Q_{rod} \quad \text{(watt)}$$

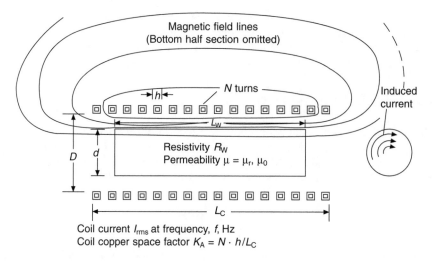

Fig. 13.14 Diagrammatic section of induction heater for round billet

where the parameters are as shown in **Figs 13.14** and **13.15**. In particular, Q_{rod} is shown in **Fig. 13.15** as a function of d/δ, and K_C, the coil shortness factor, is dependent on the ratios d/D, d/δ and L_W/L_C. **Reference 13C** gives empirical values for K_C, which tend to unity as d/D and L_W/L_C approach unity.

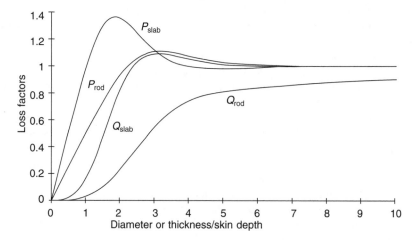

Fig. 13.15 Loss factors Q and P for rods and slabs

The power induced in a hollow cylinder of wall thickness t is calculated with Q_{rod} in the above expression replaced by an equivalent factor Q_{cyl}, which is a function of t/d, d/δ and μ_r. **Reference 13B** shows graphs of Q_{cyl} for a range of these parameters, although different terminology is used.

Similarly, for a rectangular slab of length L which has a width W much greater than its thickness, the induced power is:

$$P_W = (N \times I \times K_C)^2 \times (2W \times \rho)/(\delta \times L) \times (L/L_C)^2 \times Q_{slab} \quad \text{(watt)}$$

It is necessary to know the reactive power in the workpiece in order to evaluate the power factor of the coil. This can be calculated by multiplying the induced power by the appropriate P/Q ratio obtained from **Fig. 13.15**.

The efficiency of conversion of the electrical power supplied to the coil into thermal power in the workpiece is known as the *coil* or *electrical efficiency, η_C*. This is given by :

$$\eta_C = (1 + (1/K_C^2) \times (1/K_A) \times (S_C/S_W) \times (\rho_C/\rho_W\mu_r)^{1/2} \times (1/Q))^{-1} \quad \text{(per unit)}$$

In this formula, Q is the relevant loss factor, K_A is the space factor of the coil system and S_C/S_W is the ratio of the coil perimeter to that of the workpiece in the same plane. This efficiency can be significantly increased by the use of multilayer windings instead of a single-layer coil. Such high-efficiency coils are now commonly used for the heating of non-ferrous billets at mains frequency.

The overall efficiency of induction heating is the product of the *supply efficiency*, the *thermal efficiency* and the *coil efficiency*. Of these, the supply efficiency is typically 0.8–0.9 (per unit); it accounts for losses in cables, power factor correction capacitors and frequency conversion equipment. The thermal efficiency has typical values in the range 0.7–0.9 (per unit). This represents the thermal losses from the workpiece, and is critically dependent on the operating temperature, the thermal insulation and the method of operation of the heater.

13.6.1 Power sources

Loads which can effectively be heated at 50 Hz include slabs, large billets, cylinders and process vessels. Depending on the load rating, the power input is controlled by either an off-load tap changing transformer, or an autotransformer. Power factor correction is usually provided on the primary side of the heater supply transformer, and phase-balancing networks are used to correct the imbalance of large single-phase loads. Voltage transients on the supply network are minimized by the use of soft-start arrangements when switching large loads.

Apart from mains-frequency installations, power supplies for modern induction heaters are derived from solid-state frequency converters. Unit sizes of up to 7 MW have been installed for metal melting at 1–3 kHz, and 1 MW units are now suitable for frequencies up to 500 kHz, previously the domain of power vacuum tube triodes.

13.6.2 Through-heating of billets and slabs

Induction heating is used extensively for the through-heating of both ferrous and non-ferrous billets prior to rolling, extrusion or forging. The billets, which may be of circular or rectangular section, are either heated individually or passed in line through a series of induction coils. The frequency of the currents, the power input and the length of time in the coil are chosen to provide the right throughput rate and an acceptable temperature distribution over the cross-section of the workpiece.

Metal slabs are also heated by induction processes. One of the largest recent installations is in Sweden; this is rated at 37 MW and heats 15 tonne slabs for rolling at the rate of 85 tonne/hour. Thin slabs, which may be from continuous casting machines or at an intermediate rolling state, are heated at medium frequency.

13.6.3 Strip heating

The heating of continuous strip metals by the conventional induction method requires the use of frequencies above 10 kHz, and the resulting efficiencies are low for non-ferrous metals. An alternative technique is the *transverse flux method* by which, for example, an efficiency over 70 per cent can be achieved when heating aluminium strip at 250 kHz. The strip is passed between two flat inductors comprising windings in a laminated iron or ferrite core which forms a series of magnetic poles. The flux passes transversely through the sheet and currents are induced in the plane of the sheet. The arrangement of the windings and poles must induce a current distribution which results in a uniform temperature distribution over the sheet as the strip passes through the inductor.

Installations rated at 1.8 MW and 2.8 MW are operating for the heat treatment of aluminium strip in Japan and Belgium respectively.

13.7 Indirect induction heating

Although induction heating can only be directly effective in electrically conducting materials, it can also provide a means of heating other materials by using a metal as an intermediary, in a way that is analogous to indirect resistance heating. Examples are found in the processing of semiconductors and in the calcining of ceramic materials.

13.7.1 Semiconductor manufacture

Since the energy from the heating coil can be generated in the workpiece without any heat transfer medium, indirect induction heating can be carried out in a vacuum. This is particularly useful in the processing of semiconductor materials.

In one technique, the material is placed in an electrically conducting crucible in the vacuum space. The crucible is directly heated by induction and its heat is transferred to the semiconductor by radiation and conduction.

Semiconductors may also be heated directly by induction without the need for a conducting crucible. In this case very high frequencies up to 4 MHz are used.

13.7.2 Process heating and calcining

There is an important and growing series of applications in which induction heating is used to heat a vessel or other metal component, from which heat is transferred by conduction to a product.

Examples include:

- vessels containing a solid or liquid
- pipes through which a liquid flows
- chemical reactor vessels
- extruders
- screw conveyors
- rotary kilns
- mixer paddles

These applications often employ the coil arrangements already described elsewhere, but there are recent developments using novel techniques. Examples are the 'ROTEK' rotary kiln, the heated mixer paddle and the heated screw.

The 'ROTEK' kiln is heated by induction and is used for the drying and calcining of granular solids which flow continuously through a revolving drum. The kiln lining and flights are heated directly, and the flights also have the function of lifting or agitating the product as the vessel rotates. The heat is then transferred to the material by conduction. All internal components are made of stainless steel or Inconel, and are heated by the passage of a very large current from end to end. This current is perhaps greater than 20 kA, and cannot be fed through sliding contacts; the drum therefore forms the single-turn secondary of a ring transformer, with a bar primary which passes through its centre and returns through the outer copper conductors. **Figure 13.16** shows the general arrangement. The bar primary is fed from a low-voltage transformer and the efficiency of the device as a heater is very high. Because the ring core can be thermally insulated the operating temperature can be well above the Curie point, and with an Inconel drum, temperatures exceeding 1000°C can be achieved.

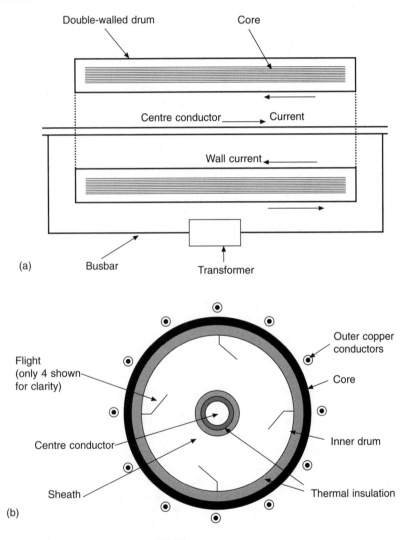

Fig. 13.16 Operating principle of a ROTEK kiln

13.8 Dielectric heating

The processing of non-metals by the conventional heating techniques of conduction, convection and radiation is often limited by the physical characteristics of the materials.

Thermal conductivity is a particular problem and since many of these methods depend on the transfer of heat through the product surfaces there is inevitably a temperature gradient between the surface and the centre. If the heat flux is raised above a certain level in such circumstances there is a risk that the surface will be damaged by overheating. The choice in many operations is between process inefficiency and product degradation.

Dielectric heating is a generic term covering radio frequency (RF) and microwave processing. RF and microwave occupy adjacent sections of the electromagnetic spectrum, with microwaves having higher frequency than radio waves. However, the distinction between the two frequency bands is very blurred; for example, some applications at about 900 MHz are described as radio frequency (cellular telephones) and some as microwave (dielectric heating). Nevertheless, RF and microwave heating can be distinguished by the technology that is used to produce the required electric fields. RF heating systems use high-power electrical valves, transmission lines and applicators in the form of capacitors, whereas microwave systems are based on magnetrons, waveguides and resonant (or non-resonant) cavities.

There are internationally agreed and recognized frequency bands which can be used for RF and microwave heating; these are known as the Industrial, Scientific and Medical (ISM) bands. At radio frequencies, the ISM bands are:

- 13.56 MHz ± 0.05% (± 0.00678 MHz)
- 27.12 MHz ± 0.6% (± 0.16272 MHz)
- 40.68 MHz ± 0.05% (± 0.02034 MHz)

and at microwave frequencies the ISM bands are:

- the so-called 900 band (896 MHz in the UK, 915 MHz in the USA and Japan, but not permitted in Europe)
- 2450 MHZ ± 50 MHz

Electromagnetic Compatibility (EMC) requirements impose severe limits on any emissions outside these bands. These limits are much lower than health and safety limits and are typically equivalent to microwatts of power at any frequency outside the ISM bands. In most countries, compliance with the relevant EMC regulations is a legal requirement, and this is covered in more detail in **Chapter 15**. Often it is the higher harmonics which present the greatest containment problems.

Industrial applications are many and varied, ranging from the drying of textiles to the welding of plastics, the tempering and thawing of meat, the heating of rubber extrusions and the firing of ceramics. Details are given in **reference 13D**.

The heating effect arises from a number of polarizations, the most commonly described being orientation polarization or oscillating dipole, as shown in **Fig. 13.17**. Although this is the principal mode at microwave frequencies, it is of relatively little significance at radio frequencies. The dominant mode in the RF range is space charge polarization, which in turn is dependent on the ionic conductivity of the material being processed. RF can therefore be regarded as a special case of direct resistance heating.

It is possible in theory to choose for any particular material the most appropriate frequency from those available in the industrial, scientific and medical bands. In reality, many products can be processed by either, and the choice between them can

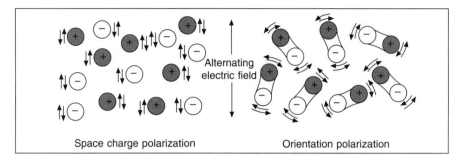

Space charge polarization   Orientation polarization

Fig. 13.17 Orientation polarization and space charge polarization in dielectric heating

be based on other considerations such as the engineering needed to make a satisfactory applicator which is compatible with process line requirements.

The heat transferred per unit volume of product, P, is given by:

$$P = 2\pi f \varepsilon_r \varepsilon_0 E^2 \quad [\text{W/m}^2]$$

where f is the frequency (in Hz), ε_0 is the permittivity of free space and E is the applied electric field strength (in V/m). ε_r is the loss factor, which is a dimensionless property of the material indicating its susceptibility to heating at the given frequency; the loss factor varies with moisture content, temperature and other factors.

13.8.1 Radio frequency power sources and applicators

Most RF power supplies in the 10 MHz to 100 MHz range are based on a class C oscillator/amplifier. The oscillator valve is most often a triode, which is built into a circuit of inductors and capacitors. When the components are rationalized this circuit can take on the appearance of a tank, which is normally fabricated from aluminium. With the increasing importance being attached to the need to avoid electromagnetic interference with other equipment, alternative generator types such as crystal-driven oscillators followed by amplifiers are being used. These so-called '50 Ω' systems have many other advantages such as controllability and the simplicity of the applicator construction.

RF applicators are essentially capacitors in which the product forms all or part of the dielectric. **Figure 13.18** shows three types in which the electric field, for the unloaded condition, is shown by the broken lines.

The simplest and most widely used of these three is the *through-field* or *parallel-plate electrode*. In addition to the product there may be in some cases a clearance which forms a series air-space capacitor. Such systems can be used in conjunction with pressure, as in the case of plastic welding, which is still the biggest use of dielectric heating. In this case there is no air gap and the field strength in the load can therefore be relatively high for a given electrode voltage. When plate electrodes are used for operations like drying, an air space is required above the dielectric to allow for the movement of product through the machine and for ventilation of the water vapour. This leads to an increase in voltage between the plates in order to maintain an adequate field strength in the product. It is therefore important to consider the relative dimensions of the dielectric and air-space capacitors so that heating is provided without the risk of an electrical discharge.

For very thin materials it may be necessary to use an alternative electrode configuration which gives a more suitable distribution. For example, to dry the

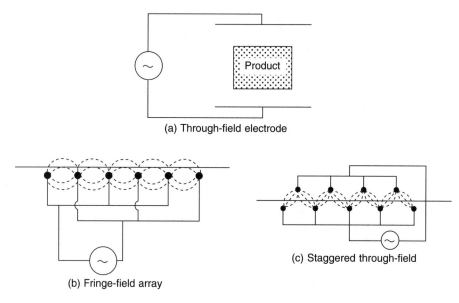

(a) Through-field electrode

(b) Fringe-field array

(c) Staggered through-field

Fig. 13.18 Radio frequency applicators

adhesive on a paper web in the manufacture of envelopes, it is necessary to use the *fringe* or *stray-field array* shown in **Fig. 13.13**, in which the electric field is arranged to be in the plane of the paper.

For intermediate thickness applications such as post-baking of biscuits it is usual to use a *staggered through-field* as shown in **Fig. 13.18**.

13.8.2 Microwave power sources and applicators

For industrial microwave heating applications the usual power source is the magnetron. This is a thermionic device which will launch electrons at a specific frequency when connected to a high-voltage dc source. At the permitted operating frequency of 2450 MHz, the highest output from a magnetron until recently was 5 kW, but 15 kW units are now becoming available. For the '900' band high-efficiency magnetrons of up to 60 kW now exist. Where higher powers are needed, a number of magnetrons may be fed into one applicator.

The most common form of industrial heating oven is the *multimode oven*, which is essentially an enlarged version of the familiar domestic microwave oven; in the case of continuous processing this will have appropriate product ports to allow the passage of product but confine the microwave within the oven. In such applicators the antenna of the magnetron may be mounted directly into the oven, but more often the microwave power is transmitted from the power supply via waveguides to the oven cavity, where it is launched into the chamber. Other options for microwave applicators include *directional horns* and '*leaky waveguides*'.

Industrial microwave heating has been used extensively in the rubber industry for curing and preheating prior to moulding. In the food industry it has been used for tempering, melting, cooking and drying. Microwave vacuum dryers have recently been developed for drying expensive, high-quality pharmaceuticals which are temperature sensitive.

13.9   Ultraviolet processes

Inks and surface coatings can be cured at high rates with ultraviolet sources. The coatings are specially formulated using monomers with photo-initiators, so that a very rapid polymerization is brought about on exposure to ultraviolet radiation. Although this is not strictly a heating process, it has much in common with infrared heating for drying or crosslinking and is very often in direct competition with infrared. The energy usage in ultraviolet is much lower than in infrared because the process requires only the stimulation of a reaction which is overall exothermic.

The active spectral region covers the wavelength range 250–400 nm in the ultraviolet, and visible wavelengths in the range 400–500 nm may also be used. The distribution can be confined to a few intense and narrow bands of wavelength. A typical output spectrum from a medium-pressure (sometimes called high-pressure) mercury vapour discharge lamp is shown in **Fig. 13.19**. Additional spectral bands can be generated by incorporating metal halide dopants in the lamp fill. The power ratings of these lamps are in the range 2–20 kW, over active lengths of 250–1800 mm.

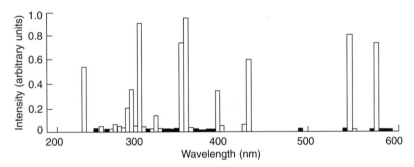

Fig. 13.19 Spectral output from a medium-pressure mercury lamp

13.10   Plasma torches

The plasma torch uses an electric arc discharge to generate a thermal plasma, which is a partially ionized gas in which the degree of ionization is linked to the temperature of the gas. Here the temperature of the gas rises above 6000°C; the gas becomes a reasonably good conductor of electricity. Temperatures in the core of an arc may reach 20 000°C or higher.

The torches that are used in electroheat applications may be broadly classified into the three families:

- *rod and nozzle electrode types.* An arc burns between the end of a rod and the internal surface of a nozzle, and gas is blown around the arc and through the nozzle. **Figure 13.20** shows the general layout. The supply is usually dc, but torches designed for ac use are also available.
- *linear coaxial-tube types.* The arc burns between the internal surfaces of tube electrodes, and gas is blown through the electrodes around the arc. **Figure 13.21** shows the arrangement. The power handling capacity is improved as the arc root motion is induced by the imposition of axial magnetic fields.
- *electrodeless types.* These include particularly induction-coupled and microwave types. An induction-coupled arc burns in a ring-shaped electric field which is

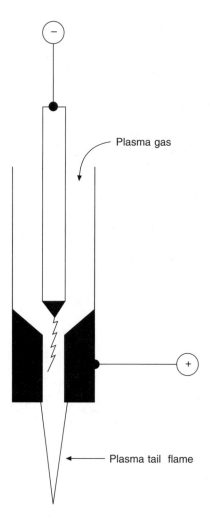

Fig. 13.20 Rod and nozzle type of plasma torch

induced within a coil-carrying current typically at a frequency of a few MHz; the arrangement is shown in **Fig. 13.22**. The arc plasma is effectively the workpiece in an RF induction heater. In microwave torches the arc is maintained by currents which are driven by microwave fields in a resonant cavity. Peak plasma temperatures in electrodeless torches are typically within the range 7000–11 000°C, which is rather lower than for the electrode types.

Transferred torches are available for industrial applications at power ratings up to about 7 MW in a single torch. Non-transferred torches are available in power ratings up to about 100 MW, but for industrial applications a maximum rating of about 10 MW is imposed by practical considerations of electrode life. The efficiency of non-transferred torches may reach 90 per cent for coaxial-tube electrode types used with high gas-flow rates, corresponding to relatively low gas temperatures of 3000– 4000°C. Rod and nozzle types are generally used for lower power ratings up to about 100 kW, and their efficiency tends to be lower, being about 60 per cent at best.

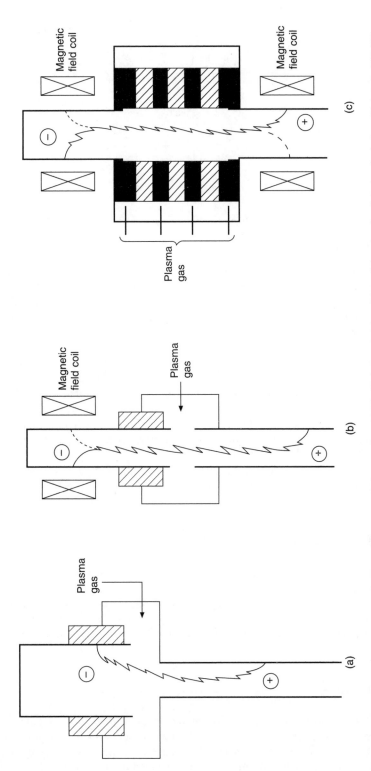

Fig. 13.21 Linear coaxial tubular electrode types of plasma torch construction: (a) The Hüls type developed from Schoenherr's original early design: (b) A typical 2 MW non-transferred torch (c) A typical stretched or segmented torch rated at 6 MW or higher

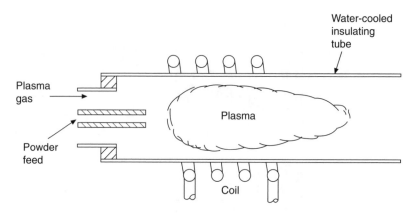

Fig. 13.22 An induction-coupled plasma torch

13.10.1   Plasma furnaces or reactors

In principle, the simplest type of furnace or reactor is the '*in-flight*' *reactor* in which the input materials are injected directly into the plasma stream or even into the torch, and the required processes occur with all the reactants suspended in the plasma stream. In-flight reactors generally require the input materials to be gaseous or finely divided because the time available for the process is short, typically of the order of milliseconds.

Transferred plasma torches are commonly used in open-bath furnaces. A return electrode is then usually built into the hearth refractories, but three-phase ac and bipolar dc multiple-torch systems have been developed to eliminate the need for the hearth connection.

Non-transferred torches are often used in shaft-type furnaces where a hot reaction zone is created close to the point of injection of the reactive plasma gases in the base of a packed shaft.

Furnaces with rotating shells are also available, the layout being shown in **Fig. 13.23**. These are used particularly for the fusion of pure refractory compounds, or when a long residence time is required, as in the treatment of some waste materials.

13.11   Glow discharge processes

Glow discharges are being increasingly used for industrial surface coating and surface modification processes.

The characteristic of a glow discharge, from which it derives its name, is that some large portion of the discharge vessel should be filled with a weakly ionized and luminous plasma. The two types of glow discharge are the dc and the low-pressure RF discharge.

13.12   Lasers

Since its discovery in 1960, the laser has found extensive application in industry for cutting, welding, material removal and heat treatment processes which require power densities above 100 W/mm^2. The principal feature of a laser as a heat treatment source is its monochromatic output, low divergence and high intensity. This enables

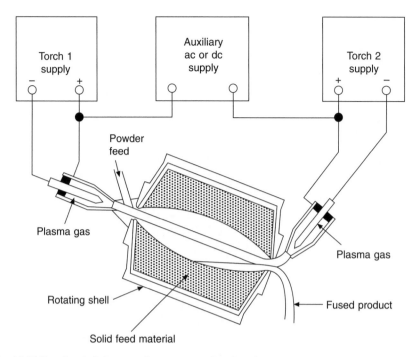

Fig. 13.23 Rotating shell furnace using a non-transferred torch

the parallel monochromatic beam to be focused to a smaller diameter than is possible with other light sources.

Industrial applications for CO_2 lasers include the dividing and perforation of substrates for integrated circuits and electronic components, perforation of elastomeric materials, the manufacture of dies for cutting cartons, ready-to-wear suits and thin glass tubes for fluorescent lamps, quartz tubes and borosilicate glass, and for the manufacture of flexigravure plates. Deep-penetration welds can be obtained at power levels above about 1 kW, and the ratio of the weld depth to the heat-affected zone using a laser can be ten times greater than that obtainable with electron-beam welding. Selective heat treatment is also possible.

13.13 Standards

Table 13.4 gives a selection of standards covering the electroheat field, with special emphasis on RF and microwave equipment and methods.

Table 13.4 Standards relating to electroheat

IEC	EN	BS	Subject	CISPR (International Special Committee on Radio Interference)	N. American
		2316	Radio frequency cables		
		2316-1	– general requirements and tests		
		2316-3	– cable data sheets		
		3041	RF connectors		
		3041-1	– general requirements and measuring methods		
		3056	Sizes of refractory bricks		
		3056-4	– bricks for electric arc furnace roofs		
		6656	Prevention of inadvertent ignition of flammable atmospheres by RF radiation		
	160200	EN160200	Assessment for electronic components' sectional specification: microwave modular electronic units of assessed quality		
	160200-2	EN160200-2	– index of test methods		
	55011	EN55011	Limits and methods of measurement of radio disturbance characteristics of ISM RF equipment	11	
			Specification for radio interference measuring apparatus and measuring methods	16	
			Determination of limits for ISM equipment	23	
240	60240	EN60240	Electric infrared emitters for industrial heating		
240-1	60240-1	EN60240-1	– short-wave infrared emitters		
519	60519	EN60519	Safety in electroheat installations		
519-2	60519-2	EN60519-2	– resistance heating equipment		
519-3	60519-3	EN60519-3	– induction and conduction heating and induction melting installations		
519-4	60519-4	EN60519-4	– arc furnace installations		
519-6	60519-6	EN60519-6	– industrial microwave installations		
519-9	60519-9	EN60519-9	– high-frequency dielectric heating installations		
1307	61307	EN61307	Industrial microwave heating installations – test methods for determination of power output		

Table 13.4 (contd)

IEC	EN	BS	Subject	CISPR (International Special Committee on Radio Interference)	N. American
1308	61308	EN61308	HF dielectric heating installations – test methods for determination of power output		
			Guidelines on limits of exposure to RF electromagnetic fields:		ANSI C95
			– recommendation (protection guide 1982)		ANSI C95.1
			– techniques and instrumentation for the measurement of potentially hazardous electromagnetic radiation		ANSI C95.2
			– recommended practice for measurement of potentially hazardous electromagnetic fields		ANSI C95.5
		NRPB GS 11	Guidance as to restrictions on exposures to time-varying electromagnetic fields		

References

13A.   Electric resistance furnaces for metal heat treatment, EA Technology Ltd, Technical Note No. EATL 1124.
13B.   Davies, E.J. *Conduction and Induction Heating,* Peter Peregrinus Ltd, London, 1990.
13C.   Orfeuil, M. *Electric Process Heating,* Battelle Press, Columbus, Richmond, Ohio, 1987.
13D.   Jones, P.L. and Rowley, A.T. Dielectric Drying, in Mujundra and Kudra (eds), *Advances in Drying,* Marcel Dekker, New York, 1996.

Chapter 14

The power system

Dr B.J. Cory
Imperial College of Science, Technology and Medicine

14.1   Introduction

All countries now have a power system which transports electrical energy from generators to consumers. In some countries several separate systems may exist, but it is preferable to interconnect small systems and to operate the combination as one, so that economy of operation and security of supply to consumers is maximized. This integrated system (often known as the 'grid') has become dominant in most areas and it is usually considered as a major factor in the well-being and level of economic activity in a country.

All systems are based on alternating current, usually at a frequency of either 50 Hz or 60 Hz. 50 Hz is used in Europe, India, Africa and Australia, and 60 Hz is used in North and South America and parts of Japan.

Systems are traditionally designed and operated in the following three groupings:

- the source of energy – *generation*
- bulk transfer – *transmission*
- supply to individual customers – *distribution*

14.2   Generation

Generators are required to convert fuels (such as coal, gas, oil and nuclear) and other energy sources (such as water, wind and solar radiation) into electrical power. Nearly all generators are rotating machines, which are controlled to provide a steady output at a given voltage; the main types of generator and the means of control are described in **Chapter 5**.

The total power output of all operating generators connected to the same integrated system must at every instant be equal to the sum of the consumer demand and the losses in the system. This implies careful and co-ordinated control such that the system frequency is maintained, because all generators in an ac power system must run in *synchronism*, that is their rotors, which produce a magnetic field, must lock into the rotating magnetic field produced by alternating currents in the stator winding. Any excess of generated power over the absorbed power causes the frequency to rise, and a deficit causes the frequency to fall. As the demand of domestic, commercial and industrial consumers varies, so the generated power must also vary, and this is normally managed by transmission system control which instructs some generators to maintain a steady output and others (particularly hydro and gas turbine plant) to 'follow' the load; load 'following' is usually achieved by sensitive control of the input, dependent upon frequency. The aim of the system controller is to run the generating plant such that the overall cost of supplying the consumer at all times is a minimum, subject to the various constraints which are imposed by individual generator characteristics.

14.3 Transmission

Many large generators require easy access to their fuel supply and cooling water, so they cannot necessarily be sited close to areas of major consumption. Environmental constraints may also preclude siting close to areas of consumption. A bulk power transmission system is therefore needed between the generators and the consumers.

Large generating plant produces output ranging from 100 MW to 2000 MW and for economic reasons this normally operates with phase-to-phase voltages in the range 10 kV to 26 kV. In order to reduce transmission losses and so that transmission circuits are economic and environmentally acceptable, a higher voltage is necessary. Phase-to-phase transmission voltages of up to 765 kV are used in sparsely populated large countries such as Brazil, the USA and Canada, but 380–400 kV is more prevalent in Europe. The standard voltages recommended by IEC are 765, 500, 380–400, 345, 275, 230–220, 185–138 and 66–69 kV.

Most transmission circuits are carried overhead on steel pylons. An example is shown in **Fig. 14.1**. They are suspended from insulators which provide sufficient insulation and air clearance to earth to prevent flashovers and danger to the public. A typical suspension-type insulator is shown in **Fig. 14.2**. Each country has tended to have its own acceptable tower and conductor design. At higher voltages, Aluminium Conductor Steel Reinforced (ACSR) conductor is used, a core of steel strands providing the required strength. A typical cross-section for an ACSR conductor is shown in **Fig. 14.3**. For voltages over 200 kV two or more conductors per phase are used. This results in lower losses because of the large conductor cross-section, and lower radio interference and corona because of the lower voltage stress at the conductor surface.

Where an overhead line route is impossible because of congestion in an urban area or for environmental amenity reasons, buried cables may be employed, but the cost is 15–20 times higher than an equivalent overhead line. On sea crossings an underwater cable is the only solution, but these are often dc, for reasons explained in **section 14.3.1.1**.

A high-voltage transmission system interconnects many large generators with high areas of electricity demand; its reliability is paramount, since a failure could result in loss of supply to many people and to vital industry and services. The system is therefore arranged as a network so that the loss of one circuit can be tolerated. This is shown in **Fig. 14.4**. In many countries, three-phase lines are duplicated on one tower, in which case a tower failure might still result in a partial blackout. Mixed-voltage systems are often carried on a single tower, but this is not the practice in the UK.

In order to achieve flexibility of operation, circuits are marshalled at substations. The substations may include transformers to convert from one voltage level to another, and switchgear to switch circuits and interrupt faults. Substations are normally outdoors and they occupy an extensive secure area, although compact indoor substations using SF_6 (as described in **section 7.5.3(b)**) have become more prevalent recently because of their improved reliability in adverse weather.

Fig. 14.1 Transmission line tower – a 400 kV double-circuit line

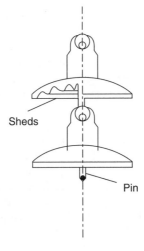

Fig. 14.2 Suspension-type insulator

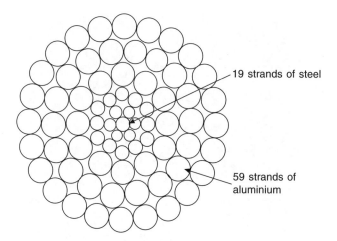

Fig. 14.3 Typical cross-section of ACSR overhead line conductor

An interconnected transmission network can comprise many substations which are all remotely controlled and monitored to ensure rapid reconnection after a disturbance or to enable maintenance.

14.3.1 Principles of design

The two main requirements of transmission systems are:

- the interconnection of plant and neighbouring systems to provide security, economic operation and the exchange of energy on a buy–sell basis
- the transport of electrical energy from remote generation to load centres

These objectives are met by selection of the most economical overhead line design, commensurate with the various constraints imposed by environmental and national considerations. The design and approval process for new lines can take many years, following public enquiries and judicial proceedings before planning permission is granted and line construction begins. Typical objections to new overhead lines, particularly in industrialized countries, are:

- the visual deterioration of open country areas
- the possibility that electromagnetic field propagation may cause interference with television, radio and telecommunications, with an increasing awareness of carcinogenic risks
- emission of noise by corona discharges, particularly at deteriorating conductor surfaces, joints and insulation surfaces
- the danger to low-flying aircraft and helicopters
- a preference for alternative energy supply such as gas and local environment-friendly generation including solar cells and wind power, or measures for reducing electricity demand such as better thermal insulation in houses and commercial premises, lower energy lighting, natural ventilation in place of air conditioning, and even changes of lifestyle

Power system planners are required to show that extensive studies using a range of

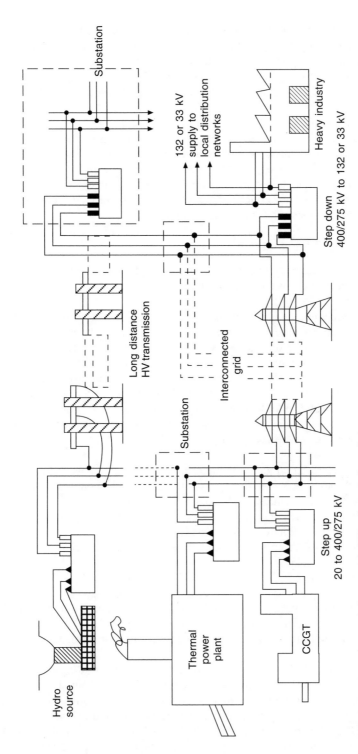

Fig. 14.4 Generation and transmission network (UK voltages and practice)

scenarios and sensitivities have been carried out, and that the most economical and least environmentally damaging design has been chosen. Impact statements are also required in many countries to address the concerns regarding the many issues raised by local groups, planning authorities and others.

Technically, the key issues which have to be decided are:

- whether to use an overhead line or underground cable
- the siting of substations and the size of substation required to contain the necessary equipment for control of voltage and power flow
- provision for future expansion, for an increase in demand, and in particular for the likelihood of tee-off connections to new load centres
- the availability of services and access to substation sites, including secure communications for control and monitoring

14.3.1.1 HVDC transmission

Direct current is being used increasingly for the high-voltage transport of electrical energy. The main reasons for using dc in preference to ac are:

- dc provides an asynchronous connection between two ac systems which operate at different frequencies, or which are not in phase with each other. It allows real power to be dispatched economically, independently of differences in voltage or phase angle between the two ends of the link.
- in the case of underground cables or undersea crossings, the charging current for ac cables would exceed the thermal capacity of the cable when the length is over about 50 km, leaving no capacity for the transfer of real power. A dc link overcomes this difficulty and a cable with a lower cross-section can be used for a given power transfer.

Where a line is several hundred kilometres long, savings on cost and improvements in appearance can be gained in the dc case by using just two conductors (positive and negative) instead of the three conductors needed in an ac system. Some security is provided should one dc conductor fail, since an emergency earth return can provide half power. The insulation required in a dc line is equivalent only to that required for the peak voltage in an ac system, and lower towers can therefore be used with considerable cost savings.

Against these reductions in cost with a dc line must be set the extra cost of solid-state conversion equipment at the interfaces between the ac and dc systems, and the corresponding harmonic correction and reactive compensation equipment which is required in the substations. It is normally accepted that the break-even distance is 50 km for cable routes and 300 km for overhead lines; above these distances dc is more economical than ac.

Because many ac systems already exist, and because the trading of energy across national and international boundaries is becoming more prevalent, dc transmission is being increasingly chosen as the appropriate link. An added advantage is that a dc infeed to a system allows fast control of transients and rapid balancing of power in the case of loss of a generator or other supply, and it does not contribute to the fault level of the receiving system; it is important here that when a short-circuit occurs on the ac system side, the current that flows can be safely interrupted by the ac-side circuit breaker.

The accepted disadvantages of a dc infeed, apart from the already-mentioned extra cost of conversion equipment, are the lack of an acceptable circuit breaker for flexible circuit operation, and the slightly higher power losses in the conversion equipment compared with an equivalent ac infeed.

With careful design of the transmission and conversion components, the reliability of dc and ac systems is comparable.

A typical HVDC scheme providing two-way power flow is shown schematically in **Fig. 14.5**. Each converter comprises rectifying components in the three-phase bridge connection, and each of the rectifying components consists of a number of thyristors connected is series and parallel. Increasingly, Gate Turn Off (GTO) thyristors are being used because of the greater control they allow. The current rating of each bridge component can be up to 200 A at 200 kV and bridges may be connected in series for higher voltages up to 400 kV or even 600 kV +/– to earth, each bridge being supplied from a three-phase converter transformer. By triggering the thyristors the current flow through the system can be controlled every few electrical degrees, hence rapid isolation can be achieved in the event of a fault on the system. Similarly, by delay triggering the current can be easily controlled, the direct voltage being best set by tap change on the converter supply transformers. The triggering of the thyristors may be by a light pulse which provides voltage isolation. *Inversion*, which enables power flow from the dc system to the ac system, depends on the ac back-emf being available with a minimum fault level in the receiving systems, so inversion into an isolated system is not possible unless devices with turn-off capability are available.

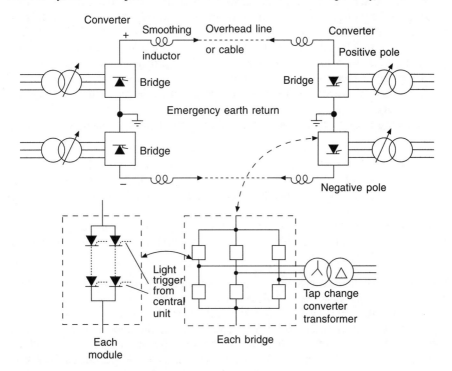

Fig. 14.5 Complete HVDC scheme, showing converters and dc link

A further feature of dc converter substations is the need for ac transmission filters to produce an acceptable ac sinusoidal waveform following the infeed or outfeed of almost square-wave blocks of current. Such filters for $6n \pm 1$ harmonics (where n is the number of bridges and substations) can add 25 per cent to the cost of a substation, although they can also be used to provide some of the VAr generation which is necessary to control the power factor of the inverter.

14.3.1.2 AC system compensation

As ac power systems become more extensive at transmission voltages, it is desirable to make provision for flexible operation with compensation equipment. Such equipment not only enables larger power flows to be accommodated in a given rating of ac circuit, but also provides a means of routing flows over the interconnected system for economic or trading purposes.

Compensation equipment consisting of fixed or variable inductance and capacitance can be connected in series with the circuit, in which case it must be rated to carry the circuit current, or it may be connected in shunt, and used to inject or absorb *reactive power* (VArs) depending upon requirements. In the same way that real power injected into the system must always just balance the load on the system and the system losses at that instant, so too must the reactive power achieve a balance over regions of the system.

Transmission circuits absorb VArs because their conductors are inductive, but they also generate VArs because they have a stray capacitance between phases and between phase and earth. The latter can be particularly important with high-voltage cables. The absorption is proportional to I^2X, where I is the current and X is the reactance of the circuit, and the generation is proportional to V^2B, where V is the voltage and B is the susceptance of the circuit. When I^2X is equal to V^2B at all parts of the circuit, it is found that the system voltage is close to the rated value. If V^2B is greater than I^2X, then the system voltage will be higher than the rated value, and vice versa. Designers therefore need to maintain a balance over the foreseeable range of current as loads vary from minimum to maximum during the day, and over the season and the year.

Compensation may be provided by the following three main methods:

- *series capacitors connected in each phase* to cancel the series inductance of the circuit. Up to 70 per cent compensation is possible in this way.
- *shunt inductance* to absorb excessive VArs generated by the circuit stray capacitance or (exceptionally) to compensate for the leading power factor of a load
- *shunt capacitance* to generate VArs for the compensation of load power factor or excessive VArs absorbed under heavy current flow conditions on short overhead line circuits

Combinations of these arrangements are possible, especially in transmission substations where no generators are connected; generators are able to generate or absorb VArs through excitation control (see **section 5.3.2**). Recently, *Flexible AC Transmission Systems (FACTS)* devices have become available. In these, the amount of VAr absorption by inductors is varied through the control of thyristors connected in series with the inductor limbs. A typical shunt controllable unit is shown in **Fig. 14.6**. This is known as a *Shunt Variable Compensator (SVC)*. Other FACTS devices are variable series capacitors, variable phase shifters and Universal Power Controllers (UPC), in which energy is drawn from the system in shunt and injected back into the system in series at a controlled phase angle by means of GTO thyristors.

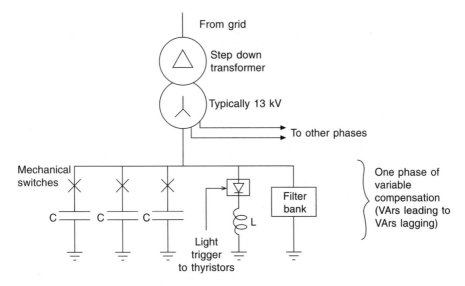

Fig. 14.6 Schematic of a Shunt Variable Compensator (SVC)

14.3.2   System operation

A transmission system may be *vertically integrated,* in which case the generating plant belongs to the same utility, or it may be *unbundled,* in which case it has only transmission capacity, with no generation plant. In either case the main tasks are to maintain a constant frequency and voltage for all consumers, and to operate the system economically and securely. Security in this context means maintaining voltage within limits, staying within a prescribed stability margin and operating all circuits within their thermal rating. This requires adequate monitoring of all the transmission components, with sufficient communication and control facilities to achieve these desired goals. Most transmission systems will, therefore, have a co-ordinating room and possibly a number of manned outstations for local or regional devolvement of responsibility.

For frequency control, some of the synchronized generators are equipped with sensitive governors which use a frequency signal rather than a speed signal. The output of these generators is dependent upon the balancing power required to achieve a steady frequency over the whole system. The control engineer, backed up by computer forecasts of load variations and knowledge of the available plant and their costs, can then instruct generators to start up or shut down (*unit commitment*) and set their output (*loading* or *dispatching*) so that over a prescribed hourly, daily or weekly period they generate energy to meet the consumer demand at the minimum overall cost. In the UK the use of a Generating Ordering And Loading (GOAL) program attempts to ensure that over a 24-hour period the total cost of generation commensurate with maintaining a secure system is the minimum that can possibly be provided. This includes the trading of energy with Scotland and France, in addition to any surplus power available from co-generation within industry and from renewable sources such as wind, hydro, waste incineration and landfill gas.

There is considerable scope for minimization of the losses in an interconnected system through the control of the compensation devices described in **section 14.3.1.2**. This control is guided by the use of optimal load flow programs, security assessments and calculations of transient stability margin. One of the main concerns is to arrange patterns of generation, including some plant which is out-of-merit, to maintain voltage despite outages of circuits and other components for maintenance, extension and repair. Safety of utility personnel and the operation of the system to avoid risk to the public are at all times paramount.

14.4 Distribution

An example of a three-phase distribution system is illustrated in **Fig. 14.7**. In the UK, voltages of 132, 110, 66, 33 and 11 kV are typically used to provide primary distribution, with a 380–415 V three-phase and neutral low-voltage supply to smaller consumers such as residential or smaller commercial premises, where 220–240 V single-phase to neutral is taken off the three-phase supply. Distribution voltages in continental Europe are typically 110, 69 and 20 kV, but practice varies from country to country. In the USA, voltages of 138, 115, 69, 34.5, 13.2 and 4.16 kV are employed.

The transformer stepping down from the primary distribution to the low-voltage supply may be pole mounted or in a substation, and is close to the consumers in order to limit the length of the low-voltage connection and the power losses in the low-voltage circuit. In a national power system, many thousands of transformers and their associated circuit breakers or fuses are required for distribution to low-voltage circuits, in contrast to high-voltage transmission and primary distribution systems, where the number of substations is in the hundreds.

It will be noted from **Fig. 14.7** that the primary and low-voltage distribution systems are connected in a radial configuration. Circuit loops between adjacent substations are avoided because these can lead to circulating currents, which increase the power losses and create difficulty in protection schemes. However, tie circuits between adjacent lines and cables are available to reconfigure the network when a portion of the low-voltage circuit is out of service for maintenance or because of failure. These tie circuits are controlled by a normally open switch which can be closed manually within a few minutes, although an increasing trend is for automation of this operation by radio or teleswitching.

In urban and suburban areas, much of the primary and low-voltage distribution system is underground, with readily accessible substations sited in cellars or on small secure plots. Industrial sites may also have a number of substations incorporated into buildings or secure areas; these may be controlled by the works engineer or they may be operated and maintained by an electricity distribution company.

In rural areas and in more dispersed suburban areas, three-phase overhead lines operating at 10–15 kV or 27–33 kV are supported for many miles on poles which may be of wood, concrete or steel lattice. The 380–415 V three-phase supply is taken from these lines through a small pole-mounted fused input/output transformer. If the maximum load to be taken is below about 50 kW, the supplies for homes or farmsteads may be derived from a single-phase 10–15 kV supply. Typically, a rural primary distribution system supplies up to 50 step-down transformers spread over a wide region. The lines in such a system are vulnerable to damage by tree branches, snow and ice accumulation and lightning strikes and it therefore has lower reliability than underground systems in urban areas. Considerable ingenuity has been applied to protection of this type of system with the use of auto-reclosing supply circuit breakers and automatic reconnection switches, which are described in **section 7.4.2(b)**.

It is now common practice in developed countries to monitor the primary distribution

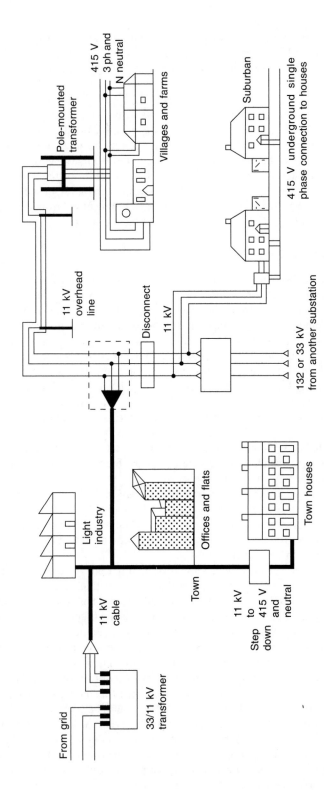

Fig. 14.7 Distribution network

system down to 10–15 kV and to display alarm, voltage and power-flow conditions in a control room, and in the event of an incident repair crews are dispatched quickly. Repairs to the low-voltage system are still dependent, however, on consumers notifying a loss of supply.

The proper earthing of distribution systems is of prime importance in order that excessive voltages do not appear on connections to individual consumers. It is the practice in the UK and some other countries to connect to earth the neutral conductor of the 4-wire system *and* the star point of the low-voltage winding on the step-down transformer, not only at the transformer secondary output but also at every load point with a local meter and protective fuse. This is known as the Protective Multiple Earth (PME) system, which is described in relation to cable technology in **section 9.3.1.3**. It is designed to ensure that all metallic covers and equipment fed from the supply are bonded so that dangerously high voltages do not hazard lives.

14.4.1 System design

It is essential that a distribution system is economical in operation and easy to repair, and its design should enable reconnection of a consumer through adjacent feeders to supply substations in the event of failure or outage of part of the system.

Copper or aluminium conductors with a cross-section of 150 mm^2 to 250 mm^2 are typically used at the lowest voltages, and these are arranged so that the maximum voltage drop under the heaviest load conditions is no more than 6 per cent; alternatively the voltage at the connection to every consumer must not rise more than 6 per cent under light load conditions. Local adjustment is achieved by off-load tap changing (usually ± 2.5 per cent or ± 5 per cent), and voltage in the primary circuits (usually 11 kV and 33 kV in the UK) is controlled by on-load tapchangers under automatic voltage control. The construction and operation of tapchangers is described in more detail in **section 6.2.5**. Reinforcement or extension on the LV network is usually arranged through the installation of a new primary feed point with a transformer, rather than by upgrading the LV network.

Primary underground networks now employ cross-linked polyethylene (XLPE) three-phase cables, as described in **sections 9.2.4** and **9.3.1.3**. The latest designs incorporate fibre-optic strands for communication purposes. Cables require careful routing and physical protection to minimize the risk of inadvertent damage from road and building works in the vicinity.

Overhead lines are usually of a simple, flat, three-phase configuration which avoids conductor clashing in high winds and ice precipitation. An example is shown in **Fig. 7.20**. Impregnated wood poles are normally used. These provide some degree of insulation to ground, and they can be quickly replaced in the event of collapse or decay. Insulators are usually of the cap-and-pin type, an example of which is illustrated in **Fig. 14.8**. Surge diverters and arcing horns provide a considerable measure of overvoltage protection, particularly where many kilometres of exposed line are fed from a primary substation. Arcing horns consist of a carefully positioned air gap between each conductor and earth. They are designed to flash over at a particular voltage when a potentially dangerous surge occurs on the system. Since there is no provision for extinguishing the resulting arc in this event, it is preferable, although more expensive, to provide a surge diverter at strategic positions in the overhead system. A surge diverter consists of a zinc oxide resistance between the live conductor and earth; this presents a very high resistance at normal voltages, but a low value at overvoltages, thereby conducting surge current safely to earth.

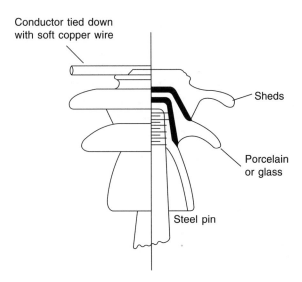

Conductor tied down
with soft copper wire

Sheds

Porcelain
or glass

Steel pin

Fig. 14.8 Pin-type insulator

Primary substations consist usually of two or three step-down tapchanging transformers, with common busbars which may be duplicated for reliability and circuit separation. Remotely controlled interconnecting circuit breakers are installed to provide operating flexibility. Such substations are continuously monitored and metered, the data being brought back to a distribution control centre for display and archiving. A modern tendency is to reduce the number of control centres and increase the geographical area covered by each through the use of computers and clever alarm handling and by analysis with powerful software which will in the future incorporate artificial intelligence.

14.4.2 System operation

The most important requirement in distribution control is good communication with district personnel and maintenance crews in order to ensure that maintenance and repair is efficiently, safely and swiftly carried out. Ready telephone access by the public and other consumers is also necessary so that dangerous situations or supply failures can be easily reported to engineers and technicians in the field.

The main task of the distribution controller is, therefore, to monitor equipment alarms and ensure that rapid and effective action is taken. Schedules of equipment outages and maintenance have to be planned and effected with the minimum of disruption to consumers.

14.5 Future trends

Power systems are continually evolving and with the increasing capability of computers and software, systems are becoming more centrally controlled for economy of operation and security.

There is a trend towards the separation of the ownership, management and control of generating plant from that of the transmission and distribution system. This is being accompanied by considerable economies in operating and technical personnel and is leading to more competitive trading.

Economists have constantly urged utilities to operate at marginal costs in order to achieve economic efficiency with the consequent downward pressure on prices, and this has led to the setting up of electrical energy markets in many countries. In such systems, entrepreneurial generators can enter with the expectation of making a profit by supplying electrical energy to any consumer over the transmission and distribution system. A charge has to be made for transport of energy over this system, but there is no reason to believe that this process will be more expensive than the previous vertically integrated system. Experience is beginning to show that such a market in electrical energy can work, as in the UK, and can produce cost savings not only in the supply of energy but also in its transportation.

References

14A.   Weedy, B.M. and Cory B.J. *Electric Power Systems* (4th edn), Wiley, 1987.
14B.   *EHV Transmission Line Reference Book*, Edison Electric Institute, USA, 1968.
14C.   *Modern Power Station Practice*, volume L, 3rd edn, Pergamon, 1991.

Chapter 15

Electromagnetic compatibility

Mr A.J. Maddocks
ERA Technology Ltd

15.1   Introduction

Electromagnetic Compatibility (EMC) is achieved when co-located equipment and systems operate satisfactorily, without malfunction, in the presence of electromagnetic disturbances. For example, the electrical noise generated by motor-driven household appliances, if not properly controlled, is capable of causing interference to domestic radio and TV broadcast reception. Equally, microprocessor-based electronic control systems need to be designed to be immune to the electromagnetic fields from hand-held radio communication transmitters, if the system is to be reliable in service. The issues covered by EMC are quality of life, spectrum utilization, and operational reliability, through to safety of life, where safety-related systems are involved.

The electromagnetic environment in which a system is intended to operate may comprise a large number of different disturbance types, emanating from a wide range of sources including:

- mains transients due to switching
- radio frequency fields due to fixed, portable and mobile radio transmitters
- electrostatic discharges from human body charging
- powerline surges, dips and interruption
- power frequency magnetic fields from power lines and transformers

In addition to having adequate immunity to all these disturbances, equipment and systems should not adversely add electromagnetic energy to the environment above the level that would permit interference-free radio communication and reception.

15.1.1   Sources

The essence of all EMC situations is contained in the simple source–path–receptor model shown in **Fig. 15.1**.

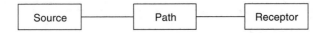

Fig. 15.1 Source, path, receptor model

Sources comprise electromechanical switches, commutator motors, power semiconductor devices, digital logic circuits and intentional radio frequency generators.

The electromagnetic disturbances they create can be propagated via the *path* to the receptor such as a radio receiver, which contains a semiconductor device capable of responding to the disturbance, and causing an unwanted response, i.e. interference.

For many equipments and systems, EMC requirements now form part of the overall technical performance specification. The EU's EMC Directive, 89/336/EEC, was published in 1992 and came into full implementation on 1 January 1996. All apparatus placed on the market or taken into service must, by law, comply with the Directive's essential requirements, that is it must be immune to electromagnetic disturbance representative of the intended environment and must generate its own disturbance at no greater than a set level that will permit interference-free radio communication. The Directive refers to relevant standards which themselves define the appropriate immunity levels and emission limits. More information on the EMC Directive and its ramifications is available in Marshman (**reference 15A**).

15.1.2 Coupling mechanisms

The path by which electromagnetic disturbance propagates from source to receptor comprises one or more of the following:

- conduction
- capacitive or inductive coupling
- radiation

These paths are outlined in **Fig. 15.2**.

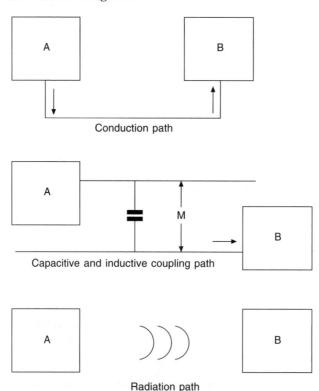

Fig. 15.2 Coupling mechanisms for electromagnetic disturbance

Coupling by conduction can occur where there is a galvanic link between the two circuits, and dominates at low frequencies where the conductor impedances are low. Capacitance and inductive coupling takes place usually between reasonably long co-located parallel cable runs. Radiation dominates where conductor dimensions are comparable with a wavelength at the frequency of interest, and efficient radiation occurs.

For example, with a personal computer, the radiation path is more important for both emission and immunity at frequencies above 30 MHz, where total cable lengths are of the order of several metres. Designers and installers of electrical and electronic equipment need to be aware that all three coupling methods exist so that the equipment can be properly configured for compatibility.

15.1.3 Equipment sensitivity

Analogue circuits may respond adversely to unwanted signals in the order of millivolts. Digital circuits may require only a few 100 millivolts of disturbance to change state. Given the high levels of transient disturbance present in the environment, which may be in the order of several kilovolts, good design is essential for compatibility to occur.

15.2 Simple source models

For many EMC situations such as coupling by radiation, effective prediction and analysis are achieved by reference to simple mathematical expressions. For nearly all products experiencing EMC problems, the equipment will work perfectly in the development laboratory and only when it is subjected to external electromagnetic disturbance do other facets of its characteristics emerge. Under these circumstances, circuit conductors are considered as antennas capable of both transmitting and receiving radio energy. The circuit can usually be assessed as either a short monopole antenna, for instance where one end of the conductor is terminated in a high impedance, or as a loop where both ends are terminated in low impedance.

For the monopole equivalent at low frequencies and at distances greater than a wavelength, the field strength E, at distance d (in metres) is given by:

$$E = \frac{60\pi \times I \times h}{\lambda \times d} \quad \text{[volts/metre]}$$

where I is the current in amps, h is the length of the conductor and λ is the wavelength (= 300 ÷ frequency in MHz), both in metres. It can be seen that the field strength is greater for shorter distances and higher frequencies (shorter wavelengths).

For loop radiators at low frequencies, the field is given by:

$$E = \frac{120\pi^2 \times I \times A \times h}{d \times \lambda^2} \quad \text{[volts/metre]}$$

where n is the number of turns and A the area of the loop. The field strength is greater at shorter distances and higher frequencies.

At high frequencies where the conductor lengths are comparable with a wavelength, a good approximation of the field at a distance can be taken if the source is considered as a half-wave dipole. The field is then given by:

$$E = \frac{7\sqrt{P}}{d} \text{ [volts/metre]}$$

where P is the power in watts available in the circuit.

15.2.1 Receptor efficiency

To estimate the degree of coupling in the radiated path, empirical data give values of induced currents of about 3 mA for an incident electromagnetic wave of 1 volt/metre. This relationship can be used to good effect in converting the immunity test levels in the standards into an engineering specification for induced currents impressed at an input port due to coupling via an attached cable.

15.3 Signal waveforms and spectra

For many digital electronic systems, the main concerns in emission control at low frequencies are associated with power line disturbance generated by switch mode power supplies. These devices switch at a relatively high rate, in the order of 30–100 kHz, and produce a line spectrum of harmonics spreading over a wide frequency band as shown in **Fig. 15.3**.

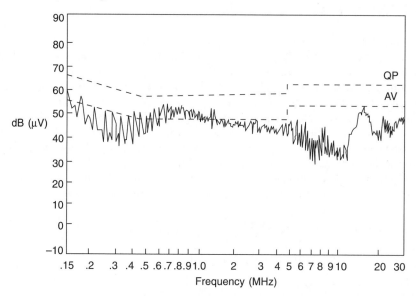

Fig. 15.3 Typical conducted emission spectrum of a switch mode power supply

With no mains filtering, the emission levels from the individual harmonics are considerably in excess of the common emission limits. Care is needed in the sourcing of these subassemblies to ensure that they are compliant with the relevant standards.

At higher frequencies, noise from digital circuits switching at very high rates (clock frequencies in excess of 30 MHz are not uncommon) couples via external cables and radiates in the VHF band (30–300 MHz) or it may radiate directly from circuit boards in the UHF band (300–1000 MHz). A clock oscillator has a waveform and a spectrum of the type shown in **Fig. 15.4**.

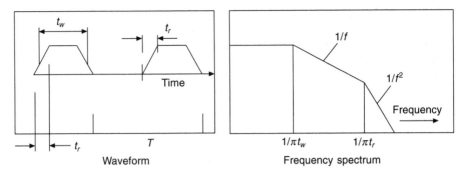

Fig. 15.4 Waveform and frequency spectrum of a digital signal

It can be seen that the turning points in the spectrum are $(1/\pi) \times$ the pulse width t_w, above which frequency the spectrum decreases inversely proportionally to frequency, and $(1/\pi) \times$ the rise time t_r. At frequencies greater than $1/\pi t_r$ the spectrum decays rapidly, in inverse proportion to the square of the frequency. Thus for longer pulse durations the high-frequency content is reduced; if the rise time is slow then the content is further reduced. For good emission control, slower clock speeds and slower edges are better for EMC. This is contrary to the current trends where there are strong performance demands for faster edges and higher-frequency clocks.

Many equipment malfunction problems in the field are caused either by transient disturbances, usually coupled onto an interface cable, or by radar transmitters if close to an airfield. The disturbance generated by a pair of relay contacts opening comprises a series of short duration impulses at high repetition rate, as illustrated in **Fig. 15.5**. As the contacts separate the energy stored in the circuit inductance is released, causing a high voltage to occur across the contacts. The voltage is often sufficient to cause a discharge and a spark jumps across the gap. This is repeated until the gap is too wide. When coupled into a digital electronic circuit, the disturbance can change the state of a device and interference in the form of an unwanted operation or circuit 'lock-up' occurs. Control is exercised by ensuring that the receptor circuit has adequate immunity to this type of disturbance.

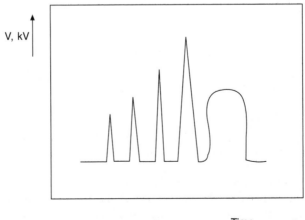

Fig. 15.5 Voltage across opening relay contacts

The radar transmission is one example of a modulated radio frequency signal, and this can often cause interference even in low-frequency electronic circuits. The radio frequency energy is rectified at the first semiconductor junction encountered in its propagation path through the equipment, effectively acting as a diode rectifier or demodulator. The rectified signal, an impulse wave in the case of radar, is thus processed by the circuit electronics and interference may result if the induced signal is of sufficient amplitude, see **Fig. 15.6**.

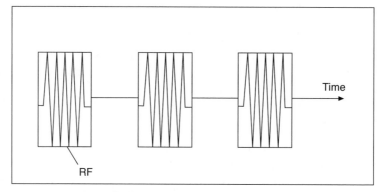

Radar signal pulse modulated RF

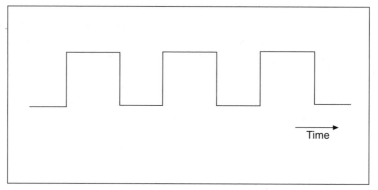

Radar signal after circuit rectification

Fig. 15.6 Radar signal

Most equipment is designed and constructed to be immune to radio frequency fields of 3 volts per metre (or 10 volts per metre for industrial environments) at frequencies up to 1 GHz. Many radars operate at frequencies above 1 GHz (for example, 1.2, 3 and 9.5 GHz) and equipment close to an airfield may suffer interference because it is not designed to be immune. Under these circumstances, architectural shielding is required, often comprising the use of glass with a thin metallic coating which provides adequate light transmittance and, more importantly, effective shielding at these high radio frequencies.

15.4 EMC limits and test levels

15.4.1 Emissions

The radio frequency emission limits quoted in EMC standards are usually based on the level of disturbance that can be generated by an apparatus or system such that radio or TV reception in a co-located receiver is interference free.

15.4.1.1 Conducted emissions

Conducted emission limits are set to control the disturbance voltage that can be impressed on the voltage supply shared by the source and receptor where coupling by conduction occurs. Limits currently applied in European emission standards are shown in **Fig. 15.7**. The Class A limits are appropriate for a commercial environment where coupling between source and receptor is weaker than in the residential environment, where the Class B limits apply. The difference in the limits reflects the difference in attenuation in the respective propagation paths.

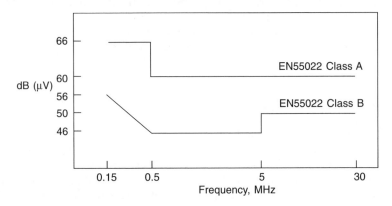

Fig. 15.7 Conducted emission limits (average detector)

The limits shown apply when using the 'average' detector of the measuring instrument, and are appropriate for measuring discrete frequency harmonic spectral line emissions. A 'quasi-peak' detector is employed to measure impulsive noise for which the limits are up to about 10 dB higher, i.e. more relaxed. The standards often require that the 'average' limit is met when using the 'average' detector, and the 'quasi-peak' limit is met using this 'quasi-peak' detector. Nearly all modern EMC measuring instrumentation contains provision for measuring using both detectors to the required degree of accuracy.

Typical limit levels are in the order of 1–2 mV for Class A and 0.2–0.6 mV for Class B. These are quite onerous requirements given that some devices such as triacs for motor speed control may be switching a few hundred volts peak.

15.4.1.2 Radiated emissions

The radiated emission limits are derived from a knowledge of the field strength at the fringe of the service area, from typical signal to noise ratios for acceptable reception, and also from applying probability factors where appropriate. Typical limits are shown in **Fig. 15.8**.

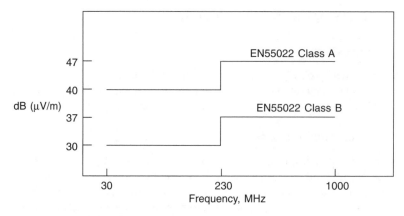

Fig. 15.8 Radiated emission limits

Below 230 MHz the Class B limits are equivalent to a field strength of 30 μV/metre at a distance of 10 metres. Many items of information technology equipment have clock frequencies in the range 10–30 MHz, for which the harmonics in the ranges up to a hundred MHz or more are effectively wanted signals, required to preserve the sharp edges of the digital waveform. This harmonic energy has to be contained within the apparatus by careful PCB design and layout and/or shielding and filtering.

15.4.1.3 Power frequency harmonics

Low-frequency limits for harmonic content are derived from the levels of disturbance that the supply networks can tolerate, and are expressed either as a percentage of the fundamental voltage or as a maximum current. Even harmonics are more strongly controlled than odd harmonics, since the even harmonics indicate the presence of a dc component. Similarly, flicker limits are set by assessing the effects on lighting of switching on and off heavy power loads such as shower heaters.

15.4.2 Immunity

Immunity test levels are set to be representative of the electromagnetic environment in which the equipment is intended to operate. In European standards, two environments are considered, the residential, commercial and light industrial environment, and the heavy industrial environment. The distinction between the two is not always clear, but most equipment suppliers and manufacturers are knowledgeable on the range of disturbances that must be considered for their product to operate reliably in the field, and can make the appropriate choice without difficulty. The key factor is whether heavy current switching and/or high-power radio frequency sources are present. If they are, the more severe 'industry' levels should be selected.

15.4.2.1 Electrostatic Discharge (ESD)

Electrostatic charge is built up on a person walking across a carpet or by other actions where electric charge separation can occur. The charge voltage is much higher for synthetic materials in dry atmospheres with low relative humidity. Although charge potentials in the order of 10–15 kV may be encountered in some environments such as hotels, the standards bodies have selected an air discharge level of 8 kV as being representative of a broad range of circumstances. The ESD event is very fast with a sharp edge having a rise time of about 1 nanosecond and a duration of about 60 nanoseconds. This generates a spectrum which extends into the UHF bands and therefore presents a formidable test for many types of equipment.

15.4.2.2 Electrical fast transients

This is the disturbance type adopted by the standards bodies to be representative of the showering arc discharge encountered across opening relay contacts. **Figure 15.9** shows the general waveform of the disturbance applied in the harmonized European standards.

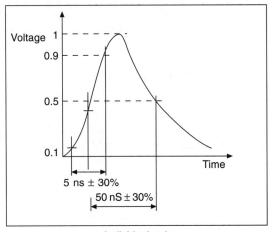

Individual pulse

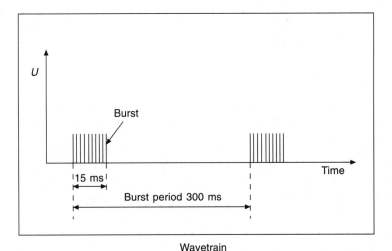

Wavetrain

Fig. 15.9 Waveform of fault transient

These transients are applied directly to the mains power conductor or to interface and input/output cables via a capacitive coupling method to simulate the effects of co-located noisy power conductors. The voltage levels applied on the mains supply are 1 kV for the residential environment and 2 kV for the industrial environment.

15.4.2.3   Radio Frequency (RF) fields

At low frequencies (below 80 MHz) the interaction of incident electromagnetic waves with receptor systems can be simulated effectively by a simple induced voltage impressed either with respect to ground via a network known as a CDN, or longitudinally via a transformer (bulk current injection). At higher frequencies (80–1000 MHz) the field is applied directly to the equipment under test, usually by setting up a calibrated transmitting antenna situated at a separation distance of about 3 m. The tests are performed in shielded enclosures in order that the RF energy can be controlled and no external interference occurs.

In the more recent standards the walls of the screened rooms are lined with absorbing material to provide a reasonably uniform field within the chamber. The applied waveform is modulated with either a 1 kHz tone to simulate speech or a pulse train to represent digital cellular radio transmission.

15.4.2.4   Dips, surges and voltage interruptions

Many other types of disturbances are present in the environment. These are the slow-speed types which are usually associated with power switching, lightning pulses and power failure. They are usually simulated by specialist disturbance generators. In many cases, immunity to these disturbances is achieved by good design of the mains power supply in the equipment under test.

15.4.2.5   Magnetic fields

Equipment containing devices sensitive to magnetic fields should be subjected to power frequency (50 Hz) magnetic fields. Typical levels are 3 amps/metre for the residential, commercial and light industrial environment and 30 amps/metre in the more severe industrial environment.

15.5   Design for EMC

15.5.1   Basic concepts

The preferred and most cost-effective approach to the achievement of electromagnetic compatibility is to incorporate the control measures into the design. At the design inception stage some thought should be given to the basic principles of the EMC control philosophy to be applied in the design and construction of the product. The overall EMC design parameters can be derived directly or determined from the EMC standards to be applied, either as part of the procurement specification or as part of the legal requirements for market entry. There are usually two fundamental options for product EMC control:

- shielding and filtering (see **Fig. 15.10(a)**)
- board level control (see **Fig. 15.10(b)**)

For the *shielding and filter* solution, all external cables are either screened leads, with the screens bonded to the enclosure shield, or unscreened leads connected via a filter. The basic principle is to provide a well-defined barrier between the inner surface of the shield facing the emissions from the PCBs and the outer surface of the

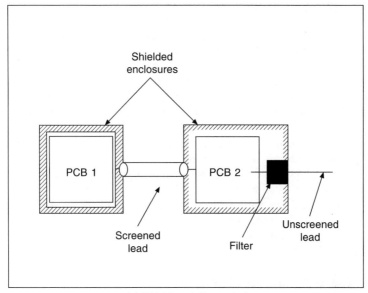

(a) Shield and filter solution

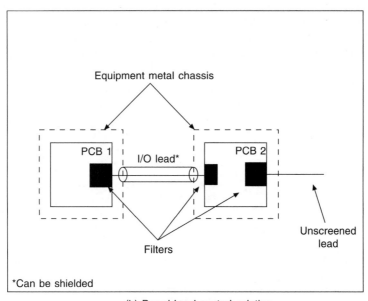

(b) Board level control solution

Fig. 15.10 Options for product EMC control

shield interacting with the external environment. This solution requires the use of a
metal or metal-coated enclosure.

Where this is not possible, or not preferred, then the *board level control* option is
appropriate, where the PCB design and layout provides inherent barriers to the
transfer of electromagnetic energy. Filters are required at all the cable interfaces,
except where an effective screened interface can be utilized.

15.5.2 Shielding

For design purposes, effective shielding of about 30–40 dB can be provided at high frequencies by relatively thin metal sheets or metal coatings on plastic. The maximum shielding achievable is associated with the apertures, slots and discontinuities in the surface of the shield enclosure. These may be excited by electromagnetic energy and can resonate where their physical length is comparable with a wavelength, significantly degrading the performance of the shield. The following basic rules apply:

- the maximum length of a slot should be no greater than 1/40 of the wavelength at the highest frequency of concern
- a large number of small holes in the shield gives better performance than a small number of large holes
- the number of points of contact between two mating halves of an enclosure should be maximized
- mains or signal line filters should be bonded to the enclosure at the point of entry of the cable

15.5.3 Cable screens termination

Ideally for maximum performance, cable shields should be terminated at both ends with a 360° peripheral (i.e. glanded) bond. This is not always possible and in some cases it is undesirable because of the associated ground loop problem, **Fig. 15.11**. Noise currents, I, in the ground generate a voltage V between the two circuits A and B which can drive a high current on the outer surface of the interconnecting screen, thus permitting energy to be coupled into the internal system conductors.

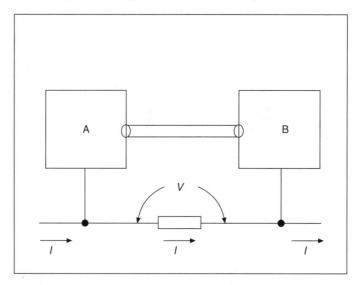

Fig. 15.11 The ground loop problem for screen terminated at both ends

In many applications involving the use of long conductors in noisy environments, the simplest solution is to break the ground loop by bonding the screen at one end only, as shown in **Fig. 15.12**. Here the ground noise voltage is eliminated but the shield protects only against electric fields and capacitive coupling. Additional or

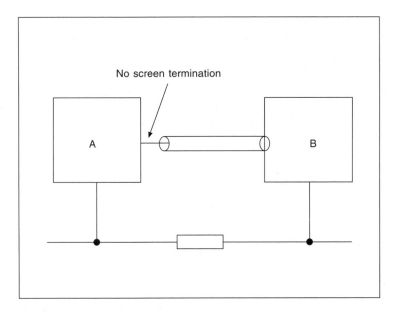

No screen termination

A

B

Fig. 15.12 Ground loop broken by termination of the screen at one end only

alternative measures such as opto-isolation are required if intense magnetic fields are present.

15.5.4 PCB design and layout

Control at board level can be achieved by careful design of the board involving device selection and track layout. As discussed in **section 15.2**, emissions from the PCB tracks may be reduced at high frequencies if the devices are chosen to have slow switching rates, and slow transition (i.e. long rise and fall) times.

Device selection can also improve immunity by bandwidth control. The smaller the bandwidth, the less likely it is that high frequency disturbances will be encountered within the pass-band of the circuit. Although rectification of the disturbance may occur in the out-of-band region, the conversion process is inefficient and higher immunity usually results.

The tracks on PCBs can act as antennas and the control methods involve reducing their efficiency. The following methods can be applied to good effect:

- reduce the area of all track loops
- minimize the length of all high-frequency signal paths
- terminate lines in resistors equal to the characteristic impedance
- ensure that the signal return track is adjacent to the signal track
- remove the minimum amount of copper on the board, i.e. maximize the surface area of the OV (zero-volt ground) and VCC (power) planes

The latter two points are generally achieved where a multi-layer board configuration is employed. These measures are highly effective at reducing board emissions and improving circuit immunity to external disturbances. Multi-layer boards are sometimes considered relatively expensive but the extra cost must be compared with the total costs of other measures that may be required with single or double-sided boards, such as shielding, filtering and additional development and production costs.

15.5.5 Grounding

Grounding is the method whereby signal returns are managed, and it should not be confused with earthing, which deals with protection from electrical hazards. Grounding is important at both the PCB level and circuit interconnection level. The three main schemes are shown in **Fig. 15.13**.

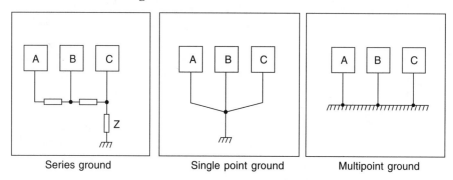

| Series ground | Single point ground | Multipoint ground |

Fig. 15.13 Three main grounding schemes

In the series ground scheme, noise from circuit A can couple into circuit C by the common impedance Z. This problem is overcome in the single point ground, but the scheme is wasteful of conductors and not particularly effective at high frequencies where the impedance of the grounding conductors may vary, and potential differences can be set up.

The ideal scheme is the multipoint ground. Generally, single point grounding is used to separate digital, analogue and power circuits. Multipoint grounding is then used whenever possible within each category of analogue or digital. Some series ground techniques may be employed where the coupled noise levels can be tolerated. Usually the overall optimum solution is derived from good basic design and successive experimentation.

15.5.6 Systems and installations

The general principles discussed above may be applied at a system or installation level. Guidance on this topic is available in **references 15B** and **15C**.

15.6 Measurements

15.6.1 Emissions

Conducted emission tests comprise measurements of voltage across a defined network which simulates the RF impedance of a typical mains supply. These Line Impedance Stabilizing Networks (LISNs) also provide filtering of the supply to the Equipment Under Test (EUT) and are also known as artificial mains networks or isolating networks. The EUT is connected to the LISN in a manner which is representative of its installation and use in its intended environment. **Figure 15.14** shows the general arrangement.

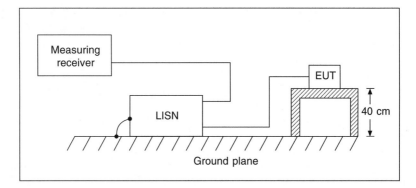

Fig. 15.14 Measurement of conducted emissions

The EUT is configured in a typical manner with peripherals and inputs/outputs attached, and operated in a representative way which maximizes emissions.

Radiated emissions are made by measuring the field strength produced by the EUT at a defined distance, usually 3 m or 10 m. The measurements are made on an open area test site which comprises a metallic ground plane, over a flat surface with no reflecting objects and within a defined ellipse.

The ground plane should cover a larger area than the test range, for example a 6 × 9 m area would be ideal for a 3 m range, and a 10 × 20 m area for a 10 m range. The EUT is situated 1 m above the ground plane on an insulating support (unless it is floorstanding equipment) and a calibrated antenna is placed at the required test distance from the EUT. At any emission frequency, such as the harmonic of the clock oscillator in a PC, the receiver is tuned to the frequency and the antenna height is raised between 1 and 4 metres in order to observe the maximum field strength radiated by the product. (The net field strength is the sum of the direct and ground-reflected waves and it varies with height.) The EUT is also rotated about a vertical axis in order to measure the maximum radiation in the horizontal plane.

Measuring instruments for both conducted and radiated emission measurements comprise spectrum analysers or dedicated measuring receivers. The spectrum analyser usually has to be modified to have a stage for preselection which prevents overload and damage in the presence of impulsive noise, and it may require additional external pulse-limiting protection when performing conducted emissions measurements with a LISN. Both instruments usually have facilities of computer control by the IEEE bus, avoiding the necessity for manual operation. When using spectrum analysers it is important to check for overload or spurious emissions by ensuring that the observed indication on the display reduces by 10 dB when an additional 10 dB RF attenuation is introduced at the front end of the analyser.

EMC measuring receivers are designed to meet the stringent requirements of CISPR (Committee International Special Perturbations Radioélectrique), a subdivision of the IEC. This sets out specifications for input impedance, sensitivity, bandwidth, detector function and meter response, such that the reproducibility of tests can be guaranteed.

15.6.2 Immunity

ESD tests are made with an ESD 'gun', set to the desired voltage which is equivalent to the human charge potential, and having well-defined charge and discharge characteristics. The ESD discharge is applied to all user-accessible parts of the EUT. The operation of the equipment is thus observed for any malfunction. Immunity to the ESD event is improved by minimizing the ESD energy that can enter the enclosure containing the electronics. The ideal solution is either a good shielded enclosure with small apertures and good bonding between sections, or a totally non-conducting surface. Generally it is difficult to design a product which completely satisfies either solution, but designers should attempt to steer towards one or the other.

Measurements of immunity to RF fields are made in a shielded enclosure, the modern types being lined with absorbing materials, such as ferrite tiles on at least five of the six inner surfaces. The EUT is subjected to radiation from an antenna situated in the near vicinity as shown in **Fig. 15.15**. The field is precalibrated to the required level of field strength specified in the appropriate standard, prior to the introduction of the EUT into the chamber. The RF frequency is swept slowly from 80–1000 MHz and any equipment misoperations noted, the performance level of the EUT having been defined prior to the start of the test.

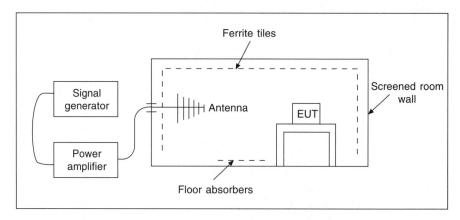

Fig. 15.15 RF field immunity-test arrangement

Transients, surges, dips and interruptions tests are performed with dedicated test instrumentation which fully satisfies the requirements of the relevant standards. Generally the tests are much quicker to perform than the RF field test, and information on the EMC performance of the EUT can be gathered rapidly.

For ESD tests and fast transient tests the equipment should carry on working after the application of the disturbance without any loss of data. For the RF field test there should be no loss of performance outside that specified by the manufacturer at any time during the test. For dips and surges etc., provided the equipment works satisfactorily, after a manual reset, it will be deemed to have passed the test.

15.7 Standards

The three types of EMC standards in current use in Europe are *basic* standards, *generic* standards and *product-specific* standards. Basic standards contain the test methods and test levels at limits but do not specify a product type. Product standards contain comprehensive details on how the product should be configured and operated during the test and what parameters should be observed. Generic standards apply in the absence of a product standard and are relevant to all products which may be operated within a defined environment. Both product standards and generic standards may refer to basic standards for their test methods. The important standards in current use in Europe are listed below.

15.7.1 Generic standards

EN 50 081-1 : Emissions; Residential, Commercial and Light Industrial Environment
EN 50 081-2 : Emissions; Industry Environment
EN 50 081-1 : Immunity; Residential, Commercial and Light Industrial Environment
EN 50 081-2 : Immunity; Industrial Environment

15.7.2 Important product standards

EN 50011 : Emissions, Industrial, Scientific and Medical Equipment
EN 50013 : Emissions, Radio and TV Equipment
EN 50014-1 : Emissions, Household Appliances and Portable Tools
EN 50015 : Emissions, Lighting Equipment
EN 50022 : Emissions, Information Technology Equipment
EN 55014-2 : Immunity, Household Appliances
EN 55020 : Immunity, Radio and TV Equipment
EN 55024 : Immunity, Information Technology Equipment

15.7.3 Basic standards

EN 61000-4-2 : Immunity: ESD
EN 61000-4-3 : Immunity: RF fields
EN 61000-4-4 : Immunity: fast transients
EN 61000-4-5 : Immunity: surges
EN 61000-4-6 : Immunity: conducted RF voltages
EN 61000-4-8 : Immunity: magnetic fields
EN 61000-4-11 : Immunity: dips and interruptions

Other product standards are emerging at a high rate which will mean that reliance on generic standards will diminish significantly in the future.

References

15A.   Marshman, C. *The Guide to the EMC Directive 89/336/EEC*, EPA Press, 1992, ISBN 095173623X, 308 pp.
15B.   Maddocks, A.J., Duerr, J.H. and Hicks, G.P. *Designing for Electromagnetic Compatibility: A Practical Guide,* ERA Report 95-0030, Leatherhead, ERA Technology Ltd, March 1995, ISBN 0700805842, 147 pp.
15C.   Goedbloed, J. *Electromagnetic Compatibility*, Prentice Hall, 1992, ISBN 0132492938, 400 pp.
15D.   Ott, H.W. *Noise Reduction Techniques in Electronic Systems* (2nd edn), John Wiley, USA, 1988, ISBN 0471850683, 426 pp.

Chapter 16

Hazardous area equipment

Mr K. Morris
CEAG Crouse-Hinds

16.1   Introduction

In areas where explosive mixtures of gas and air can be present, sparks given off by electrical equipment or hot surfaces on this equipment constitute a potential hazard, and the consequences of an explosion can be disastrous.

Combustible materials can form explosive atmospheres and can under certain circumstances cause an explosion. Such materials are used widely in the chemical, mining and other industries and even in everyday life. The concept of '*explosion protection*' of electrical equipment has been developed and formalized in order to prevent explosive accidents in hazardous areas during the normal operation of the equipment.

The coal mining industry has provided one of the main pressures for the development of special equipment, procedures, standards and codes because of its especially hazardous working environments. Davy's safety lamp for miners is an example of a piece of equipment developed specifically for use in hazardous areas. The legacy of the importance of this industry remains today through the difference in the regulations which apply to mining and to other hazardous areas. Apart from mining, hazardous areas are found especially in offshore gas and oil installations and in petrochemical complexes.

There are numerous standards and codes of practice governing the manufacture, selection, installation and maintenance of electrical equipment in potentially hazardous areas. These tend to differ around the world and despite harmonization across the European Union, the complexity can be intimidating. A summary is presented in **section 16.6**. Because of the special implications for safety, equipment and systems for use in hazardous areas must be tested and certified by approved authorities; certification is covered in **section 16.5**.

16.2   Hazardous areas

Hazardous areas are classified into zones according to the nature of the gases present in the potentially explosive atmosphere, and the likelihood of that atmosphere being present. The nature of the atmosphere is characterized by the chemical composition of the gas and its ignition temperature, and the notions of *gas grouping* and *temperature classification* have been developed in order to formalize this.

A useful concept in the consideration of how explosions occur is the *hazard triangle* shown in **Fig. 16.1**. The sides of the triangle represent *fuel, oxygen* and a *source of ignition*, all of which are required in order to create an explosion. For the purposes of this chapter, a fuel is considered as a flammable gas, vapour or liquid, although dust may also be a potential fuel. Oxygen is of course present in air at a concentration of about 21 per cent. The ignition source could be a spark or a high temperature. Given that a hazardous area may contain fuel and oxygen, the basis for preventing explosion is to ensure that any ignition source is either eliminated or prevented from coming into contact with the fuel–oxygen mixture.

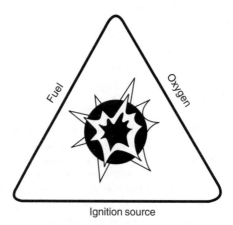

Ignition source

Fig. 16.1 The 'hazard triangle'

16.2.1 Zone classification

The zone classification defined in IEC 79 is used in Europe and most other parts of the world; it is summarized in **Table 16.1**. Various types of explosion protection are available, and their suitability for the different zones is shown in the table.

Table 16.1 IEC 79 classification of hazardous area zones

	Suitable protection
Zone 0	
Areas in which hazardous explosive gas atmospheres are present constantly or for long periods, for example in pipes or containers	Ex 'ia' Ex 's' (where specially certified Zone 0)
Zone 1	
Areas in which hazardous explosive gas atmospheres are occasionally present, for example in areas close to pipes or draining stations	Ex 'd'; Ex 'ib'; Ex 'p'; Ex 'e'; Ex 's'; Ex 'o'; Ex 'q'; Ex 'm'; Equipment suitable for Zone 0
Zone 2	
Areas in which hazardous explosive gas atmospheres are rare or only exist for a short time, for example areas close to Zones 0 and 1	Ex 'N'/Ex 'n'; Equipment suitable for Zones 1 & 0

In the USA hazardous areas are classified in a slightly different way, according to the National Electrical Code. In brief, hazardous areas are classified either as *Division 1*, where ignitable concentrations of flammable gases or vapours may be present during normal operation, or as *Division 2*, where flammable gases or vapours occur in ignitable concentrations only in the event of an accident or a failure of a ventilation system.

16.2.2 Gas grouping and temperature classification

The energy required for ignition differs from gas to gas, and the grouping of gases together with their classification by temperature is used in Europe to describe the suitability of a piece of electrical equipment for use with explosive atmospheres of particular gases.

Table 16.2 lists common industrial gases within their appropriate groups. Gas group I is reserved for the classification of equipment suitable for use in coal mines. Gas group II contains those gases found in other industrial applications, and it is subdivided according to the relative flammability of the most explosive mixture of the gas with air.

Table 16.2 CENELEC/IEC gas grouping

Group	Representative gases
I	Methane
IIA	Acetone, ethane, ethyl acetate, ammonia, benzol, acetic acid, carbon monoxide, methanol, propane, toluene, ethyl alcohol, I-amyl acetate, N-hexane, N-butane, N-butyl alcohol, petrol, diesel, aviation fuel, heating oils, acetaldehyde, ethyle ether
IIB	Town gas, ethylene (ethene)
IIC	Hydrogen, acetylene (ethyne), hydrogen disulphide

Temperatures are classified from T1 to T6, as shown in **Table 16.3**. The levels show the maximum surface temperature permitted for an equipment which has been assigned that temperature class, and the common gases for which each class is appropriate are also shown.

North American practice is to define hazardous materials in classes. Flammable gases and vapours are Class 1 materials, combustible dusts are Class 2 materials and

Table 16.3 CENELEC/IEC temperature classification

Class	Highest permissible surface temperature (°C)	Representative gases
T1	450	Acetone, ethane, ethyle acetate, ammonia, benzol, acetic acid, carbon monoxide, methanol, propane, toulene, town gas, hydrogen
T2	300	Ethyl alcohol, (-amyl acetate, N-hexane, N-butane, N-butyl alcohol, ethylene
T3	200	Petrol, diesel, aviation fuel, heating oils
T4	135	Acetaldehyde, ethyl ether
T5	100	
T6	85	Hydrogen disulphide

'flyings', such as sawdust, are Class 3 materials. Class 1 is subdivided into four groups depending on their flammability: A (including acetylene), B (including hydrogen), C (including ethylene) and D (including propane and methane). The subgroup letters are in the opposite order of flammability to the IEC groupings shown in **Table 16.2**. The North American temperature classification is similar to the IEC system shown in **Table 16.3**, but the classes are further subdivided to give more specific temperature data.

16.2.3 Area assessment procedure

Companies using flammable materials carry out an area assessment exercise in accordance with national standards such as BS 5345, Pt 2 in the UK or relevant industry codes such as those existing in the petroleum and chemical industries. In general, this assessment procedure results in a written report which identifies and lists the flammable materials used, records all the potential hazards with their source and type, identifies the extent of the zone taking into account for instance the type of potential release of flammable material and ventilation systems, and includes other relevant data. This procedure is complicated, and there are specialist companies which offer a commercial hazardous area assessment service.

16.3 Protection concepts for electrical equipment

There is a need for electrical power in hazardous areas to supply motors, lighting, control equipment and instrumentation, and a range of equipment is available which has been tested and certified to the appropriate standards. Equipment that has been designed and tested in Europe usually carries a string of codes which gives information about its suitability for use, and also carries the mark of the certifying body.

The definitions of gas group and temperature classification have already been explained. The following sections describe the basis of the *protection code*.

According to the harmonized European standard EN 50014, electrical equipment for use in explosive atmospheres can be designed with various protection concepts; these are listed in **Table 16.4**. Also listed in the table are the Ex 'N' and Ex 's' concepts which have not been the subject of European harmonization. They are not covered in EN 50014 and do not attract the EEx mark signifying certification to a harmonized European standard.

The engineer designing an electrical system has to make a choice regarding the method of protection to be used, and has to select apparatus and components accordingly.

Table 16.4 Types of protection

Protection concept	Designation
Oil immersion	EEx 'o'
Pressurized	EEx 'p'
Quartz-sand filled	EEx 'q'
Flameproof enclosure	EEx 'd'
Increased safety	EEx 'e'
Intrinsic safety	EEx 'i'
Encapsulation	EEx 'm'
Non-sparking/incendive	EXn 'N'/'n'
Special protection	Ex 's'

Fig. 16.2 The EN conformity mark for certified explosion-protected equipment

In practice, the majority of electrical equipment for use in hazardous areas will be designated according to the EEx 'd', EEx 'e', Ex 'N'/'n' or EEx 'i' concepts and these four are therefore highlighted in the following sections.

16.3.1 Flameproof enclosure – EEx 'd'

In this concept, those parts of the electrical equipment that can ignite an explosive air–gas mixture are contained in an enclosure. The enclosure can withstand the pressure created in the event of ignition of explosive gases *inside* it, and can prevent the communication of the explosion to the atmosphere surrounding the enclosure.

The rationale is the *containment* of any explosion which may be created by the equipment. The concept is therefore applicable to virtually all types of electrical apparatus, provided that the potential sparking or hot elements can be contained in a suitably sized and sufficiently strong enclosure.

The factors taken into account by equipment manufacturers and system designers include:

- arc and flame path lengths and types
- surface temperature
- internal temperature with regard to temperature classification
- distance between components in the enclosure and the enclosure wall (12 mm minimum is specified by the standard)

Some types of component are unsuitable for use in a flameproof enclosure. These include rewireable fuses and components containing flammable liquids.

A major consideration in the use of EEx 'd' enclosures is the making of flameproof joints, which must be flanged, spigotted or screwed. The maximum gaps and minimum lengths of any possible flame path through a joint are defined by the standard, which also lays down requirements for the pitch, quality and length of screw threads. The nature of the construction required is shown in **Fig. 16.3** which illustrates an EEx 'd' induction motor.

The certification for EEx 'd' equipment involves examination of the mechanical strength of the enclosure, and explosion and ignition tests under controlled conditions.

16.3.2 Increased safety – EEx 'e'

Measures are taken in this concept to prevent sparks or hazardous temperatures in internal or external parts of the electrical equipment during normal operation. The guiding philosophy here is the *prevention* of explosion from normally non-sparking/arcing or hot equipment.

Fig. 16.3 Sectional illustration of an EEx 'd' induction motor (courtesy of Invensys Brook Crompton)

No sparking devices can be used, and various electrical, mechanical and thermal methods are used to increase the level of safety to meet the certification test requirements.

One advantage of increased safety protection compared with flameproof protection is that boxes and enclosures can be made of plastics and other materials that are easier to work with. Examples of equipment commonly designed and constructed to EEx 'e' protection include luminaires, terminal boxes and motors. Examples of an EEx 'e' junction box, emergency luminaire and cable gland are shown in **Figs 16.4**, **16.5** and **16.6** respectively.

Key design considerations for EEx 'e' equipment are the electrical, physical and thermal stability of the materials and the compatibility of different materials which might be used for items such as terminations. Specialist manufacturers have been ingenious in the design of electrical terminations for EEx 'e' equipment to ensure firm, positive and maintenance-free connection of conductors.

16.3.3   Non-sparking – Ex 'n'

This is a British designation for electrical equipment, commonly known as 'Type N', which is suitable for use in Zone 2 applications, but not in Zone 1. It is not yet incorporated into harmonized European standards and does not therefore carry the EEx designation.

The concept combines certain aspects of EEx 'e' and EEx 'i' protection in its use of 'non-incendive' circuit elements and in the design for increased safety. Equipment manufactured with Type N protection includes terminal boxes, luminaires and motors. Because of the less hazardous nature of locations assessed as Zone 2 and the

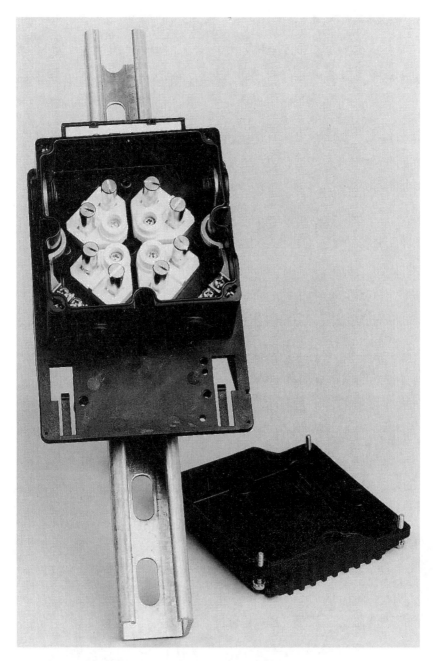

Fig. 16.4 EEx 'e' junction box and mounting plate (courtesy of CEAG Crouse-Hinds)

consequently less demanding requirements of the standard, this equipment can be manufactured more simply and at less cost than other types of explosion-protected equipment for Zone 1 use.

16.3.4 Intrinsically safe – Ex 'i'

In an Intrinsically Safe (IS) equipment, under normal operation and certain specified

Fig. 16.5 EEx 'e' Zone 1 emergency luminaire (courtesy of CEAG Crouse-Hinds)

Fig. 16.6 Deluge-protected cable gland with integral entry thread 'O' ring seal (courtesy of CMP Products)

fault conditions, no sparks or thermal effects are produced which can cause ignition of a specified gas atmosphere. Sparks or thermal effects are not produced because the energy in the IS circuits is very low. IS circuits are control and instrumentation circuits rather than power circuits.

According to the Code of Practice in BS 5345, Pt 4, in order to be defined as a 'simple apparatus' the maximum stipulated voltage that a field device can generate is 1.2 V, with current, energy and power not exceeding 0.1 A, 20 mJ and 25 mW respectively. IS circuitry exceeding these ratings may still meet the requirements of EEx 'i', but it requires certification.

There are two types of EEx 'i' protection, these being EEx 'ia' and EEx 'ib'.

EEx 'ia' equipment will not cause ignition in normal operation, with a single fault and with any two faults. A safety factor of 1.5 applies in normal operation and with one fault, and a safety factor of 1.0 applies with two faults. EEx 'ia' equipment is suitable for use in all zones, including Zone 0. An example of EEx 'ia' equipment is the pressure transmitter shown in **Fig. 16.7**.

EEx 'ib' equipment is incapable of causing ignition in normal operation and with a single fault. A safety factor of 1.5 applies in normal operation and with one fault. A safety factor of 1.0 applies with one fault if the apparatus contains no unprotected switch contacts in parts which are likely to be exposed to the potentially explosive atmosphere, and the fault is self-revealing. EEx 'ib' equipment is suitable for use in all zones except Zone 0.

Components for IS circuits contain barriers to prevent excessive electrical energy from entering the circuit. The two principal types are Zener barriers, which are used when an IS earth is available, and galvanically isolated barriers, which are used when an IS earth is not available. An IS earth must be provided by a clearly marked conductor of not less than 4 mm^2 cross-section and with an impedance no greater than 1 Ω from the barrier earth to the earth on the main power supply.

Energy can be stored in the inductance and capacitance of a cable, and this must be taken into account when designing an IS circuit. This is achieved by strict control of capacitance and L/R ratios in conjunction with Zener barriers or galvanic isolators. BS 5308 covers polyethylene-insulated cables for use in petroleum refineries and related applications and the PVC-insulated cables which are widely used in chemical and industrial applications, and cables meeting this specification may be suitable for Group II IS systems.

16.3.5 Pressurized – EEx 'p'

This type of protection uses air or inert gas to maintain a positive pressure within the enclosure or room. The positive pressure prevents the entry of flammable gas or vapour into the protected area.

An alternative method is to reduce the volume of gas or vapour within the enclosure or room below the explosive level of the gas–air mixture by dilution from a clean external source. The main features of a pressurization installation are shown in **Fig. 16.8**, and **Fig. 16.9** shows the layout of a leakage compensation unit.

16.3.6 Oil immersion – EEx 'o'

This rarely encountered protection refers to apparatus in which the ignition of a gas–air mixture is prevented by immersing the live or sparking equipment in a specified minimum depth of oil. The necessary depth is determined by testing.

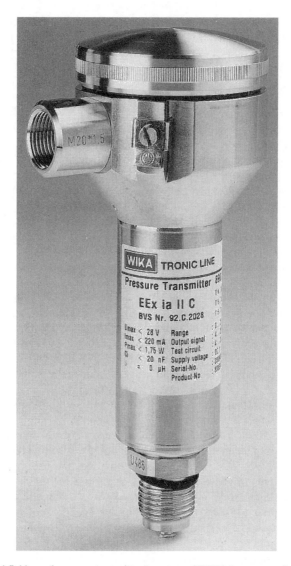

Fig. 16.7 EEx 'ia' field-cased pressure transmitter (courtesy of WIKA Instruments Ltd)

16.3.7 Quartz-sand filled – EEx 'q'

EEx 'q' protected apparatus has the live or sparking elements immersed in granular quartz or other similar material.

16.3.8 Encapsulation – EEx 'm'

In this form of protection, potential sources of ignition are encapsulated to prevent them from coming into contact with explosive atmospheres.

16.3.9 Special protection – Ex 's'

Ex 's' equipment has been shown by test to be suitable for use in the appropriate

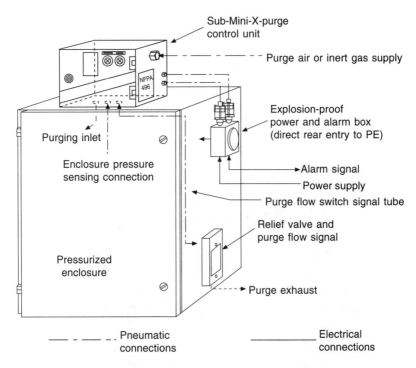

Sub-Mini-X-purge
control unit

Purge air or inert gas supply

Explosion-proof
power and alarm box
(direct rear entry to PE)

Purging inlet

Alarm signal

Enclosure pressure
sensing connection

Power supply

Purge flow switch signal tube

Relief valve and
purge flow signal

Pressurized
enclosure

Purge exhaust

Pneumatic
connections

Electrical
connections

Fig. 16.8 Main features of a pressurization installation (courtesy of Expo Safety Systems Ltd)

zone, although the apparatus does not comply with the standard of any of the established concepts previously described.

16.4 Installation, inspection and maintenance

In addition to giving information on the selection of electrical equipment for use in hazardous areas, BS 5345 also contains guidance on the installation, inspection and maintenance of equipment. The code is divided into nine parts. Part 1 gives general guidance and part 2 covers the classification of hazardous areas. Each of the remaining parts is specific to one of the types of protection concept. It contains useful general information concerning electrical work in hazardous areas, but it does not cover work in mines or areas where explosive dusts may be present, and it is not intended to replace recommendations which have been produced for specific industries or particular applications.

In general, operation and maintenance should be taken into account when designing process equipment and systems in order that the release of flammable gases is minimized. For example, the requirement for routine opening and closing of parts of a system should be borne in mind at the design stage.

No modifications should be made to plant without reference to those responsible for the classification of hazardous areas, who should be knowledgeable in such matters. Whenever equipment is reassembled, it should be carefully examined. The code gives recommended inspection schedules for equipment of each type of protection concept. These schedules set out what should be inspected on commissioning and at periodic intervals. For all equipment, the protection type, surface temperature class

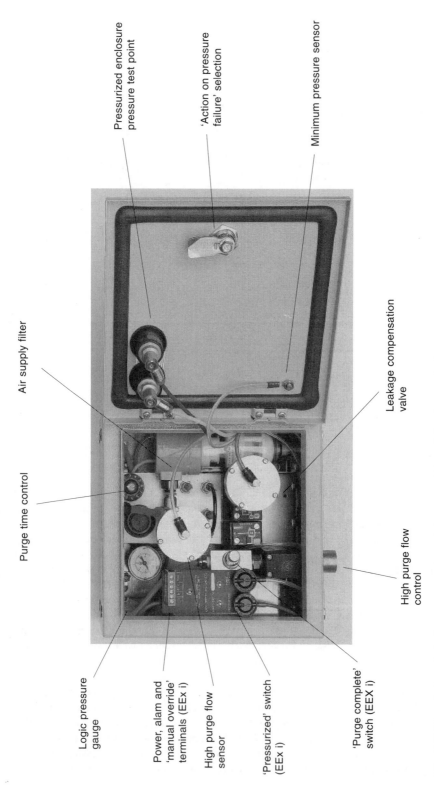

Pressurized enclosure pressure test point

'Action on pressure failure' selection

Minimum pressure sensor

Air supply filter

Purge time control

Logic pressure gauge

Power, alam and 'manual override' terminals (EEx i)

High purge flow sensor

'Pressurized' switch (EEx i)

'Purge complete' switch (EEX i)

Leakage compensation valve

High purge flow control

Fig. 16.9 Internal view of a leakage compensation unit (courtesy of Expo Safety Systems Ltd)

and gas group should be checked to ensure that the equipment is suitable for its zone of use, and circuit identification should also be checked.

Some of the areas of recommendation of the code are presented below, in particular from those parts relating to EEx 'd', EEx 'e' and EEx 'i' equipment, but these highlights can in no way replace the code itself. The electrical engineer requiring detailed and specific guidance on the installation, inspection and maintenance of electrical equipment in hazardous areas should consult BS 5345 or other codes relevant to specific industries or applications.

16.4.1 Flameproof EEx 'd' equipment

For installations making use of flameproof enclosures, the code recommends ensuring that solid obstacles such as steelwork, walls or other electrical equipment cannot be close to flanged joints or openings. Minimum clearance distances of up to 40 mm are given.

Gaps should be protected against the ingress of moisture with approved non-setting agents, and extreme care should be taken in the selection of these agents to avoid the potential separation of joint surfaces.

The code specifies the type of threads to be used for entry tappings into flameproof enclosures, and it is stressed that directions contained in the certification documents for cable systems and terminations should be followed. The types of cables suitable for use and their appropriate methods of entry are also specified. EEx 'd' equipment with integral cables where the cable terminations are encapsulated must be returned to the manufacturer if maintenance is required.

16.4.2 Increased safety EEx 'e' equipment

The section of the code dealing with EEx 'e' equipment includes a recommendation that the ratings of lamps are correct, since these may have been replaced. An appendix gives guidance on the chemical influence of combustible gases on certain mechanical and electrical properties of any insulating materials such as panels, gaskets or encapsulation.

An appendix covering cage induction motors and their associated protection equipment gives recommendations which are intended to ensure that all parts of the motor do not rise above a safe temperature.

16.4.3 Intrinsically safe EEx 'i' equipment

The part of the code dealing with EEx 'i' equipment pays particular attention to the interconnecting cables used in IS systems. It recommends, for instance:

- minimum conductor sizes to ensure temperature compliance in fault conditions
- specific separations between individual IS circuits and earth
- insulation thicknesses
- screening and mechanical properties of cables

The use of multicore cable is considered, as is the siting of cables to avoid potential induction problems. In general, cable entries should be designed to minimize mechanical damage to cables.

With all IS equipment, the need to follow the specific requirements of the certification documents is emphasized and it is recommended that during inspection attention is paid to lamps, fuses, earthing and screens, barriers and cabling. Certain specified on-site testing and maintenance of energized IS circuits is permitted inside the hazardous area, provided that the test equipment is certified as intrinsically safe in itself, and that conditions on the certification documents are followed.

16.5　Certification

The certification process involves an assessment of the equipment with regard to its conformity to the specific standard sought, an examination of a prototype to ensure that it complies with the design documents, and testing. The certified design is defined in a set of approved drawings listed in the certificate.

Within Europe there are national authorities or test houses which issue certification documents for electrical equipment in order to prove that the equipment meets a specific standard for explosion protection. **Table 16.5** lists these authorities by country. In the UK BASEEFA and SIRA are the accredited authorities. As well as being accepted throughout the European Union, certification by BASEEFA and SIRA is also recognized in other parts of the world, especially the Middle East and Far East.

Table 16.5 Certifying authorities

Country	Certifying authority
Belgium	INIEX
Canada	CSA
Denmark	DEMKO
France	CERCHAR
	LCIE
Germany	PTB
Italy	CESI
Norway	NEMKO
Spain	LOM
Sweden	SEMKO
UK	BASEEFA
	SIRA
	MECS (for mining applications)
USA	FM

Certification to the relevant standard for use in mines is carried out in the UK by MECS. The procedure is similar to the certification to standards for other hazardous areas, but in addition to explosion protection requirements, electrical equipment must also provide the high degree of electrical, mechanical and operational safety (pitworthiness requirements) demanded by the Mines Inspectorate.

Once a product has been tested and certified to a specific standard by an authorized test house, the equipment manufacturer may then certify that product under licence from the test authority. Equipment certified as meeting the European Norms carries the distinctive hexagonal conformity mark shown in **Fig. 16.2**. Individual components usually have conditions for safe use attached to their certificate. A 'certificate of conformity' for a complete piece of electrical apparatus allows installation in hazardous areas without further verification.

16.5.1　ATEX Directive

The current procedure for manufacturers introducing products to the market has been outlined in the previous section. In future, however, manufacturers will follow Directive 94/9/EC, dated 23.3.1994. This came into effect on 1.3.1996, and will become mandatory from 1.7.2003. This new Directive has a wider scope and differs from the current procedure in the following respects:

- CE marking must be applied with explosion-protected marking
- mining and surface gas groups are addressed
- electrical and mechanical equipment is covered
- equipment categories (1–3) are included
- the issue of dust is addressed

The Directive places more emphasis on continued compliance and does not allow for different interpretations.

Installers and operators will not see a great change in equipment, as most electrical equipment currently on the market will comply with the new Directive. They will see different equipment markings, with more information being included on equipment labels.

16.6   Standards and codes of practice

As in many areas of industry, European Norms (EN) exist alongside equivalent British Standards. The ENs, with equivalent BS and IEC references are shown for the different explosion protection concepts in **Table 16.6**. The origins of the ENs are in the European Directives published in 1975 and 1979 concerning electrical equipment

Table 16.6 International, regional and national standards relating to electrical equipment for hazardous areas

IEC	EN	BS	Subject of standard	N. American
IEC 79-0		5345-2	Zone classification	
IEC 79-0	EN 50014	5501-1	General requirements	
IEC 79-6	EN 50015	5501-2	Oil immersion	
IEC 79-2	EN 50016	5501-3	Pressurized	
IEC 79-5	EN 50017	5501-4	Quartz-sand filled	
IEC 79-1	EN 50018	5501-5	Flameproof enclosure	
IEC 79-7	EN 50019	5501-6	Increased safety	
IEC 79-11	EN 50020	5501-7	Intrinsic safety	
	EN 50039	5501-9	Intrinsic safety	
	EN 50028	5501-8	Encapsulation	
		4683-3	Non-sparking/incendive	
		SFA 3009	Special protection	
Of special relevance to cables:				
IEC 92-3		6883	Wiring for ships and offshore topside installation	
IEC 754-1		6425	Smoke and halogen emission	
IEC 331		6387	Fire resistance	
IEC 332		4066	Flame retardance	
		6207	Mineral-insulated cables	

for use in potentially explosive atmospheres. A separate Directive relating to mines was published in 1982.

It is important for the electrical engineer to be aware of the ENs and their equivalents in their latest editions because they determine the types of equipment available for use, and also impinge on installation procedure.

References

There are a few reference texts on this subject and the reader is referred for further detail to the comprehensive literature which is produced by reputable manufacturers. This is produced in much greater depth than ordinary commercial literature because of the complexity of specifying, installing and using equipment for hazardous areas.

Index